Insurgent Democracy

Insurgent Democracy

The Nonpartisan League
in North American Politics

MICHAEL J. LANSING

The University of Chicago Press

Chicago and London

The University of Chicago Press, Chicago 60637
The University of Chicago Press, Ltd., London
© 2015 by Michael J. Lansing
All rights reserved. Published 2015.
Paperback edition 2016
Printed in the United States of America

24 23 22 21 20 19 18 17 16   3 4 5 6 7

ISBN-13: 978-0-226-28350-0 (cloth)
ISBN-13: 978-0-226-43477-3 (paper)
ISBN-13: 978-0-226-28364-7 (e-book)
DOI: 10.7208/chicago/9780226283647.001.0001

Library of Congress Cataloging-in-Publication Data

Lansing, Michael, author.
    Insurgent democracy : the Nonpartisan League in North American politics /
Michael J. Lansing.
        pages ; cm
    Includes bibliographical references and index.
    ISBN 978-0-226-28350-0 (cloth : alk. paper)—ISBN 978-0-226-28364-7
(e-book) 1. National Nonpartisan League—History. 2. Insurgency—
United States—History—20th century. 3. Social movements—United States—
History—20th century. 4. Farmers—Political activity—United States—History—20th
century. I. Title.
    HD1485.N4L36 2015
    324.2732'7—dc23
                                                        2014049399

♾ This paper meets the requirements of ANSI/NISO Z39.48–1992
(Permanence of Paper).

For Nina, for everything

Contents

Prologue

We need fuller and richer histories of American democracy.
MERLE CURTI, "The Democratic Theme in
American Historical Literature" (1952)

This book means to make you think differently about politics. I wrote it because we live in a time when lobbyists wield inordinate power, officeholders
seem distant and corrupt, corporations command votes, bureaucrats ignore
lived realities, and citizens remain firmly cynical. Pressing problems—including climate change, poverty, mass incarceration, endless war, and discrimination—pervade our lives. Our political system seems poorly equipped
to do much about them.

We are not the first to face a failed political machinery. In the early twentieth century, farmer discontent in North Dakota grew despite rising prices
for wheat. The brutal northern plains environment made eking a living out
of the land difficult. Credit ran short. Local banks gouged farmers with ex-
orbitant interest rates on mortgages. Minneapolis-based milling companies
controlled commodity prices and railroad shipping rates. They also exerted
undue political influence in Bismarck, the state capital.

In 1915, agrarians responded to their plight by creating the starkest challenge to party politics in twentieth-century America. Their movement—the
Nonpartisan League (NPL)—deployed novel tactics that challenged existing
institutions. In an effort to empower citizens, the League drove a brief but
powerful electoral insurgency. At its peak, almost 250,000 paying members
lived in thirteen states and two Canadian provinces. As a result, the NPL dramatically shaped North American politics in the late 1910s and early 1920s.

Despite the significance of the Nonpartisan League, most today know little about its rise and fall. The few who do remember it as a radical, if doomed,
recapitulation of earlier forms of producer politics limited to North Dakota
and Minnesota. But viewing the NPL as an outdated, transitory, and local
agrarian reaction to specific economic relationships misreads the movement's

history. It also obscures the persistent potential of the League's perspective and tactics.

In fact, the NPL used the then new tools of direct democracy to insist on a moral economy premised on accumulation without concentration. It grew out of cooperative movements and insisted that government establish state-run competition in various economic sectors. It pushed for publicly owned enterprises to compete with private corporations and bring equity to the marketplace. Convinced that markets were the beating heart of the republican experiment, NPL members nonetheless insisted on state-sponsored market fairness. In their minds, the health of the nation's political democracy depended entirely on what Leaguers often referred to as "economic democracy"—an equal chance to succeed in a market, a chance that drew from interdependence to foster self-sufficiency.

These commitments seem strange to us today. Small-property holders anxious to use government to create a more equitable form of capitalism cannot be easily categorized in contemporary political terms. They do not fit on the standard political spectrum. Their movement to protect private property from corporate inroads defies easy definition. Yet the League embodied an older, alternative political economy, even as it embraced modern society. It reminded Americans that corporate capitalism was not the only way forward. Most important, it showed that the health of our democracy depended greatly on the specific economic relations we share. In a time when talking heads typically ignore moral economy in their economic analysis, the NPL's perspective deserves our attention.

A nonpartisan stance proved to be the League's most original tactic. Sidestepping deep-rooted political parties allowed the NPL to offer those without influence a chance to shape their society. As a candidate-endorsing political organization, the Nonpartisan League took advantage of the newly created direct primary to bypass entrenched politicians. They simply backed anyone who supported the League's program, regardless of party.

This bold denunciation of parties—an assumed cornerstone of modern politics—raised the ire of politicians everywhere. It proved to be one of the most important developments in an era of innovative democratic reform. Firmly committed to their affiliations, officeholders envisioned the NPL as a powerful threat. Nonpartisanship offered a way for citizens to directly fashion governance. Replacing the mediating force of a political party with a self-organized polity, these farmers invented an effective alternative to politics as usual. The rejection of parties appealed to farmers in Canada as well. Inroads in Saskatchewan and Alberta showed that the NPL's methods and ideology had a transnational reach.

Home front paranoia during World War I created an opportunity for opponents to strike back. Yet the NPL survived. In many places, it was stronger than ever. By 1920, observers speculated that it would soon transform the national political landscape. Envisioning government not as big or small but instead as the means to express the people's will, the League sought to restore equity to an unfair economy. The successful establishment of a state-owned bank, flour mill, and grain elevator in North Dakota suggested the changes to come.

But internal dissent caused fissures. Rival organizations and political opponents challenged or co-opted the NPL and its platform. Agricultural depression further weakened the League. In 1923, its national office closed. In the United States, postwar liberals laid claim to the League's legacy, even as they buried the NPL's ideology and tactics with elite-managed bureaucracies. In Canada, NPL-inflected thinking lived on in the Cooperative Commonwealth Federation.

At the very moment most American intellectuals gave up on a central role for the people in politics—ruing what journalist Walter Lippmann, in 1922, famously dubbed the "manufacture of consent"—the League showed the ongoing potency of carefully organized and platform-focused citizen politics. It offered a model (however imperfect) for citizen engagement and organizing through its application of cooperative principles to political life.

The Nonpartisan League created an antimonopolist popular politics for a rapidly urbanizing America. It generated new tactics for electoral success within the existing political system. It exemplified the push for democratic innovation during the Progressive Era. It deployed the lessons of cooperation in political venues. It established pathbreaking state-owned enterprises. It encouraged rural women to become deeply involved in electoral politics. It brought together agrarians and laborers in powerful third parties. It provided the electoral base for populist-minded US senators who influenced federal policy until World War II. Finally, it proposed an alternate future for American capitalism. By illustrating exactly how citizens could compel government to meet their needs rather than serve corporations, farmers in the NPL left their mark.

Their hoped-for future offers us new perspectives on present-day politics. For a historian like me, it is a chance to reconnoiter and to reconsider. Though the problems we face differ from those found in any other time or place, these predecessors—flawed and fully human—point us toward new ways of doing and thinking. Rooting arguments in evidence and respecting scholarly norms, historians must craft stories that shape the future as well as the past. We must reclaim alternatives. Objectivity does not demand neutrality.[1]

Though the chapters that follow trace the Nonpartisan League's rise and fall, this narrative does not follow a simple trajectory. Like any movement for change, controversy dogged the NPL. Its leadership often fell short. Its experiments sometimes missed the mark. Its history lays out the clear limits of people-centered politics in a corporate age. Nonetheless, it also points to largely forgotten possibilities. Inventive citizens grappled with established powers and challenged the status quo. Drawing on their own limited resources, western farmers seized the opportunity to remake their own lives. They transcended cynicism to create enduring—if regularly ignored—legacies that hold the potential to reshape politics today.

1

Birth

There are only two ways of making a living in North Dakota. One way is to dig the
wealth out of the ground. And the other way is to dig it out of the hide of the fellow
who digs it out of the ground.

ALBERT E. BOWEN JR., founder of the
Nonpartisan League, March 6, 1917

When the rain fell in North Dakota on June 28, 1916—primary election day—
it came down in torrents. A huge midsummer thunderstorm rumbled across
the prairie. Dark clouds, booming thunder, and gale-force winds transformed
day into night. Small streams swelled until they burst their banks. Indeed,
the "severe storms paralyzed wires and made roads impassable at points."[1]
Cloudbursts washed out bridges and turned two-rut paths into impassable
mud, keeping many rural voters from the polls.

In the midst of the deluge, journalists and politicians alike awaited election results. Initial returns suggested that the weather helped establishment
candidates of the state's ever-dominant Republican Party to survive a feisty
challenge from a new and little-understood farmer's movement called the
Nonpartisan League (NPL). Voters in major cities—Fargo, Grand Forks, Minot, and Bismarck—had all elected to keep things as they were. But as the
hours passed and the rural returns dribbled in, it slowly became clear that the
insurgent farmers had scored a major victory.

Just two months earlier, the Nonpartisan League had promised to harness "the rumbling thunder of wrathful discontent, at first low and indistinct,
but ever growing louder and louder, coming from various parts of the state,
until now it has reached the intensity of a hurricane." Their agrarian tempest
triumphed. Despite nearly impossible conditions, determined citizens made
their way to the polls. The League's endorsed candidates won nearly every
office.[2]

Though the general election was months away, the stranglehold that the
GOP had on state politics all but ensured that the Nonpartisan League would
send its candidates for governor, lieutenant governor, attorney general, and
the legislature to the state capital in November. One stunned reporter noted

that "the farmers marched in a solid phalanx . . . stuck with a vengeance and emphatically registered their demands. No state in the union furnishes a parallel instance." A farmer from Tagus, North Dakota, put things more drily: "It rained pitchforks and saw-logs but . . . the results are fine."[3]

For the first time in the state's history, citizen-farmers rather than career politicians would direct affairs. League member O. T. Haakensen celebrated the news on his farm in Barton by running up a new American flag. He believed the election meant that "the people of the state of North Dakota got their freedom and they are going to get a government by the people for the people." Recognizing that the primary victory was only the beginning, a more careful NPL member directly addressed future challengers: "Let the fight go on."[4] In an agricultural state, farmers had stuck together, voted together, and won. Agrarians across North America took notice.

The tempest fostered by the dramatic emergence of the Nonpartisan League—a candidate-endorsing political organization established by and for western farmers—rattled the foundations of American politics in the 1910s and 1920s. Born in western North Dakota, the insurgency empowered disadvantaged citizens to directly engage in electoral politics. The NPL transformed disinterested and discontented voters across thirteen states and two provinces into a movement. It defied the common assumption that agrarians could not transcend regional, ethnic, and economic differences to vote as a bloc. Instead of creating a third party, NPL members took over existing political parties from the inside. Touting a program of government competition that set up state-owned enterprises to challenge private corporations in various economic sectors, the NPL hoped to improve the lot of lower-middle-class agrarians and workers. Equally important, they applied novel electoral tactics that led to significant successes. This is their story.

*

American farmers enjoyed real prosperity from 1900 to 1920. After decades of instability in farm economies, the value of their farms tripled. Agrarian incomes doubled. The value of their global exports swelled. New technologies eased the lives of farm women and men alike. In sum, life was better across rural America than ever before.[5]

As the most agricultural state in the nation, North Dakota grew rapidly during these years. The forcible removal of indigenous peoples to reservations left broad swaths of wide-open land. The Northern Pacific and Great Northern railroads connected unbroken grasslands to regional centers such as Minneapolis and Chicago and, beyond them, a global marketplace. Boosters spoke of fortunes to be made. Thousands of European migrants joined

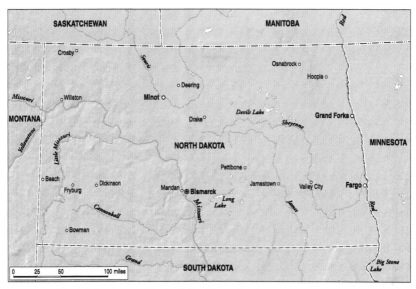

FIGURE 1. North Dakota in 1915. (Map by Philip Schwartzberg)

native-born Americans and Canadians flooding into North Dakota, claiming land and breaking the shortgrass sod west of the Missouri River. Indeed, the state proved among the most diverse in the country.[6]

Small towns chock-full of the businesses farmers needed—stores, banks, implement dealers, blacksmiths—sprouted up along rail lines. Farms spread out from them. By 1915, 637,000 people lived in North Dakota, many on homesteads smaller than four hundred acres. For most, the efforts to get and then hold on to property—a small-town shop or a small farm—defined their world. A handful of village elites and itinerant wage laborers represented the apex and the nadir of economic and social possibilities.[7]

Wheat and work demarcated the lives of these settlers. The grain accounted for 80 percent of all farm income by 1909. Though oats, barley, flax, and rye also played a role, the yearly cultivation of wheat defined the state's economy. By the 1910s, more than seven million acres in North Dakota were dedicated to wheat. The demand for it grew by leaps and bounds, especially as Minneapolis (with its prime waterpower) evolved into the world's flour-milling center.[8]

Wheat farming demanded constant labor and a strong streak of self-reliance. Spring planting and fall threshing set the rhythms. Endless chores meant farmers and their families worked long days. Aspiring to economic independence, male farmers depended on the innumerable contributions made by women and children. Men held firmly to patriarchal family structures, devaluing the essential work of women who cared for animals, sold eggs, watched the children, and cooked all the meals. Furthermore, farm women built relationships with neighbors—relationships that ensured help would be there when it was needed. The small-town proprietors that served farm families worked nearly as hard. They adjusted their schedules and strategies to serve rural customers, making their own labors intense.[9]

The rigors of establishing new farms and businesses in an extreme climate meant that most North Dakotans valued work above all else. Character-shaping labor, displayed through perseverance and dogged commitment, proffered rewards and contributed to the broader community. Hard-earned prosperity, most residents believed, would ensue. Comfortable independence correlated to the amount of work done. Civic organizations and a diversity of religious denominations further encouraged the sense that work offered a path to success. They also encouraged a culture of mutual dependence that cultivated a greater good. These ideas formed the bedrock of a middle-class culture.[10]

Yet deep-seated divides lay beneath the surface. Significant differences separated rural North Dakotans. Whether businessmen or farmers, immigrant or native-born, in small towns or on farms, Protestant or Catholic, tenants or landowners, men or women, they encountered these tensions daily. Men held tightly to their role as head of household. Furthermore, a belief that prosperity should not result in profligate wealth ensured that jealousy might accrue alongside one's accumulation of capital. Finally, the persistent unpredictability faced by farmers everywhere—in market prices for their commodities as well as erratic weather—pulsed through town and country alike. Farmers struggled more than small-town businessmen, even as both grasped for prosperity. Most agrarians found that though they shared middle-class aspirations with townsfolk, their struggle to attain that status made them firmly lower middle class.[11]

Such was the vulnerable nature of one-crop agriculture in a severe climate. Winter temperatures that dropped below -30 degrees Fahrenheit and summer heat that soared above 100 provided one challenge. Ten to twenty inches of unevenly distributed rain per year presented another. The new dry farming methods in vogue promised fixes for these problems. Even then, the margin for success proved thin. In the end, relatively brief periods of unusually high rainfall proved the only reason crops grew with any consistent success west of the Missouri River in the early twentieth-century.[12]

Agrarians searched for the type of wheat best suited for stable, large yields. In the mid-1910s, most farmers turned to Marquis hard red spring wheat. It made good bread flour and germinated early, making it less susceptible to dry summer heat and invasive fungi. Minneapolis millers valued it highly. This meant that calls for diversification often fell on deaf ears. The ongoing labor of farmers (and their animals and machines) to turn the sun's energy and the soil's nutrients into wheat alone slowly mined the soil. The lack of crop diversity and careful cultivating practices created new, human-made limits. Yields per acre slowly slipped. Yet farmers remained focused on wheat.[13]

On the northern plains in the early twentieth century, power—social, environmental, economic, and political—flowed from wheat. Grain embodied nature's energy as well as the expended bodily energies of farmers through cultivation and reaping. Control of the kernels—in aggregate—defined the powerful. Usually expressed in economic values, the power embodied in wheat involved agrarians in struggles with businesses over how to evaluate and price their product. Farmers proved especially vulnerable in these tussles over the value of nature.

In mid-nineteenth-century Chicago, grain traders and buyers had developed systems of evaluating crop quality in order to confer economic value on wheat. Assigning standardized grades to kernels of wheat—regardless of their place of origin—allowed businesses to store and sell large quantities. It also gave rise to futures trading—a way of making money from abstractions. When Minneapolis and northern plains hard spring wheat overtook Chicago and midwestern soft winter wheat in the 1890s, these systems persisted.[14]

Indeed, the Minneapolis Chamber of Commerce (the city's central grain exchange) created the same stranglehold on the wheat trade that the Chicago Board of Trade had pioneered decades before. In fact, the Minnesotans' position proved stronger than their predecessors' in Chicago, because they harnessed the power of falling water at Saint Anthony Falls and invented a new milling process that made the primary product of wheat—bread flour— more valuable. At harvest time, the local representatives of Minneapolis grain traders and buyers—not farmers—did the actual grading of kernels. Grain agents looked to their own vested interest in assessing the wheat. This left agrarians at the mercy of outsiders.[15]

At its best, the ability to mix the rivers of wheat that entered grain elevators every fall to create new grades provided legitimate ways to increase profits at the farmer's expense. At its worst, the system encouraged outright fraud. One North Dakota farm woman remembered the local "elevator man who said papa brought in 23 sacks of wheat in the wagon" though he always carefully "hauled 26 sacks to a load." When "three sacks full of wheat just disappeared inside the elevator," her father erupted in anger but had little recourse. Values and even quantities determined by others rarely worked in the middling agrarian's favor.[16]

Thus multiple dangers loomed amid presumed prosperity. A farmer from outside Jamestown, North Dakota, wrote in 1921: "I came here in 1880. I am no kid-glove farmer. Few men have in that time done more actual work. I've done straight farming—wheat, hogs, and cattle. Now at the end of over forty years I solemnly affirm that *there has been nothing in it.*" Middle-class ideals could not surmount a system that simultaneously encouraged family farming and undercut its viability. Economic divisions within the rural middle class— between townspeople and farmers—emerged.[17]

The ongoing volatility in wheat values, yields, and corresponding farm income threatened the very stability of the property-owning agrarian's world. Even as prices for wheat and land remained high, many farmers borrowed money in order to stay on their land or to make their operations big enough to be viable. Sometimes they even lent small amounts of money to each other. With large infusions of cash coming only at harvest, credit came to define local economies. All this threatened agrarians' already tenuous middle-class status.[18]

By 1910, over half of all North Dakota's farms were mortgaged. By 1920, banks owned a lien on 70 percent of the state's farms. Tenancy rates grew, as fewer and fewer farmers proved able to hold on to their land. In 1921, US Senator Edwin F. Ladd (R-ND) noted that in 1915 "more than two-thirds of the farms of North Dakota were mortgaged and more than one-fourth of the farmers of North Dakota were tenants on the land they tilled."[19] The ratio of mortgage debt to farm values also exploded during this period—from 14.8 percent in 1890 to 22.4 percent in 1910 to 28.5 percent in 1920. One contemporary observer noted that "farms, abandoned because of mortgage foreclosure, were astonishingly common."[20]

With the rise of indebtedness grew the fear of foreclosure. Magnified by one's proximity to neighboring farmers losing their land, anxieties over property loss rose dramatically in the 1910s. Because agrarians still imagined themselves as the heirs of Jeffersonian democratic ideals, land ownership proved crucial to farmers' self-identity. First, their entrepreneurial success depended on it. Second, in a culture that valued hard work, foreclosure meant that one's moral economy could be questioned. Losing one's farm signified not only economic failure but also social failure. It unmanned farmers, whose sense of self depended on the outward display of self-reliance and independence embodied in farm ownership. Third, it represented the loss of even lower middle-class status. In fact, foreclosure signaled a complete undoing not just of a farmer's economic future but of the most basic assumptions agrarian families held about their world. Meanwhile, because land values remained high, foreclosures enriched a handful of small-town bankers and real estate agents, even as they hurt farmers and the merchants who served them.[21]

In their efforts to exploit the soil, farmers found themselves exploited. The dependence they felt took on structural forms. Minneapolis millers purchased and milled most of their wheat, and Minneapolis companies often owned their mortgages. Saint Paul was the region's primary rail center, and railroad companies there set the costs for transporting wheat from nearby small towns. The Twin Cities played a disproportionate role in North Dakota's economic and cultural life. Living in these cities' hinterland encouraged farmers to believe that urbanites gained the real prosperity earned through farmers' sweat and sacrifice. Others controlled their destiny—and their profits. The power that they brought into the world flowed to someone else.

Despite the assumption that hard work led to a bright future, by the early 1910s conditions left these farmers feeling more powerless than ever. Reluctance to consider alternatives to wheat monoculture left them with less-productive land. They could not control bad weather or wheat rust. They could not control how their grain would be valued. In some cases, they could not even hold on to their property. Those who could feared losing what they had. Together, these factors fostered a culture of anxiety and dependence dramatically at odds with agrarian expectations. The corresponding anger provided fuel for the coming storm.

✶

Unrealistic expectations, the exacting environment, and the uneven economic relationship with grain buyers made for an unstable atmosphere. In this context, pockets of grievances blew up into widespread discontent. The seeming powerlessness of individual farmers encouraged calls for action. Though the position of middling agrarians gradually deteriorated, they responded as best they knew how. They knew the significance of both cooperative action and the ballot box.

Agrarians in early twentieth-century North Dakota drew on an older populist tradition of facing down the newly emergent corporate America to confront these challenges. They insisted on responsive governance. They recognized that industrial society seemed to be leaving them behind and anxiously argued for inclusion. They tried to fight off feelings of dependence by working even harder. Indeed, farmers added political, economic, and community organizing to the list of productive labors that enveloped every waking hour. Farmer-backed Populists had seized state government in 1892. They hoped to curb the power of elevator and railroad operators. Though they appropriated $100,000 in 1893 for a state-owned terminal elevator, the poorly conceived bill never made it to the governor's desk. Political defeat soon followed.[22]

Their pyrrhic victory ensured that politicians in North Dakota failed to recognize farmers' slowly growing anger. A Republican political machine run by Alexander McKenzie, whose coziness with Twin Cities corporations disturbed even the most cynical observers, held a firm grip on state politics from the territorial period until the early 1900s.[23] In 1906, a Democratic candidate for governor, John Burke, defied the McKenzie machine. With help from like-minded colleagues from both parties, he instituted political reforms such as the direct primary and the referendum. He also worked to undo the crony capitalism that defined state politics under the still-active McKenzie. Neither tack directly addressed lower-middle-class farmers' specific problems. Indeed, Burke found most of his support among small-town proprietors and those disgusted by machine politics. Many farmers remained apathetic about his good-government crusade.[24]

Recoiling from electoral politics, North Dakota agrarians responded by creating a state branch of the American Society of Equity. Conceived in Indiana in 1902, the society pushed agrarians to collectively hold their crops from the market, set prices themselves, and beat the commodity monopolists at their own game. When organizers entered North Dakota in 1907, more than ten thousand farmers pledged to work together and keep their wheat until told to sell. In a society where credit was king, the campaign failed.

But Theodore Nelson, a young farmer from Mayville, North Dakota, became the head of the national equity society's grain growers. Early experiences with farmer-owned stores and elevators—vestiges of 1880s and 1890s populist movements—gave him energy and example. He immediately began lobbying the US Congress to create federal standards for grain grading. Meanwhile, in 1908, his companions at home created their own wheat holding company—the Equity Cooperative Exchange—to compete in the Minneapolis grain market.[25]

Exchange organizers looked north for guidance. In Saskatchewan and Alberta, wheat also dominated the landscape. Many Americans emigrated there in the early 1900s. Branches of the Canadian Society of Equity soon followed. Facing down Winnipeg-based monopolies, Canadian agrarians formed the Saskatchewan Grain Growers' Association in 1902, the Grain Growers' Grain Company (which traded grain in the Winnipeg market) in 1906, and the United Farmers of Alberta in 1909. All successfully pushed provincial governments to directly support farmer-owned cooperatives. They offered a compelling example. In 1911, Theodore Nelson even tried to organize a federated body of Canadian and American wheat farmers. Though the effort failed, the attempt illustrated multiple connections and overlapping concerns. Canadian farmers continued to exchange ideas and advice with their exchange colleagues in the United States.[26]

Knowing that advice was not enough, the Equity Cooperative Exchange began recruiting wheat farmers across western Minnesota, eastern Montana, and South Dakota. After all, grain traders in the Minneapolis Chamber of Commerce purchased most of their grain as well. Chock-full of young agrarians dedicated to cooperative work, the Equity Cooperative Exchange incorporated in 1911 and began operating that same year. It mimicked private commission firms on the Minneapolis market. It also proved these lower-middle-class producers' penchant for capitalism of a different sort. The Equity Cooperative Exchange represented democratically controlled, mutually organized, collaborative self-help in order to ensure private economic benefit for individual farmers in a competitive marketplace. It embodied a moral economy as well as an economic enterprise.[27]

Searching for effective strategies, the organization turned to George Loftus for new leadership in 1912. Loftus, a campaign manager for Robert La Follette's 1912 presidential bid, brought fiery oratory and showmanship to the job. One farmer and Equity Cooperative Exchange organizer remembered that Loftus "showed them how the elevator man could raise No. 1 wheat on his mahogany desk!" Beloved by farmers but short on business skills, Loftus provoked an immediate response from the established grain traders.[28]

Angered by Loftus's heated rhetoric and fearing the Canadian example of state-sanctioned farmer controls, Chamber of Commerce members actively and sometimes illegally undercut the farmers' organization. Equity organizers made their plight public, ensuring that expansion followed on the heels of persecution. The intensity of the Minneapolis millers' resistance to farmer participation in the wheat market stemmed as much from their weakened grip on flour production as their oft-cited greed. The rise of electric-powered mills—factories untethered from direct proximity to waterfalls—threatened their water-derived power. Buffalo, New York–based companies already directly challenged their milling monopoly. Indeed, they would soon seize the mantle of flour production from Minneapolis altogether.[29]

Besides joining the Equity Cooperative Exchange, which traded more wheat every year, agrarians began organizing locally. By 1914, North Dakota farmers had created more than 350 cooperatively run elevators. These directly competed with existing elevators for wheat. Constantly attacked by corporate competitors with more capital, local elevators struggled to organize, let alone survive. Building such a cooperative movement, remembered one agrarian, took "sweat and blood." In the early 1910s in New Rockford, North Dakota, farmers woke up one morning "and found [their] elevator, full of grain, had been set a fire." They lost thousands of bushels of wheat. But as Ole Olson recalled, "We started right away, ordered lumber, and built a new one." Even when they allied with the Equity Cooperative Exchange, their long-term future promised little more than constant struggle.[30]

Some farmers lived on such a slim margin that they had to patronize the elevator giving them the best price, regardless of affiliation. Others disliked the confrontational approach taken by the Equity Cooperative Exchange. The exchange's failure to move beyond the market and engage entire farm families in educational and community-building efforts at the local level likely limited its effectiveness. New farm organizations sprang into the breach.

A nascent Farmers Union began organizing in central North Dakota in 1913. Their program mimicked that of the American Society of Equity but added cooperative purchasing, self-education, and calls for agrarian brotherhood. Local chapters opened their meetings by singing songs about cooperation. Socials, picnics, musicals, and plays created community not just for farm men but also for farm women and children. When it came to selling their crop, locals looked to build or purchase elevators in their communities. These Farmers Union elevators chose to sell their crop to long-standing firms rather than the Equity Cooperative Exchange. For farmers skeptical of the exchange, the Farmers Union offered a welcome alternative. As a third alternative, brand-new chapters of the Patrons of Husbandry—better known as the Grange—began sprouting up in 1912.[31]

Though the North Dakota Society of Equity, the Equity Cooperative Exchange, the Farmers Union, and the Grange all eschewed electoral politics, they and their members pressured local institutions to verify their sense of injustice. Innumerable investigations substantiated their concerns. As early as 1906, US Senator Porter McCumber (R-ND) publicly attacked Minnesota grain merchants for cheating North Dakota's farmers out of three to five million dollars a year.[32]

Follow-up analysis by the North Dakota Bankers Association—whose small-town membership shared a middle-class culture with farmers and were anxious for capital to stay in the state—and conservation-minded agricultural scientists at the North Dakota Agricultural College (NDAC) concurred. Both offered concrete examples of economic injustice even as they chided farmers for overlooking concerns about fertility, diversification, and better business practices.[33]

In 1906, the Bankers Association investigated grain marketing, noting that one terminal elevator in Minnesota shipped out thousands more bushels of No. 1 and No. 2 wheat than it had recorded receiving.[34] The next year, the state

legislature instructed NDAC chemist Edwin Ladd to build a small mill in order to study wheat processing. In a series of reports in the mid-1910s, he determined that lower-grade wheat easily made high-quality flour—representing
a real profit for millers, who gave farmers less for grain that worked almost as
well as top-quality kernels. Ladd also determined that grain graded at a small-
town North Dakota elevator as second- or third-rate suddenly became top-
quality wheat upon arrival in Minneapolis. Indeed, by 1916, he proved that
the state's wheat crop earned farmers $5 million less than its actual worth.[35]

Despite consistent confirmation of the agrarians' plight, it took continual
pressure to prod the North Dakota state legislature into action. In 1909, a
legislative commission resurrected the old Populist plan and recommended
that the state own and operate its own terminal grain elevator in Minnesota
or Wisconsin. A state-owned terminal elevator would empower farmers by
ensuring they could help determine the value of North Dakota's most important export. Yet it would take a constitutional amendment to create it.

Later that year, and again (as the cumbersome state statutes dictated) in
1911, the legislature approved such an amendment. In 1912, 74 percent of all
voters ratified it. A revised version—one that called for the terminal elevator to be located in North Dakota—was approved by legislators in 1911 and
1913. In 1914, 73 percent of state residents voted for the new amendment. The
complicated constitutional process and the dramatic challenge the terminal
elevator posed to Twin Cities business interests (who lobbied hard against
the change) ensured slow movement. Finally, late in 1914, a legislative committee expected to produce a concrete plan of action to move forward with a
state-owned elevator instead prepared an unfavorable report.[36]

Even though it identified the problem and presented a much-agreed on
solution, the state's political system offered little relief. More farmers lost
their farms. Their spirits flagged. To the average voter, government seemed to
stand in the way rather than respond to the will and needs of the people. The
limits of organizing outside electoral politics grew more visible. The elevator
fight suggested that dramatic change could only come by yoking together
economic and political action. Frustration and failure, however, fed agrarians' sense of dependence and sapped their strength. Little did farmers know
that the huge amount of energy they had invested in change would soon elec-
trify not only the state but also the entire nation. The storm was building
toward a breaking point.

★

The silver-tongued Albert E. Bowen Jr. moved to tiny Bowman, North
Dakota, with his wife and two children in February 1914. The handsome

FIGURE 2. Albert E. Bowen, creator of the Nonpartisan League, circa 1915. (Minnesota Historical Society, Por 18119 r1)

twenty-eight-year-old grew up on a farm in Dickey County, North Dakota. As a young man he took classes at the University of North Dakota and taught in a one-room schoolhouse outside Grafton, North Dakota. In 1912, he ran for governor on the Socialist Party ticket. Bowman, in the southwest corner of the state, exemplified the wheat frontier. Recently settled farmers hoping to cash in on the boom faced real hardships. Bowen, however, did not move there to grow grain. He arrived in Bowman with a reputation as a fine organizer and orator, intending to grow the local Socialist presence.[37]

Socialism in the United States in the early twentieth century—a sizable and vigorous movement—focused on the ills of industrialized America. Drawing as much from indigenous traditions of independence, the dignity of manly labor, and democracy as from theoretical Marxism, the Socialist Party emerged in 1901. It grew slowly over the next decade. Representing diverse perspectives—from the social gospel to the cooperative commonwealth to Marxist revolution—the party's leaders generally looked to urban laborers for support in their struggle against corporate capitalism. Electoral success typically came in local elections for posts such as justice of the peace, sheriff, and alderman. In 1912, however, Socialist presidential candidate Eugene Debs captured 6 percent of the popular vote.[38]

Across the northern plains, small pockets of Socialists ensured that the party remained truly national. In North Dakota, dedicated comrades consistently agitated for change. Socialism attracted few adherents but remained quite visible. Nonetheless, Socialists in agricultural North Dakota struggled with the national party's focus on urban wage workers and its reputation for radicalism. Middling agrarians neither earned wages nor lived without property. As one contemporary put it, these "farmer-proprietors are at once capitalists and laborers." Anxious to prosper in the wheat economy, they proved suspicious of revolutionary talk. Furthermore, it took ideological and rhetorical gymnastics to connect agrarians to the mainstream of Socialist thought. Calls for collectivization of farms, for instance, fell on deaf ears. Socialists also wanted to organize seasonal agricultural laborers. In contrast, most farmers envisioned the migrants as a necessary evil. As a result, socialism got little traction in rural towns and farms.[39]

Arthur LeSueur, a lawyer in Minot, served as president of that city's commission (effectively, the mayor) from 1909 to 1912 and became North Dakota's most prominent Socialist. Under his guidance, the 1912 Socialist Party state platform represented a shift in strategy. Muting the unpopular notion of collective farming, it instead called for state-owned elevators, banks, and packinghouses and state-sponsored insurance programs. Through these commitments, the party hoped to claim a space in the electoral mainstream and prove the worthiness of its cause to agrarians. At the same time, the party continued to publicly critique the Equity Cooperative Exchange and other farmer organizations as stopgap remedies that would ultimately fail.[40]

As he spent more time with agrarians like those he grew up with, Albert Bowen began to challenge this perspective. One colleague remembered that "he wasn't like the city socialists. He saw more from the farmer's standpoint."[41] At the January 1914 state Socialist convention, for instance, Bowen argued for "the creation of a state department of agriculture" to run industries that processed farm products. And even as he spent the spring of 1914 visiting farms across Bowman County and nearby Golden Valley County, Bowen published an article in the state's Socialist newspaper that broke with others in the party. "The farmer is not a capitalist," he argued. Because they did not control the markets for their product, agrarians were "absolutely dependent upon others for the very means of life." This meant that "the exploitation of the farmer is now complete." In identifying the growing dependence of lower-middle-class agrarians, he took their feelings of powerlessness more seriously than most Socialists.[42]

FIGURE 3. Arthur C. Townley, leader of the Nonpartisan League, circa 1915. (Minnesota Historical Society, Por 25668 p3)

Meanwhile, Bowen started spending time with a new colleague, Arthur C. Townley. Born in western Minnesota in 1880, Townley grew up on a farm. After graduating from high school in Alexandria, Minnesota, he taught school for two years. In 1904, he moved to Beach, North Dakota, and began wheat farming with his brother. Two years later, he went in on a scheme to grow wheat on thousands of acres outside Cheyenne Wells, Colorado. The plan—farming on a large scale—failed. He gave up on wheat, returned to Beach, and became a flax farmer. By 1912, he had eight thousand acres in cultivation. That year he overextended himself, going into debt to buy machinery and land. Flax prices plummeted. He hired out his thresher in an effort to stem the losses, but by early 1913, Townley lost his farm and owed creditors thousands of dollars.[43]

Soon thereafter, Townley drifted into the Socialist Party. In March 1913 he sponsored a series of public debates in Beach, challenging a local attorney to suggest a better alternative than socialism. His reputation as a convincing public speaker grew. Yet he needed money to live. The party hired him in

November 1913 to organize in McKenzie, Mountrail, and Divide Counties. As a farmer, he "confined his labor almost exclusively to the country districts" and took advantage of the less labor-intensive winter months to convince fellow farmers to join the Socialist Party. One agrarian who encountered Townley claimed that he gave "by far the most effective talk to the farmers, for he is one of them, having gone thru the mill." Indeed, the farmer noted, "his arguments are all based upon personal knowledge that no one can contradict."[44]

In March 1914, Bowen and Townley started working together. Both found themselves on the Socialist Party ticket to represent the area in the statehouse. Together, they filled the Beach Opera House and sold forty dollars' worth of Socialist literature. Then Townley headed to Bowman County for three weeks to help Bowen organize there, organizing a debate on taxes at the end of the month.[45] April and May saw the two ranging across western North Dakota, working alone or as a pair.[46]

Sometime in June, the entrepreneurial Townley grew tired of walking, biking, and hitchhiking through remote rural districts. Realizing that an automobile could help him organize more efficiently, he approached the Socialist Party state office with a plan. If he could sign up three hundred farmers to pledge one dollar a month to a new subdivision of the party called the "Organization Department," a single district could support a salary and a Ford automobile for the organizer. This way, even "the non-socialists cannot escape hearing the message of Socialism." It also ensured that Townley and other organizers could support themselves with a regular wage.[47]

The state office, noting his successful efforts in Golden Valley, Billings, and Bowman Counties, backed the plan. Farmers proved willing to join the new branch because it promised them action in face of political stalemate without the taint of full-fledged socialism. It was an immediate success. When Townley left to spread the organizing into Williams and McKenzie Counties in July, Bowen joined him.[48] Another Socialist organizer, Leon Durocher, excitedly joined the work. By the end of December 1914, six ardent Socialists in five automobiles ranged across most of western North Dakota, raising over $11,000 and selling six hundred dollars' worth of Socialist pamphlets, books, and subscriptions. Townley's idea proved more successful than anyone could have imagined.[49]

Even as the Organization Department grew, Bowen recognized that the successful new effort depended on downplaying the standard Socialist Party line and instead providing farmers with the feeling that they could do something. For some time, Bowen had envisioned a new organization without the stigma of socialism. He sought a group that would bring North Dakota's middle-class agrarians together with an ideology-free platform—state-owned elevators, mills, and packing plants—for action. He started openly discussing the idea with Durocher in September 1914.[50]

Just days after moving his family from Bowman back to eastern North Dakota (in the wake of the November 1914 elections, in which the Socialists scored no major victories), Bowen publicly articulated his new idea in the state party's newspaper, the *Iconoclast*:

> The farmers are being robbed. Farmers raise wheat. Wheat must pass thru elevators and mills, it must pass over railroads. . . . These are not owned by the farmers. . . . We are going into politics as farmers and re-write the rules governing the game we are compelled to play. . . . Only by political action can we establish peoples [*sic*] banks, terminal elevators, co-operative purchasing associations and selling agencies, state hail insurance, etc., and by political action we will establish these things in the next two years. On to Bismarck.

In his call for action, Bowen drew from the success of Townley's "Organizing Department" and his own frustrations with fellow Socialists.[51]

On the same day Bowen's essay appeared, the Equity Cooperative Exchange's newspaper, the *Co-operator's Herald*, echoed Bowen's call for action. In an editorial, the journal noted that "now is the time to begin the agitation for the short ballot and nonpartisan nomination and election of state district and county officials." In fact, the paper had expressed an interest in nonpartisan politics almost a year earlier, telling readers that "farmers have had sand thrown into their eyes by partisan politics long enough and it is time for them to exercise freedom of thought and action and stand together as a class." In October 1914, the same newspaper opined: "The only way you can advance the interest of your class is to work, vote, and if need be, fight for your class." After all, "so many farmers and laboring men in all parties vote party instead of men and principles. . . . If you have a man in office who has shown his interest and allegiance to the interests of farming community tie to him and stand back of him so long as he remains true blue." Ongoing agrarian frustration suggested organizing around shared experience as the only remaining solution. The future of the middle-class aspirations held by most North Dakota farmers depended on it.[52]

By late 1914, Equity Cooperative Exchange leaders as well as a small circle of Socialist organizers realized that the established parties offered little succor for agrarians. Responding to the powerful updraft of farmer discontent, they started to flirt with more novel options. Most important, they argued for a class-based politics derived from the middling culture of rural farmers. Nonetheless, it took Albert Bowen—with the help of his colleague Arthur

Townley and existing farmer organizations—to make the thunderhead visible on the horizon.

★

Days before Albert Bowen moved away from Bowman and his failed bid for the state legislature, the executive committee of North Dakota's Socialist Party decided to pull back from its commitment to the Organization Department. Many feared that "it was superficial, and it omitted the essential element of the Marxian movement—class consciousness." Opponents to the initiative rightly worried that few of the newly enrolled farmers experienced profound transformations in their politics. Yet the work proved successful. So Townley, Durocher, and others kept at it. Because of their disloyalty, on December 15, the party formally released Townley from the work.[53]

The next day, Bowen wrote to the state Socialist Party secretary, announcing his intention to meet with M. P. Johnson, president of the North Dakota branch of the American Society of Equity and a leading Equity Cooperative Exchange member. Exploring his options outside the Socialist Party, Bowen sought out new partners. Johnson represented thousands of already organized agrarians. Both men recognized that after years of struggle, only direct political action remained.[54]

Then, on January 5, 1915, the state legislative committee charged with examining the feasibility of the overwhelmingly approved state-owned terminal elevator made its report public. Arguing that "any investment made in terminal elevators . . . would be a waste of the people's money as well as a humiliating disappointment to the people of the state," the committee came out "strongly against the expenditure by the state of any money for the erection of new terminal elevators." Angry farmers noted that the committee sought advice from railroad and milling interests in Minnesota. For years, lower-middle-class farmers had pursued change through established political and legal processes. Now, at the moment of victory, the state legislature refused to confirm the clearly expressed will of the people.[55]

Anxious to capitalize on this turn of events, Albert Bowen arrived in Minot in late January to attend the Socialist Party's state committee meeting. He, Townley, Durocher, and a few others tried to persuade their fellow Socialists that the Organization Department deserved a second chance. They recognized that the rejection of the terminal elevator would spark a political storm. But the state committee did not budge. The state's Socialist leaders remained tone-deaf.[56]

Frustrated, Bowen used his new relationship with equity leaders to secure an invitation for himself and Durocher to speak to a district convention of

equally exasperated farmers meeting in Minot on January 29. While Durocher focused his speech on the limits of economic action without political action, Bowen easily persuaded them of the necessity of a nonpartisan farmer's
political organization to put agrarians in the statehouse. Both pointed to the
rising number of mortgage foreclosures in nearby towns as just one example
of the need for agrarians to enter politics as a bloc. Impressed, the farmers
in Minot invited him to speak at the Society of Equity's state convention in
Bismarck, just four days later.[57]

At the state convention in Bismarck on February 2, 1915, more than five
hundred agrarians decided to march on the state capital. They sat in the gal-
leries as the society's leaders reprimanded members of the house of representatives. After a bitter debate with legislators opposing the state-owned
terminal elevator, the Society of Equity retired to a nearby hotel for heated
speeches.[58]

Bowen arrived in Bismarck that day primed to seize the moment. He
even drew up membership forms for the new organization. When Bowen addressed the now thoroughly riled crowd of farmers, he "discussed the program of the Non-Partisan Political organization which will be perfected here
during the life of the convention for the purpose of demanding a system of
state-owned power-plants, stock yards, flour mills, hail insurance, single tax,
and rural credits at cost." The "splendid address made a remarkable impres-
sion on his audience." In its wake, he hurriedly signed up convention members anxious to join the cause.[59]

Witnessing a ready constituency for his idea, Bowen resigned from the Socialist Party the next day. Admonishing his longtime colleagues, he declared:

> I believe the time has come for those men in all parties who desire real prog-
> ress (and no party has a monopoly on these men) to stand together in favor
> of a program that shall be so specific it will appeal at once to the fair minded
> in all parties. And realizing the difficulty connected with useless antagonism
> along the paths of partisan ship, I propose to teach FIRST a program and then
> follow this with investigation.[60]

Launching into the unknown, the ambitious Bowen recognized that he could
not do the work by himself. So he turned to his colleagues in the rejected
Organization Department for help.

As early as December 1914, Arthur Townley may have considered joining
Bowen in the creation of a new organization. Townley contemplated the option, but only if he could be the absolute boss. In the wake of the passionate
meeting in Bismarck, Townley saw the potential of both Bowen's idea and the
pent-up agrarian anger it confirmed.

Bowen immediately started on a speaking campaign across the eastern part of the state. As a "lecturer for the Non-partisan League," he signed up members after rousing lectures. One small-town newspaper editor, after attending Bowen's speech, wrote that "the common people will give their first demonstration and fight their first battle as men and NOT as partisans." Appeals to self-determined manhood fell on fertile ground. Indebtedness, the fear of foreclosure, and the loss of economic independence struck at the very core of male agrarian self-conception. Still, Bowen knew he needed a better way to animate the already organized farmers.[61]

So Bowen turned to Arthur Townley. Townley's proven worth as a mobilizer stood out. At some point in late February or early March, he and Bowen cut a deal. Townley "was to be given full credit for originating the League, in return for which Bowen was to be the League's principal speaker and when the Leader started publication he was to receive $100 per week from its earnings." They also began distancing themselves from their previous Socialist affiliations.[62]

While Bowen continued to lecture across the state, Townley immediately turned to Francis B. Wood, a former member of the board of directors for both the Equity Cooperative Exchange and the North Dakota Society of Equity, for help. The two likely knew each other, since Wood had also been an active member of the state's Socialist Party. Seizing on the long-simmering terminal elevator issue, the two met with a handful of sympathetic agrarians on Wood's farm outside Deering. The assembled decided on a meeting in the hall used for equity gatherings in town on March 16, 1915. There, Townley laid out the possibilities for a nonpartisan farmers' organization to an excited audience.[63]

A formal platform soon emerged. The platform combined long-standing demands of both the Socialist Party and the equity movement, first articulated by Bowen during his speech in Bismarck. It called for

state ownership of terminal elevators, flour mills, packinghouses, and cold-storage plants;
state inspection of grain and grain dockage;
exemption of farm improvements from taxation;
state hail insurance on an acreage-tax basis;
rural credit banks operated at cost.

Each offered a solution for voters' long-standing grievances. A committee for "the Non-Partisan Political Organization League," made up of local farmers John Hagan, William A. Hayeman, and Francis Wood's son, Howard, soon thereafter advanced Townley money to purchase automobiles for organizing.[64]

By May, Townley and Bowen—busily recruiting farmers—created an executive committee for what they now called the "Farmer's Non-partisan Political League." Townley served as president, Francis B. Wood became vice president, O. S. Evans (a prominent Socialist from Bowman, North Dakota, who met Townley and Bowen in 1914) became the secretary, and T. C. Nelson and A. B. Bowman joined as board members.

In early summer 1915, prominent Fargo attorney William Lemke—whose previous work for the Equity Cooperative Exchange gave him real credibility among farmers—also joined the board. Together, the leadership quickly decided that only farmers could be members. Drawing directly from his Organization Department experience, Townley hired proven former Socialist Party organizers, allowed farmers to pay for their membership with postdated checks, and promised a newspaper subscription to members.[65]

Reports further noted that by late May the organization already claimed almost six thousand adherents. While some critics dubbed the new members "six dollar suckers" (by mid-summer 1915 a League membership cost six dollars), others recognized that a wide swath of lower-middle-class agrarians saw the League as the logical evolution of years of hard work. Indeed, Townley later admitted that the "very vigorous Equity movement" explained much of the League's initial success.[66]

Growing out from its base among Society of Equity men, membership numbers exploded throughout the rest of 1915. League organizers and lecturers fed on middling farmers' feelings of dependence. The politicians who killed the voter-approved terminal grain elevator bill proved their point. The promised insurgency gave farmers newly found confidence. Finally, the large harvest that fall permitted agrarians to pay their postdated membership checks.

At least one rural newspaper noted that "for the first time in ten years the farmers are taking a real interest in politics and doing their own thinking." George White, a farmer from Whitby, North Dakota, put it more bluntly: "This organization is an automatic double-barreled breech-loading fool killer." The storm clouds gathered strength. Soon, the whole state would hear the farmers' thunder.[67]

*

Familiar with nonparty political organizations, lower-middle-class agrarians joined the Nonpartisan League in droves. Because voluntary associations were a typical locus for civic improvement, this proved normal. Numerous "leagues" popped up across North Dakota in the early 1900s, all reflecting efforts to transform the major parties. In 1904, Norwegians in the Republican

Party organized a Scandinavian Republican League. As an ethnic bloc within the existing party, it focused on replacing boss-ridden party conventions with an open primary. The next year, a group of native-born Republicans came together to fight cronyism under the banner of the Republican Good Government League. In the 1914 elections, two leagues emerged. The Progressive Republican League drew together those who backed Theodore Roosevelt in 1912. The second, the North Dakota Personal Liberty League, grew out of a faction wanting to repeal statewide prohibition and keep women from voting. Neither reached out to the broader citizenry, and neither lasted longer than two election cycles.[68]

Similarly, the notion of nonpartisan politics grew out of Progressive Era reform initiatives. Large municipalities all over the country turned to nonpartisan elections in local races, thinking that unaffiliated bureaucrats better fit the needs of citizens in the modern, industrialized city. Other efforts to craft a nonpartisan politics came from the impulse to weaken political machines. Minnesota's state legislature, for instance, became an officially nonpartisan body in 1913. In a few cases, voting blocs broke with parties and supported specific candidates regardless of their affiliation. In 1912, following an initiative by the Conference of Progressive State Granges to identify and support profarmer politicians in Washington, DC, activists founded the Farmers' Non Partisan League. As a coalition of different organizations for farm lobbying, it focused on influencing Congress. In 1918, to avoid confusion with the movement emerging in the North American West, lobbyists dropped the name. Finally, agrarians in Idaho tried to vote along nonpartisan lines in that state in 1916, though without intense organization.[69]

Nonetheless, the Nonpartisan League represented something new. While many members brought their experiences from other farm organizations such as the Equity Cooperative Exchange, the Farmers Union, or the Grange to bear, the burgeoning insurgency depended on the specific anger over politicians' ongoing refusal to respond to the needs of a large segment of North Dakota's voters. Energized by frustration and empowered by a small group of former Socialist organizers, the farmers who signed up for the NPL embodied a politics that did not fit conventional continuums or categories. They proved neither entirely radical nor merely reform oriented. The combination of their platform and innovative tactics represented a new option for electoral politics in America.[70]

Until the League emerged, middling agrarians in North Dakota envisioned the economy and politics as separate, if overlapping, realms. Early twentieth-century farmer associations, for instance, worked on the former but expressly avoided the latter. They concentrated on cooperative selling, shipping, and marketing, leaving politics to politicians. Though agrarian middle-class culture in North Dakota encouraged service to others in local, interdependent communities, political engagement—beyond voting on election day—did not define agrarian citizenship. But Bowen's idea and Townley's tactics empowered farmers to imagine themselves as more expansive agents of political change. Instead of being acted upon by politicians, they would themselves act to simultaneously transform both the wheat economy and politics itself.

The insurgency in North Dakota in 1915 stemmed from agrarians' growing feeling of economic—and thus social—inferiority. Urban America seemed to be leaving them behind, even as it thrived on their products. As one commentator put it, farmers in North Dakota were "cautious by nature and look with suspicion on the city man. . . . Above all, farmers will rise in revolt against the supposition that they are in any sense benighted and inferior." Prosperity denied through unfair market practices as well as a search for self-respect animated their protest.[71]

Thus, joining the NPL did not represent or require an ideological transformation. In fact, lower-middle-class farmers remained thoroughly committed to capitalism and deeply respected private property. Equity lawyer James Manahan described their position in a speech in 1918: "The market-place—why, the market-place should be a holy place and the light should shine through it every hour. It should not be a place of avarice and greed. The general welfare of humanity is dependent upon the purity, honesty, and fairness of the marketplaces of the world quite as much as upon the integrity of its courts." He finished by reminding the crowd of Leaguers that "degradation and decay must always result from monopoly." With this in mind, agrarians looked to the NPL for what one farmer called "protective legislation" to safeguard their property and their tenuous hold on a lower-middle-class status.[72]

Furthermore, however dire their situation, mortgage-ridden agrarians did not look to overthrow the government. Through the early 1910s, they gradually came to realize that their ongoing movements—such as the Equity Cooperative Exchange and Society of Equity, the Farmers Union, and the Grange—were not powerful enough to change the economy. So they settled on electoral politics as the appropriate vehicle for creating the changes they sought.

The farmers created a political organization to improve their chances in the marketplace. Hard experience taught them that direct participation as a political bloc was their only remaining option. Individual initiative—so highly prized in their worldview—would not be enough. The unfair advantages of others in the wheat markets could only be balanced by cooperative political action. Agrarian property and pursuits would be better protected once the state embodied their specific interests alongside those of corporations.

What one Leaguer called "government competition" would result in the "restoration of competition" in the marketplace. Better able to pursue their own prosperity, farmers then could fulfill the middle-class ideals fostered by their communities.[73]

Even so, envisioning the Nonpartisan League as merely a reform organization also proves problematic. Reform-oriented progressive politicians in North Dakota had failed farmers as badly as politicians involved in the Alexander McKenzie machine. The former did not envision wheat markets as the primary concern of the state's mostly agrarian citizenry. Instead, they followed national trends, looking outside the state for good government and democracy initiatives that failed to address the entirety of local concerns. Furthermore, the NPL sought to transform the wheat market through the establishment of state-owned industries. Few reformers proved willing to use the state to directly challenge corporate power. Meanwhile, the latter engaged in corrupt practices that alienated voters. Their cronyism embodied the way in which corporations seemed to exert an unlimited—and decidedly unfair—influence on politics. Neither camp provided farmers with a more even playing field.[74]

Because it rejected existing political parties and failed to fit into easy categorizations such as radical or reformist, it is tempting to characterize the Nonpartisan League as a straightforward reiteration of late nineteenth-century Populism. Some historians believe that the People's Party, which emerged in the early 1890s from the Farmers' Alliance movement, was the last best chance for farmers to influence the trajectory of the nation's development. This assumption—that Populism was the last chance farmers had to shape the political future and improve their own economic position—makes all the farmer-centered political movements that followed it derivative and doomed to fail. The Nonpartisan League is the most prominent victim of this way of thinking. In fact, this explains the general neglect of the NPL, even among scholars of rural America.[75]

To be sure, the general antimonopolistic slant of Populism lived on in the Nonpartisan League. A vision of economic cooperation nurtured first by the Grange and then by the Farmers' Alliance and the People's Party, lived on in the Society of Equity and the Equity Cooperative Exchange, which deeply influenced future NPL members. Many farmers who joined the Nonpartisan League—like their forebears in the Farmers' Alliance and People's Party— also looked forward rather than backward. They wanted to fully participate in modern American society.[76]

But there the resemblance ended. Indeed, an NPL organizer told farmers that they should not "think the Farmer's Non-Partisan Political League is a revival of the Populist Party. It isn't. It's the modern product of a modern, economic and industrial and governmental need."[77] In crucial ways, the Nonpartisan League veered from the trajectory mapped out by Populism. First, it did not plan to operate as a third party. Instead, it worked as an organization willing to endorse any candidate in any party that supported its program. Second, it focused on local and state politics rather than the national stage. Third, it did not advocate changes in monetary policy. Fourth, the language of occupational specificity outweighed broad-based appeals to "the people." Farmers needed to come together, as an occupational class, to effect needed change in markets and politics. Finally, the Nonpartisan League's program articulated the need for farmers to embrace techniques invented by corporations not only in agrarian cooperative ventures but also in politics. In so doing, it offered a more complex vision of state intervention in the economy than the Populists of the 1890s.

Instead of using government to craft the "cooperative commonwealth," the NPL would simply utilize the state to foster another competitor—albeit one directly controlled by the democratic process—in a competitive marketplace. Farmers who joined the League believed that state intervention in the wheat economy would increase, not decrease, market competition. The organization promised to reorient government to ensure that the self-interest of farmers found a voice alongside the self-interest of corporations. Leaguers carefully articulated their belief that in no way would such a tack handicap "individual initiative and enterprise." Agrarians involved in the NPL envisioned the organization's plans as their version of the collaborative work that grain-trading companies and millers did to further their collective interest. In fact, the adoption of an organizational vision derived from business models represented a deepened commitment to a competitive market, not a wholesale critique of it. By democratizing property as well as politics, the NPL's vision offered middling agrarians a fighting chance in corporate economies of scale that threatened farmers' continued existence.[78]

The proprietary capitalism—a stark alternative to corporate capitalism—that animated North Dakota's middling farmers depended on direct political intervention for survival. As one contemporary put it, "The men of North Dakota are not a race of socialists. They are sturdy individualists, with a stake in the soil and stern notions of property rights." These farmers called on the agrarian past, even as they imagined themselves as small businessmen. As "capitalists against capitalism," League members simultaneously laid claim to an old tradition in American economic life and proffered a new conception of popular, antimonopoly politics. They reminded Americans in an era of corporate growth that there was more than one way to envision capitalism.[79]

Throughout the nation's history, the middle class had challenged the un-checked accumulation of capital even as it remained committed to capital-ism. The small producers who backed the People's Party in the 1890s were only the most famous example. In the NPL's twentieth-century understand-ing of agricultural economies, "the money power" that animated Populists in the 1890s was replaced by "big business." This new menace threatened not the late nineteenth-century "cooperative commonwealth" but general "economic democracy." Leaguers called for equitable accumulation in order to preserve small-property ownership. Historians note that this alternative vision of cap-italism served as a wellspring for popular democracy well into the twentieth century and may even explain the ultimate failure of socialism in America.[80]

To be sure, the NPL's call for state-run mills, elevators, banks, and in-surance made it easy for both contemporary critics and future historians to envision the Nonpartisan League as socialist. Yet the League's hope to cre-ate state-owned competition to protect private enterprise did not represent the adoption of socialist ideas. The NPL drew on but did not emerge from the broad tradition of American socialism. Only a handful of members envi-sioned themselves as part of that ideological trajectory. Even then, flexibility marked their commitment. In contrast, most agrarians hoped to embed a co-operative model—democratically run and mutually beneficial business col-laborations aimed at promoting individual success—in state-level politics. They did not want North Dakota's government to control the entirety of the agricultural economy. Leaguers simply wished for institutions accountable to voters to participate and compete in it. In their minds, only state govern-ment could support the scale of enterprise necessary for farmers to compete fairly with corporations. They believed a public option in the wheat economy would buttress the marketplace by creating more equity in it.[81]

Furthermore, North Dakota's middling farmers fiercely resisted the so-cialist label. They cared little for the urban working class. Agrarians remained hostile to calls for collectivization, the organization of migrant agricultural workers, and the effort to foster social revolt. Confirming the gulf between the NPL and socialism, the national Socialist Party quickly turned on the NPL. It deemed the League insufficiently revolutionary.[82]

Though the NPL consistently rejected what one of its leaders called "im-practicable radicalism," the discourse used to describe its outlook did display a socialist influence. Experiences with the Socialist Party provided NPL or-ganizers with the discursive tools needed to politicize the cooperative vision crafted by Equity Cooperative Exchange organizers. Previous iterations of agrarian populism drew on broad and ambiguous definitions of "the people." They pointed to the agrarian past to claim republican values. In contrast, the stark language of class-based specificity emanated from the League. As one organizer put it, "The farmer has at last awakened . . . to the fact . . . that he is no longer a unit, that he must become a part of his class, and help carry on the struggle for the rights of that class." Indeed, organizers in the early months of the League often insisted that "the farmers have a common object and a common interest in politics in this state. They can never get what they want from the state government unless they organize as farmers to get it." They emphasized the discrete grievances of farmers and created an organization defined by occupation. This separated agrarians from other middling producers. Accentuating differences, they excluded townspeople from membership. This further fractured the middle-class culture of rural North Dakota, even as it firmed up a distinct agrarian consciousness.[83]

Many of the League's leaders and initial employees moved directly from the Socialist Party into the new farmer organization. Bowen and Townley refused to be chained by socialist ideology. One report noted that the two possessed "no political philosophy; that they are solely political pragmatists, experimenting in an effort to free the people from what they believe to be monopolistic controls." But other former Socialists did not give up their dream of a wholesale transformation of American society. They put hope before understanding, mistaking the League's discourse, self-conception, and mobilizing as a first step toward a socialist state rather than an end in itself. These former Socialists failed to understand most farmers' rejection of ideology beyond their own middle-class vision of independence, prosperity, cooperation, and self-sufficiency.[84]

The NPL, even in its aspirations, rejected monopolistic accumulation yet thoroughly committed agrarians to a market-based economy. The farmers who joined the Nonpartisan League in 1915 and 1916 joined the dramatic insurgency to strengthen their economic and political position—not to foster a thoroughgoing remedy for broader social injustice. As the *Nonpartisan Leader* put it, the average agrarian became a member "not because he loves his fellow farmer" but "for self protection." Though socialist discourse proved useful for their task, agrarians remained dedicated to self-preservation. On the back of the 1915 League program, for instance, prospective members found a pledge that included "North Dakota farmers *lose millions of dollars each year.* We can *save* as much of this as our Organization compels the *middlemen* to *give up— The stronger we organize the more we can keep.*"[85]

When a committee of farmers in Nelson County, North Dakota, published its own "conception of the League" in June 1916, they laid out both their commitment to capitalism and their rage over the unfair concentration of capital in Minneapolis-based milling and railroad companies. Farmers, "the dominon [sic] power in the state," took "all the risks and do all the hard work, and in return are not allowed even a manager's salary." Meanwhile, "the Industries we ask the State to engage in, are controlled by Corporations who control the machinery of Government; and farmers will never get the State Legislature to consider their just demands until farmers themselves control the Legislature." They demanded "Economic Justice; and to get Justice they are going to take over the Government and help to write the laws under which they have to live and do business."[86]

For these men, the League, "an active, efficient farmers' organization fighting the corrupt Political Machine that has dominated the Government of this State in the interests of Trust Magnates," represented farmers finally realizing "the power they possess thru solidarity in voting—thru collective action at the polls." Nominating "brother farmers" to political office, they believed that "the League . . . means better conditions for every person having legitimate business in the State.—Farmers, laborers, merchants, tradesmen will alike benefit." In fact, "the only persons to lose anything by our program are the beneficiaries of the most unjust industrial system the world has ever known—the organized looters who take sixty cents out of every dollar paid by the consumer for farm products." All this would ensure "that the will of the people is the law of the land." For these men, democracy meant equity in both politics and the economy. Each depended on the other.[87]

Significantly, ethnicity and religion mattered little in the formulation of this outlook. Often ascribed to a Scandinavian predisposition for socialism, North Dakota farmers' initial interest in state-owned industries as proposed by the Nonpartisan League came from a commitment to the prosperity promised by middle-class ideals. It grew out of local circumstances—not imported ideas.[88]

Thousands of Norwegian migrants arrived in North Dakota in the late nineteenth and early twentieth centuries. By 1910, almost 20 percent of the state's residents came from Norway, had two Norwegian parents, or spoke Norwegian as their first language. Voting records in both Norway and North Dakota, however, suggest no innate socialism. They instead intimate that Norwegian Americans voted like other North Dakotans, responding to economic needs above all else. Even a photograph of some of the first League members, published in 1915, featured only three Norwegian Americans. In fact, the leading Norwegian-language newspaper in the state became one of the League's earliest opponents.[89]

Furthermore, despite the state's reputation as a Scandinavian outpost, North Dakota proved more diverse. Era Bell Thompson—an African American—remembered a Norwegian neighbor encouraging her father to join the League in early 1916. Milling interests, said the friend, "take our money, dey cheat us wid our grain. Ve do dere verk; dey sit behind big desk an' smoke big cigar wid our money." When Thompson's father wondered aloud about what "a bunch of green farmers" could do, the friend responded: "Dey can yust be 'onest! . . . Dat's vot dey can do!" Her father joined the League soon thereafter.[90]

Wide-ranging communities across the state responded to the League's call for change. A small pocket of Jewish farmers in Burleigh County joined the NPL in 1915. Harry Lashkowitz, a young Jewish lawyer in Fargo, envisioned the League as a powerful antidote for the corrupt political machine in Bismarck. He later handled court cases for the organization. The landowning men in a small colony of Lebanese Muslim farmers outside of Ross, North Dakota—part of a community that, in 1930, built one of the first mosques in America—faced episodic discrimination. But they followed their Norwegian-speaking neighbors into the organization in 1915. The more sizable population of German Russians in south central North Dakota tended to avoid politics altogether. But they too got behind the NPL. Though they voted against one protemperance Nonpartisan League candidate in 1916, they expressed their general preference for the NPL program in large numbers in 1916 and 1918. In each case, the economic concerns of middling farmers trumped ethnic affiliation. And in trumping ethnic affiliation, the League also transcended religious differences.[91]

Creating occupational cohesion across ethnic and religious divides showed the viability and novelty of the League's tactics, which proved—alongside the renewal of an alternative capitalist vision—to be the League's most significant contribution to American political life. The membership-based approach fostered an emergent consciousness among North Dakota's lower-middle-class agrarians.

One admiring farmer later noted that Bowen and Townley "recognized the futility of trying to remedy conditions that oppressed the farmers through a political party. No matter which of the established parties might win at the polls, the real control remained the same. The influences that combined back of the parties were the real government, and the secret of power lay in that party control." Furthermore, by creating a membership-based organization, the League fought "money with money. . . . Payments, like regular union dues, would not only help to hold the organization together but would create greater loyalty on the part of the members."[92]

In its ability to focus on one expression of the public's will, the NPL proved unmatched. By participating in an organization that remained outside political parties but worked within them, farmers gave themselves tremendous influence. The League deliberately stood "apart from every political party, every political machine and free from every political boss" and would "put men in office that will legislate in the interests of the members." It created an institutional space in which disempowered agrarians engaged directly in electoral politics. Members believed that "politics can only be purified by democracy." The approach merged populist sensibilities and antimonopolistic sentiment with an electoral method that simultaneously renewed citizen agency and offered a path to power.[93]

North Dakota's adoption of a primary system for nomination in political parties in 1907 created the opening for such an organization. League leaders took full advantage of it. With a program-based politics instead of a party-based politics, farmers offered their support in the open primary to any candidate in any party who reciprocated. League members knew that "it is almost an impossibility to get a third party success." Instead, voters needed to "pick out a nonpartisan ticket in the present parties."[94]

Besides providing flexibility, nonpartisanship made the League more attractive to potential members. By working within the system—and within the existing parties—the NPL offered lower-middle-class farmers an instrument for rapid political change without surrendering the social respectability so crucial in their communities. Farmers could still be good Republicans or Democrats and get the change they wanted. They did not need to engage in what many of their rural neighbors might deem fringe or even radical politics in order to advance their cause.

That League leaders derived their program from political goals generated by farmers themselves gave the organization integrity. It directly reflected what agrarians wanted—what they had already expressed in their votes for a state-run terminal elevator, in their farmer-run equity movement, and in the rapid growth of the Socialist Party's Organization Department in 1914. One booster rightly suggested that the League's platform was the "accrued product of several years of discussion." In its first months, the NPL responded directly to the desires of the people it brought together and thus found widespread support.[95]

Finally, the Nonpartisan League not only promised to provide agrarians with access to power but also ensured that they would transform the understanding of how political power worked in the state. Indeed, the success of the League depended entirely on the average farmer's participation in the formation of a new politics. When Nelson County agrarians spoke of their "brother farmers" replacing longtime politicians, they referred not just to a redistribution of resources but a redistribution of political power. Imagining people just like themselves in the state capitol, they hoped to become more directly involved in governance. They argued that economic relations were political relations and vice versa. Clearly representing self-empowerment in both the economy and politics, the organization embodied the hopes and aspirations of its lower-middle-class agrarian membership.[96]

All told, the combination of a popularly derived program and innovative tactics shaped a powerful storm. The insurgent democracy fostered by the League generated civic agency. As one farmer put it, the NPL meant "emancipation for the common people." It looked to transform dependency into empowerment through the promise of success at the ballot box. Agrarians saw the League as an opportunity to articulate their vision of accumulation without concentration. It even offered them a way to skirt their own complicity in a system in which they both exploited and were exploited. They forged a path that rejected radicalism even as League opponents suggested that socialism had captured the imagination of North Dakota's farmers. In fact, the farmers bent talk of working-class revolution into a vehicle for lower-middle-class protest like grass in a storm-driven wind.[97]

✷

Through the summer of 1915, the updraft of discontent drew in farmers by the hundreds. By July, the League comprised ten thousand members. The inrush taxed the young organization. Through that first summer, a former bank clerk "carried the records and books of the entire Nonpartisan League organization in the back seat of his Ford car." Nonetheless, organizers continued their work in rural districts with aplomb.[98]

By September 1915, twenty-two thousand stood with the organization.[99] Townley moved the League's headquarters first to Valley City (in the east central part of the state) and then to North Dakota's largest city, Fargo. With the infusion of cash from new members, he founded the promised Nonpartisan League newspaper. For that work, he enlisted the help of O. M. Thomason, a fellow ex-Socialist and longtime contributor to that party's newssheet.

Townley and Bowen took Thomason to a farm south of Cleveland, North Dakota. Bowen "walked out in the field and introduced himself" to the farmer. After pulling "some papers out of his pocket . . . he pointed to the items he was explaining with the pencil. He dropped the pencil. The farmer picked up the pencil." Then Bowen handed the farmer the membership form. "It wasn't long until he had signed, and made out his check." Thomason, stunned by Bowen's skill, found himself charged with organizing that entire area the next day. When Thomason returned to the office in Valley City with fifteen paid-up memberships, Townley said: "Goddam—that's the stuff—that's the goddam stuff—one, two, three—fifteen—everyone cash. . . . You didn't believe it before. You thought we were stinging you. Now you have faith in it—now you can go to Fargo and start the paper."[100]

Thomason turned to H. D. Behrens for help. Townley also wrote to prominent Socialists such as Seattle's Joseph Gilbert and New York's Charles Edward Russell, asking them to join the effort. In Fargo, Thomason and Behrens published the first issue of the *Nonpartisan Leader* on September 22, 1915. It trumpeted that it belonged "to the farmers of the Northwest." Russell contributed a column encouraging farmers to "not believe anything you read . . . unless you read it in your own journal." Meanwhile the editor assured agrarians that the *Leader* would make entrenched interests and politicians "duck and dodge" in the interest of the "farming class." Only farmers could represent farmers.[101]

With the newspaper up and running, League leaders began preparing for the June 1916 primary election. The savviest, William Lemke, looked for potential NPL candidates in his vast network of contacts. They included friends in the state's Republican Party. For instance, the League arranged a private meeting with leading gubernatorial contender Usher Burdick. A terminal elevator backer, he sympathized with the farmers' plight. But Burdick underestimated the NPL. He believed the League needed him more than he needed the League. Among prominent Republicans, only Lemke knew the true strength of the NPL. As early as July 1915, for instance, Lemke plied Morton County district attorney William Langer with assurances he would be the League's candidate for attorney general. By November 1915, Lemke assured Langer that one especially prominent Republican would back the latter, though he "seemed to be doubtful as to the outcome. Of course he does not know the strength of the revolt."[102]

As bitter cold settled over the state in December 1915, the NPL took advantage of the postharvest quiet. Without advance publicity, they scheduled a series of rallies. They hoped to cement farmers' adherence to the League. The first gathering, in tiny Pettibone, North Dakota, featured Bowen, Townley, and the new editor of the *Nonpartisan Leader*, D. C. Coates. The speakers whipped the large crowd—many members of which had defied below-zero temperatures to drive from "as far as twenty five miles across the prairie to attend"—into a frenzy. Similar efforts in January and February 1916 proved equally successful. A farmer from Edmore, North Dakota, recalled that his local rally "was a continuous performance—short, hard jabs by the speaker and then wild cheering, clapping of hands and stamping of feet." In response to the charge that state-owned enterprises meant socialism, orators claimed that "big business" sought to "make you farmers divide up, to take your farms away from you, to even share your wives with you." Adding fuel to the fire, one speaker even noted that "the Minneapolis millers and bankers have already taken most of our farms, and our wives are so worn out working on farms that they don't want them."[103]

Again and again, the organization appealed to agrarian manhood. As early as September 1915, the *Nonpartisan Leader* crowed that bankers' organizations allowed financiers to "take better care of their wives and children," even as the wives and daughters of agrarians "spend long hours at hard unremunerative labor" and farm boys "stay out of school to plow." Their claim that the persistence of property ownership and patriarchal families depended on political organization resonated with audiences.[104]

The speaking tour garnered attention across the state. Though it remained suspicious of the League, the *Fargo Forum* dubbed the NPL "the greatest political power in North Dakota today." It noted the class orientation of the organization, calling it "an uprising of the farmers of North Dakota ALMOST TO A MAN." The League immediately looked to confirm the *Forum*'s confidence in its power. Organizing its members by precinct, NPL leaders called for local conventions. They anxiously sought to measure their power and cement the organization's standing with agrarians. Every member received careful instructions directly from Arthur Townley. On George Washington's birthday, February 22, they were to gather with others in their voting precinct to elect a delegate to the upcoming League convention in Fargo. Farmers would reassert themselves and redefine the state's politics on the holiday that celebrated the father of their country.[105]

Only members were allowed to attend. League staff were ineligible to serve as delegates or candidates. Newspaper reporters were barred. The results would remain secret. Most important, the delegates selected needed to be "MEN WHOM BANKERS, MIDDLEMEN AND POLITICIANS CANNOT INFLUENCE." Anyone who lobbied for office was ineligible. Only under such conditions would the "farmer have his turn." Finally, Townley reminded members that success depended on them, "not upon the men at headquarters."[106]

On the appointed day, over twenty-five thousand farmers met across the state. They followed Townley's instructions to the letter. Across North Dakota, precinct after precinct chose "a good strong farmer" to represent them in their legislative district meetings. Often, local nominees at first refused their neighbors. They made it clear that they were not politicians. Others pressed them. They insisted on exactly that point. As night fell, the League had introduced thousands to a broader vision of citizenship. One prominent member told an opponent, "The leaders have simply undertaken to gather the farmers under a common banner and . . . enable them to vote jointly and intelligently in their own interest." Farmers themselves made the democratic insurgency real.[107]

Delegates to the legislative district caucuses met in early March. They nominated men to serve in the statehouse and as formal delegates to the state NPL convention in Fargo at the end of the month. In most cases, the top vote getter among legislative candidates also became the delegate. Though they avoided the precinct meetings, League organizers often attended the district assemblies. NPL staff did not participate in votes but instructed delegates on procedure and answered questions. This secured smooth meetings with clear results.[108]

All this activity took place outside the major parties and the established open primary process. Confusion ensued. Some members wrote to the *Nonpartisan Leader*, wondering how they should register for the state primary in June. In response, the NPL insisted that members should enroll in local primaries as they always had. Newspapers that opposed the League then charged the organization with the subversion of state primary laws. But Leaguers broke no law. They merely endorsed candidates to run in the primaries of whichever party the aspiring officeholder preferred.[109]

On March 29, 1916, the League gathered its delegates for a statewide meeting. Theodore Nelson, one of the founders of the equity movement in North Dakota, attended. He "recognized names of many men whom [he] had known of in Equity days" and happily "renewed acquaintances with many." The many years of cooperative organizing made for an exhilarating, atmosphere. As old friends found each other, the forty-five legislative district delegates met in closed session.[110]

With input from League leaders—who shared their own insights and suggestions but did not cast votes—the assembled emerged with a startling slate of candidates. For governor, Leaguers picked Lynn Frazier, a relative unknown from a farm outside Hoople, North Dakota. After meeting him in Walsh County, organizer Beecher Moore proposed him as the perfect gubernatorial candidate. A graduate of the University of North Dakota, Frazier gave up his dreams of pursuing a professional life when his father died and his mother begged him to return to the family farm. Moore insisted that Frazier, "an actual honest-to-God, on the ground, farmer of land, of unimpeachable character" would "go along and work in harmony with the organization." The devout Methodist and rock-ribbed Republican also offered an impeccable defense against charges of socialist agitation.[111]

Other nominees included William Langer for attorney general, S. A. Olsness for commissioner of insurance, Thomas Hall for secretary of state,

James E. Robinson for the supreme court, Carl R. Kositzky for state auditor, M. P. Johnson for one of the three railroad commissioners, John Hagan for commissioner of agriculture and labor, and P. M. Casey for state treasurer. In a firmly Republican state, the slate included only one Democrat. But the legislative candidates chosen by League members in their districts proved more thoroughly nonpartisan. They included twenty-one Democrats and two Socialists. Each candidate publicly pledged to "vote and work for those measures and amendments that will assure justice to the farmers and ALL THE PEOPLE OF THE STATE."[112]

The debt to previous farmer movements was clear. Many candidates—themselves mostly farmers—held direct links to cooperative movements. Johnson and Casey served as president and vice president, respectively, of the North Dakota Society of Equity. Olsness was among the first group of North Dakotans to join the equity society back in 1907. John Hagan, himself a longtime equity society member, joined the League during the first month of organizing. Even Lynn Frazier, who served on the board of his local farmer-owned elevator in Hoople, quietly ran for a position on the Equity Cooperative Exchange board of directors in 1914.[113]

Two days later, the NPL opened its first convention in Fargo. More than two thousand farmers descended on the city. Farmers excitedly approved the chosen slate of candidates. Speeches punctuated the brisk spring air. On the last night, a fireworks show riled the crowd before a nighttime march through the streets of Fargo concluded the convention. Farmers carried banners saying: "Privilege knows no party; why should the people be fooled?," "It's do it ourselves or go broke," "Let the grafters holler; the people have the votes," and "Everybody's organized; why not the farmer?" All told, the meeting was a rousing success.[114]

Yet more took place behind the scenes than met the eye. One of Lemke's close friends wrote to him soon after the convention. He saw "the fine Dutch hand of Bill Lempke [sic] written large on every page." Indeed, the friend wrote, "that list of candidates looks like the Muster Roll of your personal friends." Lemke denied it, claiming that "the thirty seven thousand farmers [sic] members of the League selected those candidates." Both were right. A month before the meeting in Fargo, Lemke reassured a nervous William Langer that he would be the League's candidate for attorney general. Almost simultaneously, Lemke acquired a list of Equity Cooperative Exchange members. Lemke also told his law partner, James E. Robinson, that he would find himself on the slate for supreme court justice. Even Lynn Frazier, initially identified by other Leaguers, played football and roomed with Lemke at the University of North Dakota.[115]

Only a handful of people knew Lemke well enough to see the personal connections he held with so many League statewide candidates. In turn, the *Nonpartisan Leader* responded to charges of bossism by noting the free and open nature of the selection process. Suggestions had been made, but the voting delegates in the room—not Townley, Lemke, or others—settled on the candidates of their choice.[116]

Those delegates took real pride in a ticket of their own devising. O. T. Rishoff, a small-town newspaper editor in Bottineau County, observed firsthand the "tremendous wave of public indignation" that gave the League its power. In a letter to his friend John Gillette, a pioneer in the study of rural sociology and a professor at the University of North Dakota, he declared that "membership in this league means a great deal more to the average farmer than being a member of one of the old political parties. It is a religion with them." Plainly visible on the horizon, the looming storm soon sent establishment politicians running for cover.[117]

✶

In response to the show of force at the convention in Fargo, at least one newspaper asked: "Who is Frazier, and is Hoople a place or a disease?" These sharp attacks confirmed the big-city and even small-town disdain for agrarians that so frustrated them. Farmers put faith in one of their own. The day after the convention, O. M. Thomason took gubernatorial candidate Lynn Frazier on his first campaign trip. Planning to "drive quietly through the country, meeting any farmers who happened to be near the road in their fields, or barnyards," the two men soon got stuck. The car's wheels spun uselessly in the mud. Frazier got out, "and placing his heavy shoulder against the car, grunted to give her the juice." When he "was plastered with flying mud . . . he laughed and said clean dirt didn't hurt a farmer." When Thomason introduced Frazier at a rally that night, he shared the story. While "Frazier blushed and the farmers applauded," it became clear that the candidate from Hoople "was their kind of a man."[118]

The incumbent governor, Louis B. Hanna—a Republican associated with the McKenzie machine—took notice. By May, League successes convinced him that the droves of farmers signing up for the organization were "misguided and badly advised, but otherwise honest citizens" led astray. Nonetheless, "a certain percentage of them will stay by the league in this fight," making the NPL "dangerous." He decided not to run again.[119]

Hanna echoed the concerns of many. A lawyer in Park River, North Dakota, believed that "there are hundreds of merchants, politicians, and others, who are afraid of the so-called Non-Partisan League, and fear prevents them from speaking." Nonetheless, he believed that "the strength of the Non-Partisan League has been grossly over-estimated, and I am sure that the League is now on the wane." Fear of a farmers' movement could not surmount the disdain many professionals held for the aggrieved agrarians.[120]

Divided Republicans—members of the McKenzie machine as well as their reform-minded opponents—dithered over what to do. Animated farmers meant trouble for the former. A private meeting between leaders of both factions in early April went nowhere. Even Alexander McKenzie proved unable to broker a deal. Besides the League-endorsed man, there would be two other Republican candidates for most of the statewide offices and legislative seats.[121]

To be sure, farmers in the League faced their own struggles. Some imagined the organization as a balm for all their ailments. Overblown expectations led in a straight line to potential disappointment. One farmer penned a ditty for the *Nonpartisan Leader*, limning the reason he joined: "The gophers eat the farmer's wheat / The bees, they get his honey / The loan sharks haunt him in his sleep / And the grafters get their money." Their faith that the NPL could fix all the farmer's problems encouraged overestimations of its power. A few agrarians in Bottineau County even believed that "if the League could put their program through . . . land would jump from $10 to $25 per acre."[122]

Others struggled with their handpicked candidates. After the state convention, the NPL-endorsed candidate for the legislature from Nelson County began to openly attack the organization. Disappointed that the League did not choose his fellow reformer and progressive Republican Usher Burdick for governor, the NPL candidate denounced "forty or more of the cleanest and clearest thinking farmers of the State as mere puppets." In the ensuing confrontation with local NPL members, he lost the endorsement. Though local agrarians expected "slanders and misrepresentations from the Myrmidons of the Gang we are fighting," one committee wrote, "we do not expect to see them in print in reactionary newspapers over the signatures of our candidate." They scrambled to find a new candidate for the June primary.[123]

League employees spent that month organizing mass meetings and picnics. In some cases, thousands of farmers came together to hear speakers such as Albert Bowen, Arthur Townley, and Lynn Frazier. Many of the speeches focused on the retention of manly independence. Townley, for instance, told audiences that if the NPL lost in the primary, "heavy mortgages, high interest, low prices will force long hours of toil." More important, "wives, mothers and sisters will work in the fields . . . all will be debt and drudgery." Acknowledging the entirety of a farm family's labor, he went on: "We will yield up tens of millions of the earnings of our wives and daughters and mothers and fathers and brothers to the greedy masters of trade and finance in the East—millions that they do not need and cannot use."[124]

Never before had North Dakota's agrarians gathered in such numbers. The fair-like atmosphere at some all-day gatherings transformed politics into a festival. Even as farmers celebrated their newfound solidarity, workers handed out sample ballots, campaign literature, and buttons. Again and again, orators insisted that if farmers stuck together, they would win the primary.[125]

Celebration turned to concern in the last week before the primary. Treacherous weather forecasts—crucial in rural districts where two-rut paths far outnumbered graded gravel roads—moved the *Nonpartisan Leader* to implore its readers to "brave the thunder and lightning if it comes." After all, it noted, "Farmers are not fair weather fighters."[126]

Agrarians responded with vigor. As one NPL organizer from LaMoure County remembered, on primary election day "people turned out to vote in droves." In fact, "some who had not voted since they left the old country many years ago were there." Despite the bad weather, "they got thru the mud in Model T Fords." In other parts of the state, farmers braved otherwise impossibly swollen streams by tying "their clothes on their backs" and literally swimming to reach the polls.[127]

League leaders trumpeted the victory as a conquest over the long-standing political machine in the state's Republican Party. The NPL rejected McKenzie's allies even as they ejected the Progressive reformers who worked for years to dethrone bought-and-sold politicians. This left many angry, encouraging opposition to the League in the general election. Ongoing rumors that members were liable for the unincorporated organization's debts lingered. Whispered accusations that the NPL represented an attempt by Socialists to take over state government continued to circulate. The *Nonpartisan Leader* directly responded to these charges, debunking each one. League leaders also consolidated their close relations with the equity movement, the Grange, and the Farmers Union to ensure farmer unity.[128]

*

In the wake of victory, a few farmers reconsidered their decision to join the NPL. Charlie Barrer, from Carrington, North Dakota, wrote the League office before the June primary. He wanted to resign from the League. Barrer's "wife didn't like it when [he] joined and she has a weak heart and worries so over this that it makes her sick." After receiving a letter assuring him that the League was primed for victory, Barrer wrote back. He admitted that his "wife got to bothering her head" about the question of shared responsibility for NPL debts. "I am a poor man and haven't got any money," he wrote. Barrer was "willing to be a member of this great movement" but worried about any liability he might be taking on. Another letter from League headquarters finally assuaged his doubts.[129]

Ensuring success in the November 1916 general election depended not only on shoring up the membership but also political maneuvering. First, in September, the League took over the Republican Party's central committee. Making William Lemke chair of the Republican Party, NPL members seized the reins of the state's majority party and drew on its extensive campaign resources. Second, in the weeks immediately before the election, the NPL purchased one of the state's major daily papers. Ownership of the influential *Fargo Courier-News* offered another outlet for a pro-League stance. Finally, $20,000 quietly provided by the aging kingpin of North Dakota's crony politics—Alexander McKenzie—allowed the League to cement its support. McKenzie hoped the NPL would continue to neuter the reform-minded in the state Republican Party. He got more than he bargained for.[130]

The seizure of the state's most powerful party created complications for the NPL. Determining the League's stance in national political races proved thorny. William Lemke, state Republican chair and member of the NPL board of directors, publicly supported Charles Hughes, the Republican challenger to President Woodrow Wilson. Lynn Frazier, the League's candidate for governor on the Republican ticket, did the same. The organization itself, however, officially refused to endorse any candidate for national office. Both major parties took out large ads in the *Nonpartisan Leader*, further confusing the matter. All this befuddled farmers anxious to "stick" with the League.[131]

In the end, the NPL won an overwhelming victory. The November elections saw Frazier earn 80 percent of the vote for governor—the biggest majority in the state's history. The other statewide candidates, with one exception, also triumphed. Farmers sent almost every NPL-endorsed legislator to the statehouse. Yet North Dakota's voters gave Woodrow Wilson a slight edge in the presidential race. Despite Lemke and Frazier's efforts, farmers clearly seemed more committed to nonpartisan voting than some of the movement's leaders. Anxious to keep the United States out of World War I, they cast their ballots for the incumbent, who promised exactly that.[132]

While the *Nonpartisan Leader* trumpeted the "peaceful revolution" of the League, farmer politicians readied themselves for a stint in the capitol. A month before the session opened, Arthur Townley, William Lemke, and other NPL leaders rented the Northwest Hotel in Bismarck. There Townley, Lemke, and Bowen met with the newly elected men daily. They tutored the new representatives in legislative procedure, drafted laws, and sequestered the men from prying reporters. The quest for privacy went so far as to require League legislators to carry membership cards certifying their status as "a member of the N.P.L. Legislative caucus." Admittance to meetings could be gained only after showing one's card.[133]

When the session finally opened in January 1917, NPL members threw conventions aside. Governor Frazier's pietistic stance on drinking and dancing led to the cancellation of the traditional inaugural ball. Some of the newly elected men "had never been inside a State capital." Frazier hired an African American—Tony Thompson, father of Era Bell—to be his personal messenger because of his previous work as a cloakroom attendant in the Iowa statehouse. The "farmer-politicians sought after him" for tips on everything from formal dining to parliamentary procedure. Sometimes, even the new governor asked for his advice.[134]

With the League in full control of the state's house of representatives, the governor's mansion, and the state supreme court, only the state senate remained a stumbling block for the NPL. Even there, the League held a strong minority—eighteen out of forty-nine seats. In the house, they elected Howard Wood—son of one of the League's earliest members—as speaker. Equally important for procedural matters, Albert Bowen became the chief clerk. "Determined that public money shall not be frittered away with useless extravagance," the farmer-dominated legislature quickly got down to business.[135]

As they took office, skepticism spread. A January 1917 editorial in the *New York Times* snidely noted that the League's plan for North Dakota was "not class legislation" but "nonpartisan social equity for farmers." Closer to home, *Saturday Night*, a Saint Paul–based weekly, drolly insisted that "in the unregenerate rural commonwealths, they elect rustics to the most exalted in the gift of the state—and seem to be proud of it." In Bismarck, the suspicious *Tribune* nonetheless welcomed League politicians and noted that if they gave "the farmers of the state a freer and more competitive market at terminal points" they might earn the "solid support of the cities, as well as the farmers."[136]

The central goal of the NPL caucus soon became clear. Calling for a new state constitution to pave the way for the state-owned grain elevator and flour mill outlined in their platform, League legislators came up against fierce opposition. House Bill 44—which soon became the object of intense debate— not only created a straightforward path for state ownership of industries but also raised the state's debt limit, provided for four-year terms for state office (rather than two-year terms), limited the power of the state supreme court to render legislative acts unconstitutional, and broadened recall and referendum powers for state voters. Demanding power for the people, Leaguers claimed a mandate derived from their overwhelming electoral victories.[137]

League opponents headed an unrepentant house minority that, in their own words, "organized the fight, not to defeat the maesre [sic] but for time for the Senate to prepare to defeat it." Representative Arthur Divet, the leader of this effort, angrily denounced the newly elected NPL members as novice legislators with no business in government. During a floor debate, Divet told one NPL member of the house that he was "so devoid of intelligence that if he were cast bodily into the womb of the Goddess of Wisdom he would not beget a single idea." In response, Leaguers threatened to use their majority to expel Divet and other leaders of the opposition. Cooler heads prevailed.[138]

Despite attempts to stall, in late January, the NPL-controlled house of representatives overwhelmingly approved the bill. But in the senate, where a number of legislators elected before 1916 held out on the grounds that House Bill 44 endangered the state's constitution, opponents narrowly defeated the legislation. Then League enemies in the senate passed a bill designed to divide the NPL. Senate Bill 84 called for the establishment of a state terminal elevator in Minnesota to be paid for from direct taxation. Once built, the venture would be turned over to the Equity Cooperative Exchange.[139]

NPL opponents crafted the legislation to drive a wedge between the Equity Cooperative Exchange and the Nonpartisan League. They knew the crucial role played by equity members in the formation and sustenance of the League. They also knew that in the previous month, a disagreement about packing plants planted seeds of discord between the two overlapping groups.[140]

Ultimately, League leaders rejected the bill, arguing the elevator should be in North Dakota, operated by the state, and paid for by bonds. Tensions between the Equity Exchange and the NPL grew. J. G. Crites, general manager of the exchange, warned NPL legislator Ole H. Olson that "there are a great many Equity people who are going to be very bitterly disappointed if a legislature composed of farmers will turn down a bill for which the Equity has been fighting all these years." In the face of fierce internal debate, Governor Lynn Frazier vetoed Senate Bill 84. Sore equity members did not bolt the League, though some began to worry that the NPL would not live up to its promises.[141]

Despite the divisive battles over House Bill 44 and Senate Bill 84, the League-dominated legislature created a state grain-grading system, devised a state bank-deposit guarantee program, established a state highway commission, tripled state moneys for education, and crafted new railroad regulations.

Equally important, the legislature approved women's suffrage for presidential and municipal elections.[142]

The League's victories in North Dakota grabbed national attention. Even those who critiqued the NPL—such as the *New York Times*—admitted that the NPL-dominated legislature would "be observed with interest." Indeed, farmers, social reformers, and radicals demanded more information about the Nonpartisan League. Requests from Michigan, Oklahoma, Minnesota, Idaho, Missouri, South Dakota, California, Nebraska, Arkansas, and Kansas for League organizers and publications flooded the NPL's headquarters in Fargo.[143]

The famous North Carolina Populist (and former US senator) Marion Butler even wrote directly to Arthur Townley, the acclaimed leader of the movement. Anxious to bring the League to his native state, Butler insisted that farmers there were "ripe for revolt against the Democratic Party." Victory created unanticipated notoriety and opportunity. NPL leaders carefully considered their next move.[144]

The storm—created and sustained by middling farmers—that struck North Dakota in 1916 shocked the political establishment of that state. Property-owning agrarians reclaimed their political agency and looked to shape brighter futures for themselves and their families. Their articulation of antimonopoly capitalism as well as their organizing drew on a rich tradition of middle-class insurgency in American political life. Newly invented methods all but ensured their victory in North Dakota.

Soon, the farmer-fueled thunder would sweep across much of the country. While NPL leaders eyed the potential in North Dakota's neighbors, enterprising farmers took matters into their own hands. Seeing potential in the League's model for popular politics, one farmer who visited old friends in North Dakota in 1916 decided to take the idea back home with him to Canada. The Prairie Provinces—like the rest of the northern plains, much of the Pacific Northwest, and parts of the Midwest—would not be spared from the storm.

2

Expansion

Find out the damn fool's hobby and then talk it. If he likes religion, talk Jesus Christ;
if he is against the government, damn the Democrats; if he is afraid of whisky, preach
prohibition; if he wants to talk hogs—talk anything he'll listen to, but talk, talk, until
you get his God-damn John Hancock to a check.

ATTRIBUTED TO ARTHUR TOWNLEY, 1918

With a cold wind howling outside, hundreds of farmers assembled in a hall
in Calgary, Alberta, to hear Louise McKinney—the first woman elected to a
legislative body in the history of the British Empire—speak. Standing "on the
platform of the Farmers' Non-Partisan League," McKinney pointed to her
election to Alberta's parliament in June 1917 as evidence that the NPL had
come to Canada to stay. As the "spontaneous expression of the people," she
told listeners on that chilly night in February 1918 that the League "is not a
shadow of the moment, not a freak, not a spasm" but "a movement that furnishes the channel through which the thought of the people can be expressed
to-day."[1]

Days later, in a newspaper article published in the Alberta NPL's *Alberta
Non-Partisan*, McKinney continued to claim the significance of the nascent
Canadian populism embodied by the League. "Organized primarily with the
idea of bettering the economic conditions under which the farmer is working," she believed the League also offered agrarians in the province a way to
fight for "the greatest movement of the age—the fight for a wider democracy." "Evils that have been fastening themselves upon our political life for
half a century cannot be uprooted in a day," McKinney explained. It took
"time to get people awake to the fact that these evils exist." But she remained
confident. Knowing that "we may through compromise be switched from
our original purpose," she implored readers to always remember that "the
Government must be controlled by the people and run in the interests of the
people."[2]

McKinney, a former resident of North Dakota living as a farm wife in
Claresholm, drew on years of experience. As a longtime Women's Christian
Temperance Union (WCTU) organizer and a social gospel adherent, she

FIGURE 4. Louise McKinney, the first woman elected to a parliamentary body in the British Empire, in her official Legislative Assembly of Alberta portrait, 1917. (Glenbow Archives, NA-5395-4)

worked hard to bring women the right to vote. Like other suffragists, she hoped to transform Canada in multiple ways and saw the League as a ve- hicle for doing so. When approached by the brand-new organization about running for office in early 1917, the Alberta WCTU president initially refused the nomination. But McKinney soon changed her mind because, in her own words, the nonpartisan thrust of the NPL "appealed to me very strongly." Driven by the desire to reform society through politics, McKinney saw the League as offering possibilities for wide-ranging change denied by the exist- ing parties. This propelled her into the movement.[3]

The manifestation of the Nonpartisan League in western Canada—which sported a different political system than North Dakota—showed the wide- ranging appeal of the model created by Albert Bowen and fostered by Arthur Townley. The dizzying pace of NPL success in North Dakota ensured that agrarians across North America kept tabs on the farmer movement in that state. In some cases—as in Alberta—farmers simply appropriated the League model and started organizing themselves. Elsewhere, agrarians invited the

NPL to send organizers. Encouraged by this outpouring of interest, League leaders who initially envisioned the organization as a vehicle for voter empowerment in one state seized on the opportunity to expand. They hoped to create, in their own words, "a 'solid West,' built on the farmers' and workers' demands for justice."[4]

They soon relocated the NPL headquarters in Saint Paul, Minnesota, and announced a national expansion effort. Visiting the new offices, University of Montana professor Louis Levine interviewed Arthur Townley and other League leaders for the *New York Times* in March 1917. Levine told readers that the NPL's "leadership has been recruited from all the reform elements of the country," making for "a composite movement in which no one single theory dominates." Indeed, the professor saw that the NPL, "animated" by a "spirit of compromise," produced a "lesser regard for theory" than most political insurgencies. It was not driven by "abstract justice or . . . a comprehensive radical program." Instead, the NPL looked to "utilize the sentiment and ideas which have already penetrated the hearts and minds of the people."[5]

Knowing that outside the rural West, farmers made up only one collective interest, Townley told Levine that the League hoped "for . . . co-operation in politics between all organized forces of reform" on a national scale. Aspiring to become the primary organization representing farmers in American politics, the NPL could then "hold counsel with other organized groups of the country and formulate a platform acceptable to all. The idea is that of a combination in which all elements remain independent in so far as they pursue different interests, but combine for common purposes." This meant "total indifference to partisan politics and all that has ever gone with it." Levine accurately reported that this plan was "the extension of the co-operative principle"—so central to the self-conception of lower-middle-class agrarians on the northern plains—"from economics to politics."[6]

In other words, Townley and other NPL leaders recognized that the dramatic victory of the League in North Dakota in 1916 stemmed from shared self-interest in the wheat economy. The consistent rejection of the people's will in that particular state primed already organized agrarians to join a new effort to bring farmers into politics. The conditions that birthed the movement in North Dakota did not necessarily exist elsewhere. In fact, outside that intensely agricultural state, the task proved more difficult. Enlisting voters from large cities and extraction-based industries would be challenging for the agrarian-oriented NPL. Moreover, each state and province had its own peculiar local politics.

But all this did not dissuade League leaders from expansion. It just meant that League employees, who began working outside North Dakota in 1916, needed not just to mobilize already willing farmers to take over a state government but instead organize them into a cohesive and unified political group that could bring the cooperative model into American politics. The difference proved significant. Effective mobilizing depended on identifying voter desires, honing a pitch, and seizing the moment. Effective organizing, on the other hand, involved education, consciousness raising, and building solidarity both within agricultural communities and with other organizations. The former offered short-term possibilities. The latter demanded long-term commitments.[7]

In North Dakota, the organizing had already been done. Equity movements, a resurrected Grange, and an emerging Farmers Union all created agrarian cohesion in the face of economic inequity and tin-eared politicians. There, the NPL effort largely consisted of mobilizing already united farmers. In other states and provinces, however, transcending ethnic, religious, and political differences in a wide range of contexts depended on careful organizing. In 1916 and 1917, as the NPL expanded into Minnesota, South Dakota, Montana, Idaho, Washington, Colorado, Nebraska, Kansas, and Wisconsin—and as Canadian agrarians adapted the model to Saskatchewan and Alberta—it confronted this tension.

To meet the challenge, as one contemporary put it, "the Non-Partisan League adopted methods of propaganda and organization which have no parallel in former farmers' movements." The League also developed a powerful new iconography that embodied the NPL's ideology. It convinced many agrarians to join and then remain loyal to the League.[8]

But other issues mounted. Dependence on postdated membership checks made the League as susceptible to crop failure as farmers themselves. League representatives began serving in the North Dakota statehouse in 1917. This gave opponents and allies alike an actual political record to critique or praise. America's entry into World War I put the divisive organization on the defensive.

The Nonpartisan League's expansion beyond North Dakota pointed toward tremendous possibilities, as well as the opportunity to transform an event into a movement. Creating the capacity for growth demanded immense energy and foresight. For the membership-based organization, everything turned on finding farmers willing to join League. Reaching and holding on to constituencies both inside and outside the North American wheat-growing belt presented real difficulties. This made the agitated farmers in Canada, who were the first to use the NPL's ideology and tactics outside the borders of North Dakota, especially significant. These agrarians made it clear that even farmers in another country faced economic and political problems that, in their mind, a new nonpartisan and popular movement might solve. Unable to wait for organizers to come to their communities, they began organizing themselves.

⋆

Even as Arthur Townley, Albert Bowen, and other Nonpartisan League leaders began considering the possibilities of a national farmers' organization based on the NPL model, a seventy-two-year-old widower from Saskatchewan named Silas Haight hatched an ambitious plan. Born in Ontario, Haight migrated to the Dakota Territory in 1881 with his family. There he established a farm. In 1896, he gave up his land and moved into the nearby town of Osnabrock to work as a blacksmith. Tiring of life in town, in 1906, Haight returned to farm life and to Canada. He bought acreage in southwestern Saskatchewan. But like so many agrarians whose lives transcended the border, Haight stayed in touch with his old friends in North Dakota.[9]

In early 1916, Haight visited his former neighbors to observe the burgeoning Nonpartisan League. During his visit, he became a member and carefully studied the organization's methods and platform. Envisioning rich opportunities for political ferment in the Prairie Provinces, Haight returned home to Canada a few weeks later. Soon thereafter, he began touring the local countryside and enrolling agrarians into a new membership-based organization patterned after the NPL.[10]

Even while living in Osnabrock—itself a mere thirty-five miles south of the international border—Haight kept close tabs on the economic and political situation facing wheat farmers in his native Canada. Like so many on the northern plains, he lived in a transnational world. From the 1870s to the 1920s, similar processes enveloped both sides of the border. The dispossession of indigenous peoples and agricultural settlement resulted from analogous policies. Canadians moved south into North Dakota and Montana even as thousands of newly arrived farmers came to the Canadian prairies from the United States. Grain trading companies in Minneapolis carved out a powerful presence on the Winnipeg grain-trading market. Farmers traveled across national boundaries to visit friends and family. People, goods, and ideas moved freely over a border that often meant little to those of European descent.[11]

As late as 1916, the number of American-born immigrants in Saskatchewan stood at 13 percent. The same year, 18 percent of Albertans claimed American origins. Mixed with British, Norwegian, Swedish, German, and eastern European immigrants, these US-born agrarians rooted new lives in marginal lands. Even some of those born elsewhere had lived in the United States for many years before migrating north.[12]

This demographic reality—coupled with a shared grassland ecology—resulted in a common economic and cultural experience across the northern plains. Dry farming techniques invented in the United States came north with American farmers. Indeed, without them, the otherwise marginal lands of the Prairie Provinces might not have attracted farmers. In turn, the search for a spring wheat to match the difficult climate resulted in a Canadian's creation of Marquis wheat in 1911. Despite its susceptibility to rust, by 1920, the variety accounted for 90 percent of all the wheat grown in the Prairie Provinces. It also became the favored grain grown in North Dakota, South Dakota, Montana, and Minnesota.[13]

Silas Haight's initial focus on organizing fellow agrarians in and around Swift Current, Saskatchewan, proved especially appropriate. Most agrarians in southwestern Saskatchewan owned farms between 200 and 320 acres in size. Many carried mortgage debt on their land. Spring wheat made up almost 80 percent of all the crops harvested—more than anywhere else in the Prairie Provinces. Nowhere else in North America could one find conditions so similar to those in the League's birthplace, western North Dakota.[14]

In fact, wheat monoculture, farm ownership, middle-class ideals, and community bonds defined life on the Canadian prairies, just as they did for farmers in North Dakota. Agrarians in Canada worked as small landowners who valued independence and private property. Farming demanded ownership, self-sufficiency, and participation in the market system. Nearby small towns offered crucial services, and social tensions between townspeople and farm folk often remained submerged in day-to-day relations.[15]

On both sides of the border, difficulties faced agrarians at every turn. Georgina Binnie-Clark, a British woman who migrated to Saskatchewan and bought a farm in 1906, noted that "the farmer"—enveloped by both markets and the difficult climate—"is at the mercy of every wave of ill-tide. . . . It is to be remembered that the small farmer is invariably pressed for small money." A young man whose family homesteaded in rural Saskatchewan in 1908 recalled privation. Despite all the hard work, he said, "we were hard up and for years suffered discomfort and deprivations." Troubled by their dependence on both markets and climate, agrarians looked to better business methods as a means for individual and communal survival.[16]

Ethnic and religious diversity—as well as gender—complicated relationships within this middling world. British- and Canadian-born whites worried about the many different nationalities that found their way to the prairies. Indeed, concerns about American mores traveling north, coupled with racial and religious anxieties about persistent indigenous peoples and Mormon immigrants, produced federal policies intent on peopling the land with married couples. Farm women's labors provided the community bonds that survival demanded and middle-class culture encouraged. In fact, the work of women often kept farms afloat. Furthermore, family farms gave native-born Canadians social and economic security that soothed their fragile sense of national identity and promised prosperity.[17]

Those bonds survived in the face of monopolistic practices in the grain trade that mimicked those found in the United States. Fueled by the high prices garnered for Canadian wheat on hungry global markets, companies eagerly capitalized on the soaring values for farmers' crops. By the early 1900s, large companies controlled the Canadian wheat market. They bought out smaller enterprises and reduced competition. Their agents cheated farmers when loading, grading, and buying their grain.

These practices flourished because Minneapolis milling interests implemented tactics used south of the border. They invested heavily in Winnipeg, the burgeoning center of the Canadian wheat trade. By 1911, Minnesotans owned almost half of the grain companies in the city. In 1916, John C. Gage, son of J. Edward Gage (a prominent member of the Minneapolis Chamber of Commerce), became president of the Winnipeg Grain Exchange. In turn, by 1918, Winnipeg-based firms held the largest out-of-town presence in the Minneapolis consortium.[18]

Meanwhile, eastern Canadians controlled not only national politics but also the supply of credit. Even as World War I sent wheat prices skyrocketing, a precarious combination of wheat monoculture, mortgaged land, and corporate power created the conditions for foreclosures. As late as 1922, the *Grain Growers' Guide*, the agrarian newspaper for the Prairie Provinces, noted that "from the standpoint of economic life it is an imaginary line that is drawn between Canada and the United States."[19]

*

As in North Dakota, farmers in Saskatchewan and Alberta recognized and responded to the power of corporations and established political parties in the wheat economy. For over a decade, Canadian agrarians organized to confront the challenges faced when outsiders graded, priced, distributed, and processed their crop. Committed to the market, their organizations sought to improve farmers' position by using cooperative ventures to ensure the equitable distribution of economic power. They also sought to reform an entrenched party system.[20]

American migrants brought the Grange and the Society of Equity to the Prairie Provinces. But Canadian farmers soon created their own institutions, such as the Saskatchewan Grain Growers' Association and the Alberta

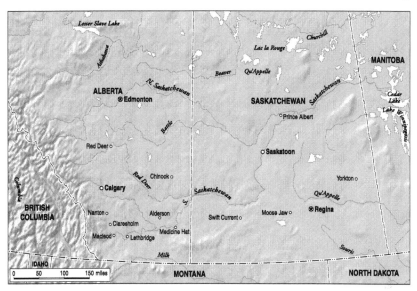

FIGURE 5. Alberta and Saskatchewan in 1916. (Map by Philip Schwartzberg)

Farmers' Association. The push for economic cooperation led provincial grain-grower organizations to join together and form the Grain Growers' Grain Company in 1907. It represented agrarians in grain trading and hoped to establish a farmer-owned terminal elevator. A year later, Manitoba's parliament passed a law opening the Winnipeg Grain Exchange to farmers' cooperatives. Simultaneously, the Alberta Society of Equity and the Alberta Farmers' Association merged to form the United Farmers of Alberta. The *Grain Growers' Guide*, published in Winnipeg, became the primary newspaper serving all these organizations. It consistently touted cooperative work and encouraged farmers to imagine themselves as a distinct occupational class.[21]

Agrarians found politicians at the Canadian provincial and federal levels more responsive than those in North Dakota. The Canada Grain Act of 1912 offered the most specific grain regulation found in any country. It provided for public terminals and elevators and even authorized the dominion to provide a government-owned terminal. That same year, the Grain Growers' Grain Company built its own elevator in Fort William, Ontario. Meanwhile, Saskatchewan created a provincial grain-elevator company. Agrarians in Alberta sought similar legislation.[22]

Given the relative success of their cooperatives, Saskatchewan's farmers did not seem to need an organization such as the Nonpartisan League. Yet despite many years of organizing, the rapid expansion of wheat growing depended on outside capital. That, in turn, increased farmers' dependence on others for their futures. Squeezed between an unforgiving climate and tight credit, they sought to succeed as respectable agrarians and property owners. In the early twentieth century, taking on debt to buy more land and machinery, agrarians stretched themselves thin. By 1913, the number of mortgaged farms throughout Saskatchewan ranged from 50 to 90 percent, depending on the area. On average, the typical farmer in the province owed $1,500.[23]

Even as it noted the "too exclusive devotion to grain growing" among the province's agrarians, a 1913 commission investigating the Saskatchewan mortgage crisis focused on "the dependence of our economic life on outsiders." Their worst fears came true in August 1914, when a financial crisis in London paired with drought to push many Canadian farmers over the edge. Foreclosure loomed for thousands. As many as 75 percent of the farmers around Medicine Hat, Alberta, found themselves on provincial debt relief, and many more were working government jobs to buy food for their starving families. Only a bumper crop and high prices in 1915 discouraged a full-scale exodus from the region.[24]

The episode taught western farmers two crucial lessons. First, government-aided cooperation in grain grading and marketing did little to solve their credit problems. Second, farmers held little sway in provincial and federal politics. Others seemed to be in charge of the major decisions affecting their lives. This growing sense of dependence challenged the notion of independence so crucial in the middle-class vision of self that they cherished. For years, the *Grain Growers' Guide* had illustrated the necessity of cooperating as an occupational class. In the wake of the mortgage crisis, many farmers began to agree. Nonpartisan political organizing seemed to be the answer to many of their problems.

Their desire for nonpartisanship emerged in a political world that differed from the American one in crucial ways. Canada's ruling elite imagined themselves as the North American expression of the values of Great Britain. The very act of creating a united nation out of disparate parts—one begun by English-speaking Canadians with the 1867 British North America Act—imposed a fragile nationalism on the country. Distinct colonial experiences and deep-seated cultural, social, and economic tensions fostered sectional politics. Regional differences loomed large. The primary means of effecting unity—the National Policy, which guided federal governments from the 1870s into the 1910s—fostered continental growth at the cost of encouraging antagonistic sectional politics. With its focus on financing a transcontinental railroad, settler colonialism west of the Great Lakes, industrialization in Ontario, and high tariffs, the National Policy left many unhappy.[25]

In Saskatchewan and Alberta, agrarians became suspicious of eastern intentions and motives. Even though the National Policy promoted western growth and made prosperity possible, resentment among prairie farmers grew. Anger over the firm commitment to protectionist tariffs joined with the sense that political elites in far-off Ottawa cared little for agrarian opinions. Commitment to the National Policy also meant that politicians in the Prairie Provinces needed to straddle the desires of their local constituents and their parties' federal commitments. This often left them in difficult straits.[26]

Furthermore, in these newly created provinces, the inrush of migrants from all over Canada, Europe, and the United States kept party loyalties in flux. The parliamentary system left no room for primaries and allowed those who governed to call for elections at opportune times. It also allowed for multiple parties. But in practice, the Liberal and Conservative parties dominated. Seeing the push in the United States for direct democracy—primaries, direct election of senators, referendums, and the recall—some Canadian farmers called for transformations in their own system. World War I brought together Liberals and Conservatives in a unity government. Suspicious of wartime corporate profiteering and corruption, western agrarians looked upon both parties with disdain. Free to create new parties, many farmers preferred to envision politics without them. Thus, nonpartisanship appealed not only to agrarians' immediate economic interests but also as a vehicle for provincial and national political reform. They believed that equity and equality depended on rejecting parties altogether.[27]

Canadian farmers also drew from a burgeoning cooperative impulse as they imagined what the *Grain Growers' Guide* called a "well-balanced Canadian yeomanry." They put forward an alternative form of capitalism, one premised on accumulation without concentration. Veterans of populist struggles in the United States—such as John W. Leedy (former Populist governor of Kansas)—lived in the Prairie Provinces. They emigrated with suspicions about monopolies and a faith in farmer cooperation. The significant presence of Americans in the region ensured that less well-known participants from these popular movements knew and looked up to these men. Limited commitments to Canadian political parties further primed them for nonpartisan action. Finally, a wide range of American magazines and newspapers made their way north of the line, outlining potential progressive reform programs for Canadian readers.[28]

Prairie farmers, however, derived their populist inclinations not just from agrarian currents imported from the United States but also from currents for change in Canada. Rejecting farmers as capitalists, Socialists attracted little

attention in rural ridings across the Prairie Provinces. But some British settlers belonged to trade unions or agricultural cooperatives before arriving in
Canada. They retained their familiarity with the discourses and practices of
popular politics. Meanwhile, rural Ontarians moving west brought years of
farmer organizing and agitation with them.[29]

The social gospel preached across Protestant Canada, however, was the
most significant source of popular politics. At least one NPL member in Alberta noted that "we found our origin among religious people." The middle-
class Christian response to the strife spawned by industrialization focused
adherents on ameliorating the human costs of rapid economic growth. The
simultaneous emergence of biblical criticism in theological debates created
an opening for sociological analysis. Many adherents hoped that through
their social critique, churches might become more relevant to both debt-
ridden farmers and the working class. As a compelling call to collective action
consistent with democratic principles, the social gospel resonated powerfully
with agrarians in western Canada.[30]

In fact, some social gospel adherents envisioned the Prairie Provinces as
the best place to create a society that reflected Christian values of equity and
empowerment. In a 1913 speech, J. S. Woodsworth, a prominent Methodist
minister, went so far as to publicly pray, "Thy kingdom come as in Heaven, so
in Regina and Moose Jaw, Saskatoon and Winnipeg." Two years later, Woodsworth wrote that on the prairie, the firmly middling efforts to make "good
roads; the getting rid of weeds; the improvement of stock; the providing of a
ball ground; the higher education of the young people; a square deal for the
stranger; better laws and better administration of law—all these are essentially religious, all are surely part of the work of bringing in the kingdom of
God in your home district."[31]

Wesley College, a small Methodist school in Winnipeg, proved to be a hotbed for social gospel thinking in the early 1900s. Students and faculty there
envisioned service in the Prairie Provinces as mission work. Seeing the straits
of prairie farmers as structural dispossessions, they blurred the boundary
between sacred and secular. Practical Christianity, in their mind, demanded
what Wesley College professor Salem G. Bland called "devotion to the common people." This demanded a response to a wide range of social and political ills, including what one of Bland's students called "*public* vices" such as
"partyism." Innumerable men and women connected to Wesley later became
significant leaders in Canada's agrarian protest movements—including the
Nonpartisan League.[32]

Finally, as nonpartisan populist sentiments developed in Saskatchewan
and Alberta, women made room for themselves. In 1914, a columnist in the

Grain Growers' Guide noted that "the State-made economic conditions of to-day" compelled the average farm wife "to crucify herself daily. . . . The average mother has no past, no present, no future. Her personal identity is lost in a clamor for bread." In response, agrarian women envisioned themselves as a powerful part of the farm struggle.[33]

Animated by social gospel thought, Protestant farm women actively crusaded for reforms such as temperance and suffrage. They also devoted themselves to building local institutions such as libraries and community halls. Participation in voluntary associations allowed them to break free of monotonous daily routines. Regional and provincial meetings helped create strong bonds between otherwise isolated farm women. In the process, they insisted on their own vision of agrarian politics, one rooted in civic connections, shared labors, and attention to their own plight. Because of their efforts, Saskatchewan's women earned suffrage in March 1916. Voting for women arrived in Alberta one month later. Almost simultaneously, agrarian women became full-blown members of both the Saskatchewan Grain Growers' Association and the United Farmers of Alberta. This meant that women played a significant role in the emerging prairie populism embodied by the NPL.[34]

★

The first stirrings of dissent in the Prairie Provinces came in 1913. At a meeting of the Saskatchewan Grain Growers' Association (SGGA), a small group of farmers called for the creation of the No-Party League. Disgusted by both the Liberal and the Conservative parties and hoping to obtain further agrarian legislation, the organization planned to "secure through Direct Legislation and Taxation of Land Values the rescue of the natural resources and public utilities from private control and their administration for the benefit of all the people." Poorly organized, the organization quickly fizzled out. Yet it showed that a membership-based nonpartisan political organization might appeal to western Canada's farmers.[35]

Silas Haight understood that North Dakota's Nonpartisan League offered a better example of exactly how to marshal farmers' desire for nonpartisanship and translate it into political power. Enlisting supporters in the summer of 1916, Haight, dissatisfied with the SGGA, avoided "all unnecessary publicity" to avoid opposition "until the League was strong enough to fight back." Then, on July 12, 1916, in Swift Current, recently recruited members of the Farmers' Nonpartisan League of Saskatchewan quietly met and elected him president. Focused on what they envisioned as a " 'new way out' for the farming class," the men devised a platform calling for provincial terminal elevators, flour mills, stockyards, and packinghouses; direct legislation; provincial inspection of dockage and grading; and rural credit banks. Their platform echoed the North Dakota NPL program. For these Canadians, "the main issue" was "'the political supremacy of the common people.'"[36]

Responding to popular sentiment, NPL members in Saskatchewan soon crafted a more expansive and ambitious agenda to meet the specific needs of Canadian farmers in the complicated politics of the broader dominion. This platform called for the nationalization of banking, credit, railroads, and telephones; women's suffrage; a graduated income tax and a graduate inheritance tax; the recall; national old-age insurance; ensuring that no court could overrule Parliament; and the abolition of the Canadian Senate. To do all this, the League planned to "elect farmers to office who, by their past records have proven themselves to be 'fighters for their class.'"[37]

Hammering home the notion that "the only place to start real cooperation is at the ballot box," the League leaders spoke directly to agrarian concerns. They told Canadian farmers that "there is another 'class' who sometimes call themselves farmers, but the only thing they have been known to 'farm' is the farmer and the wage worker of the cities." In fact, "through this kind of 'farming' the wage worker has become a 'wage slave' and the farmer has become a 'mortgage slave.'" Responding to such appeals, in just three months more than twelve hundred agrarians joined the organization. Just like their brethren to the south, Canadian members received a newspaper. Many, already members of the Saskatchewan Grain Growers' Association, led an attempt to take over that organization and push it into politics. Though they failed, by February 1917 the Saskatchewan NPL counted more than two thousand adherents.[38]

They even earned the attention of Leaguers in the United States. North Dakota's NPL sent its counterpart in Saskatchewan a triumphant telegram in November 1916, after sweeping the former's general elections. The Canadians used it to implore agrarians across the province "to do your share." Soon thereafter, an American who "was busily engaged in the organization work in North Dakota" became one of the top NPL organizers in Saskatchewan. He believed "the farmers in this country have more to kick about than the farmers across the line, and they are quite prepared to register the kick by signing up with the League."[39]

The organization also caught the eyes of agrarians in Alberta. A young American-born farmer from Alderson, Henry Johnson, found himself in Regina on business in fall 1916. There he encountered members of the Nonpartisan League. Upon his return, Johnson "interested some of his farmer friends, and with his usual vigor and wholeheartedness" formed the Farmers'

Non-Partisan League of Alberta. In an effort to "destroy partyism; to obtain a government more representative of an agricultural country, and to secure for the farmer a fairer share of what he produces," Johnson and four others met in Calgary to form the organization on December 16, 1916.[40]

A Wesley College graduate named William Irvine became the secretary of the group. Two years before, the young Scotsman had helped farmers in the tiny hamlet of Emo, Ontario, form a cooperative. It cost him his job as minister of the local Methodist Church, where he "was not preaching to get people into heaven but . . . was much more interested in getting heaven into people." He moved to Calgary soon thereafter, accepting a call from the First Unitarian Church. Throwing himself into both farm and labor politics through his newspaper, *The Nutcracker*, Irvine quickly became a well-known presence across Alberta.[41]

Irvine and Johnson proved to be a powerful combination. Johnson signed up farmers, and Irvine wrote newspaper editorials and gave lectures. The former spoke with experience and energy, and the latter's fervent oratory animated the movement. Under Irvine's influence, the League immediately sought an alliance with organized labor in Calgary. In fact, he used connections to Calgary's trade unions to help establish a complementary Labour Representation League. At the same time, the Albertans kept in touch with their colleagues in Saskatchewan.[42]

Reaching out to farmers east and south of Calgary, by February 1917, the organization had recruited two hundred members. In April, the League absorbed a small independent farmers movement organized in and around Chinook, Alberta, adding more members and momentum. Within a year they counted almost three thousand adherents. Irvine changed the name of his newspaper to the *Alberta Non-Partisan*, and it became the organization's official newssheet.[43]

The top of the editorial page made the animating ideology of the movement clear: "Politics is the business of the people." Only when the people removed political parties from the parliamentary system would government work. Albertans seeking citizen-led efficiencies needed to follow the lead of North Dakota. "Let the farmers of the prairies and the workmen of the cities strive together for a Non-Partisan government to elect which, men and women, instead of Grits and Tories, shall vote and citizens should take the place of politicians in the legislature," Irvine's newspaper crowed. Irvine even began giving speeches touting the Nonpartisan League as the "firstborn child of the new day in our political world." Another prominent League member in Alberta later referred to nonpartisanship as "merely organized independence." Combining an antiparty and anticorporate message with a call for participatory civic work, the Alberta NPL offered a new vision of democracy in the Canadian context.[44]

The spectacular growth of the Nonpartisan League forced farmers across the Prairie Provinces to pay close attention. Roderick McKenzie, a prominent farm leader and newspaper editor, chronicled the rise of the League on both sides of the border in the *Grain Growers' Guide* in September 1916. After laying out the origins and trajectory of the movement, he argued that "conditions here are similar to the conditions that impelled the farmers of North Dakota." In fact, McKenzie believed that "the request for reforms by the prairie farmers are receiving similar consideration to those which drove the North Dakota farmers to desperation." In the same issue, the newspaper's editors opined: "Western Canada will remain the stamping ground for the privileged interests until that day arrives when the farmers of this country with one accord are willing to forget that they are Liberals or Conservatives and will remember only that they are men with responsibilities to themselves, their families and their country." Such sentiments suggested a bright future for the League in Western Canada.[45]

However unintentionally, electoral strategies devised by North Dakota farmers to solve local problems in an American state offered Canadian agrarians a method for confronting federal issues in a parliamentary system. Nonpartisanship held a special appeal to many in the Prairie Provinces. The shared economic and social experiences of farmers on both sides of the border transcended political boundaries and differences. Indeed, the Nonpartisan League concept—an agrarian-derived anticorporate ideology paired with an attempt to transform electoral politics—became a transnational lynchpin. Farmers in western Canada deployed the NPL model to crystallize a distinctly independent agrarian populism that came to define Prairie Province politics for decades to come.[46]

✶

Even as Canadian farmers saw and seized on the potential in the Nonpartisan League model, a more intentional expansion stirred south of the border. Arthur Townley and other NPL leaders—altogether ignoring Canada—understood that farmers in states outside North Dakota also desired change. Fueled by success, Townley decided to formally expand the League's reach into Minnesota, South Dakota, and Montana. The NPL leader knew "that these three states and North Dakota controlled the production of hard spring wheat of the United States." In fact, "by dominating the politics of these four commonwealths," Townley "visualized an alliance that would control the flour production of the United States." Given the centrality of flour in the American diet and economy, national political power would follow.[47]

For Townley, the natural extension of the movement followed the landscape of the spring wheat–growing system. Like agrarians in North Dakota, wheat farmers in the southern and western part of Minnesota felt powerless in the face of corporate interests based in the Twin Cities. The state included not only wheat farmers but also Iron Range miners, urban workers, and entrenched corporate power. Agrarians could not claim to represent the entirety—or even the majority—of the state's population. Even the farms in Minnesota proved more diverse than those in North Dakota. Dairy, corn, and poultry operations stood alongside farms focused on wheat.[48]

Nonetheless, the League set out to organize Minnesota's agrarians. The Equity Cooperative Exchange counted thousands of Minnesotans among its membership. Appealing to that base as it had in North Dakota, the League hoped to grow rapidly. Given the ongoing efforts in North Dakota, this meant hiring more employees. "Several Grand Forks men"—including Walter Quigley, a lawyer and longtime political operative—"signed up as organizers." Knowing that early reverses in a neighboring state might hurt League efforts in North Dakota, Townley warned the new employees "to take it easy in Minnesota and leave nobody sore."[49]

On July 5, 1916, Quigley, accompanied by Townley and others, visited Arne Grundysen's farm on the North Dakota–Minnesota border. Already familiar with the League's programs, the Minnesotan "was all prepared to join when we arrived and agreed to assist" Quigley "in that locality." Indeed, a number of Minnesota farmers already subscribed to the *Nonpartisan Leader*. In just a few weeks, "at least 95% of the farmers in several townships" in Polk County joined the NPL.[50]

These initial efforts bore real fruit. By late summer, more than ninety organizers canvassed Minnesota. The NPL's most experienced employees kept signing up farmers well into 1917. One, who lived on a farm outside of Warroad, announced that "eight dollars of my money will be gladly given every year to see the time when the necessities of life will be taken out of the hands of men who have elected to make themselves millinaires [sic] out of farmers' toil and the consumer's misery."[51]

In December 1916, Ray McKaig, a leading Nonpartisan League member and director of the North Dakota State Grange, spoke at a meeting of the Minnesota State Grange. By then, most of the state's farmers wanted to learn more. McKaig described the NPL program to the more than 150 delegates, creating "much inthusiasm [sic]" and "a spirit of welcome." Unfortunately, McKaig found the state's Grange leadership less welcoming. In particular,

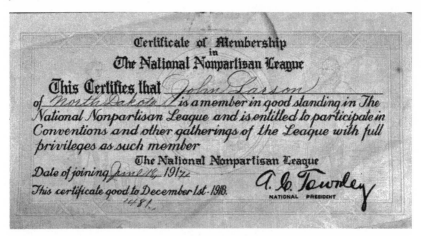

Certificate of Membership
in
The National Nonpartisan League

This Certifies that *John Larson*
of North Dakota is a member in good standing in The
National Nonpartisan League and is entitled to participate in
Conventions and other gatherings of the League with full
privileges as such member

The National Nonpartisan League
Date of joining June 16, 1917
This certificate good to December 1st-1918.   *A. C. Townley*
                                                          NATIONAL   PRESIDENT

FIGURE 6. Nonpartisan League membership certificate. (State Historical Society of North Dakota, A5623)

Charles Rier, the state master, struck McKaig as "not too friendly" to the NPL. Luckily for the League, the Grange proved to be a rather weak organization in the North Star State.[52]

Despite geographic proximity and environmental similarities, nearby South Dakota differed from both Minnesota and North Dakota in crucial ways. In the late 1800s, the absence of transcontinental railroad lines in South Dakota—as well as the presence of the Great Sioux Reservation, which covered the western half of the state—produced dense agricultural settlements east of the Missouri River. A rush for profit in territorial politics punctured settlers' more grounded sense of middle-class republican virtue. The resulting Populist movements played a prominent role in shaping South Dakota politics in the 1890s. When settlers crossed the Missouri River into newly opened indigenous lands in the early 1900s, the middle-class values of wheat farming, property ownership, community, and hard work traveled with them.[53]

The initial League effort focused on northeastern South Dakota, a region of the state with a long history of Populist politics and an active Socialist presence. Albert Bowen directed the initial work. Organizers fanned out to enlist wheat farmers, but unlike NPL efforts in Minnesota, they paid particular attention to known members of the Socialist Party and the Farmers Union. For instance, in Bristol, South Dakota, a League employee immediately sought out a local Socialist. The Leaguer admitted to the South Dakotan that the NPL "had the lists of all Socialist Party members and they first went to the Socialist Party members in order to get help to ride with the league organizer to see farmers." As in North Dakota, many Socialists left their party and took up the League cause. Meanwhile, influenced by the North Dakota

Farmers Union president—"a strong League man"—the state Farmers Union resoundingly endorsed the NPL program in March 1916. The former Socialists and ongoing Farmers Union members formed the core of the almost twenty thousand farmers who joined the NPL in South Dakota.[54]

The South Dakotans who joined the Nonpartisan League that fall and winter of 1916 imagined the NPL program as an opportunity to both increase farm profits and force established parties to pay more attention to agrarians. It also appealed to many former Populists. D. N. Bowrey, a sixty-four-year-old farmer, chronicled his experiences in the Grange, Farmers' Alliance, and Populist Party. He joined the NPL "for [his] sons' sake." Indeed, when surveyed in 1940, more than thirty former NPL members still held strongly to their desire for "a fairer share of the national income." They envisioned "improvement in marketing facilities and cheaper supplies for the farmer" as crucial to their survival. Like the farmers outside of Doland, South Dakota—where a young boy named Hubert Humphrey listened to men discuss politics in his father's drug store—they "took to it like a duck to water."[55]

Nearly as many South Dakotans looked north in fear. League organizers targeting rural precincts across the eastern part of the state soon learned that "the going was tougher in South Dakota" than it had been in North Dakota or Minnesota. Often contemptuously referred to as carpetbaggers, the NPL men faced stiff opposition. Suspicious small-town professionals quizzed the newly arrived Leaguers. One organizer tried to convince an apprehensive banker that "he was selling washing machines." When the South Dakotan finally exposed the true nature of his visit, the NPL employee feebly "defended himself by maintaining that the League was a machine to wash political linen."[56]

Such arguments sometimes fell on deaf ears. Rural South Dakotans' long-standing ability to influence local politics kept farmer discontent more tamped down than in neighboring states. The Republican Party prevailed in state elections from 1906 to 1916 by offering agrarians progressive taxation, curbs on corporate power, and a push for good roads. So even though farm organizations swelled and cooperative elevators took root, not all farmers felt impelled to directly involve themselves in the nonpartisan politics of marketing and selling wheat as a class.[57]

Like George Dixon, leader of the South Dakota State Grange, some believed that "too many Farmer Organizations" weakened the economic chances of agrarians. Indeed, the Grange resisted the League at every turn. Others remembered "how badly they were stung by the old 'Farmers' Alliance' movement as soon as it went into politics." Even famous South Dakota Populists resisted League entreaties. After a December 1916 meeting with Arthur Townley, former US senator Richard Pettigrew criticized the NPL as a moneymaking machine that served Townley's private interests. Clearly, the League would struggle to organize South Dakota's agrarians.[58]

That struggle immediately manifested itself in the November 1916 elections. Even as the NPL began organizing in the state, South Dakota's voters approved a constitutional amendment calling for a state-run rural credits system for farmers. They also elected the man who proposed it—Peter Norbeck—governor. Norbeck, a well driller from Redfield, perfectly understood farmers' grievances. His political career centered on a sensitive understanding of agrarian needs. Entering office in January 1917, Norbeck not only pushed a rural credits program through the state legislature but also called for state hail insurance and suggested the creation of a state-owned terminal grain elevator. He became the central figure opposing the spread of the Nonpartisan League in South Dakota.[59]

Norbeck directly responded to the League's challenge. He was not surprised when a friend wrote him soon after he took office declaring that the "Non-Partisan crowd is coming down from the north like a swarm of grasshoppers." In fact, while campaigning in 1916, Norbeck told voters that "he . . . favored the same things that the NonPartisan League did." Looking ahead to the 1918 election, a number of colleagues advised Norbeck to "get the N.P.L. on the hip, by espousing the planks of their platform which are not objectionable . . . and yet make no mention of the NPL." He hoped to undercut the growing League presence by rallying his progressive Republican base of small-town businessmen, avoiding criticism of the NPL, and co-opting the League platform, which appealed to many of the state's farmers. During the 1917 legislative session, he and his allies did exactly that. Nonetheless, NPL organizers crisscrossed South Dakota throughout that year, signing up farmers.[60]

League employees also looked west. They fanned out across eastern Montana in the summer of 1916. A homesteading boom there in the mid-1910s took advantage of unusually heavy rainfall and bumper crops. Anxious to participate in the wheat economy, farmers and the professionals who served them created community by establishing libraries and schools. As the newly landed, they looked forward to booster-promised prosperity in the dry grasslands.[61]

Like farmers elsewhere in the region, they soon faced an uphill battle for that prosperity. They responded by organizing. By 1916, six thousand farmers claimed membership in the Montana Society of Equity, which mimicked similar organizations in neighboring states. The influence of newly arrived settlers from the Prairie Provinces further pushed the group to envision legislation like the Canada Grain Act of 1912 for American farmers. Unlike organizations such as the Grange and the Farmers Union, their orientation toward a government-aided, cooperative plan for public elevators and terminals en-sured that equity members in Montana joined the NPL in droves.[62]

In turn, League leaders appreciated the Montana Society of Equity's ef-forts to forge alliances with the state's unions. The Anaconda Company—a copper-mining monopoly in western Montana—had controlled the state's politics for decades and alienated many residents. Fierce labor battles punc-tuated state politics from the beginning. The legacy of 1890s pro-silver Popu-lism ensured that both working men and farmers shared a general populist orientation. Recognizing that control of Montana depended on alliances with labor leaders in Butte, Anaconda, and elsewhere, the equity movement offered connections crucial to the NPL's future in the state. It helped that the Anaconda Company directly alienated farmers. In 1917, for instance, the corporation deep-sixed agrarian-backed prohibition as well as range-fencing laws. One correspondent tantalizingly asked: "Could the Non Pats beat it?"[63]

The rapid spread of the NPL, however, initially exacerbated tensions be-tween Montana's farmers and other institutions. Montana Agricultural Col-lege extension director F. S. Cooley, for instance, defended the two-party system challenged by the League. Though the state's small Farmers Union backed the League, a small number of equity-affiliated legislators who won office in November 1916 still opposed it. Once their "new-found Moses, B. C. White," and other equity leaders got "a little more experience in the Legisla-ture," they "changed their attitude" and joined the NPL. In the end, the Mon-tana Farmers Federation approved the "plans, purposes and program" of the NPL, "providing the funds collected in Montana are to be spent in Montana under the direction of Montana men." Flush with victory, the League ham-mered away at what it called "the big copper company" in Montana. It tried to show agrarians that they suffered greatly from a nonpartisan corporate influence in state government. The NPL argued it could only be countered by nonpartisan cooperation among farmers and workers.[64]

The large number of employees required by organizing efforts in states adjacent to North Dakota left League leaders scrambling to cover their grow-ing obligations. When the American Sociological Society requested a speaker on the League for its 1917 meeting, no one could be spared. Because of the "new work of organizing Leagues in Minnesota, South Dakota, and Mon-tana," University of North Dakota sociologist John Gillette reported that the NPL did "not feel that they have a man immediately among their promoting force at present who could take up the movement." Stretched thin, the League needed to ensure effective organizing and expand the membership. It began hiring new organizers and developed a strategy to train them. Teaching new employees the techniques honed by Townley in 1914 became crucial to the

League's effort to establish a broader presence among western farmers. "This work of enrolling members," he told farmers, "is a business function, requiring specialized training." The NPL's successful expansion beyond North Dakota depended on it.[65]

⋆

Some of the new League employees joined the organization because of their firm commitment to the cause. When George Griffith, director of the NPL effort to take over Minnesota, visited Blue Earth County in 1918, he brought years of organizing experience with him. Griffith usually bantered with newly met farmers to break the ice. A "new arrival in heaven sees an angel with ball and chain on his leg," one joke began. Griffith slyly continued: "Hello, Saint Peter, what's this? I thought all angels were free to come and go." Finally, he laid on the punch line: "That one's from North Dakota where they have the League. We got to put a ball and chain on him to keep him from going back." Prospective members reveled in the tale.[66]

Griffith's colleague George Cronyn reported that, once signed up, a new member "will shout League propaganda from the housetop . . . pester his neighbors, jeer at the opposition, and *get the vote to the polls. . . .* And he will follow the light of that $16.00 into the jaws of hell itself." The organizer even overheard one farmer say: "I was raised poor as dirt. I love my wife and my kids, and I ain't agoin' to breed into poverty." After encountering such sentiments, Cronyn sometimes even believed "that this is not a campaign, it's a *Crusade.*"[67]

His idealism belied many difficulties. When first approaching a farm, NPL organizers often encountered "suspicion, distrust, hostility," and an "incurious, hard stare." After all, the average agrarian "had to deal with many a 'slippery customer,' insurance agents, stock investment solicitors, dealers in 'good propositions,' clever and oily talkers," and "sly inveiglers of the credulous." Only after the League employee—himself working on commission—broke through "the crust of old disillusionment, of better experience, paid for by hard-earned dollars and notes that fell due and mortgages called in" could he "restore some degree of faith to the farmer."[68]

In both the United States and Canada, NPL organizers followed consistent formulas to enroll new members. Townley, the leader of the Nonpartisan League, depended on his experiences in North Dakota's Socialist Party to direct League organizing and expansion. Townley's turn to automobile-dependent and commission-based organizing in mid-1914 stemmed largely from his own struggle to support himself as a Socialist Party organizer. Solicited memberships produced monies for the organization and a regular salary and covered transportation costs. "The truth, a Ford automobile, and $16," he told a crowd of NPL members in 1918 "—that is all it takes to win a political victory." In reality, it took much more.[69]

Townley's organizing method depended on a businesslike structure highly dependent on the sale of memberships. The *Kansas City Star* noted that the NPL "bears none of the earmarks of the old Populist movement so far as the methods of organization are concerned." Indeed, at least one close associate asserted that the NPL leader was "capitalist minded." As a journalist put it, "The League has not been merely 'taken up' by the northwestern farmers. It has been sold to them." The NPL's protest of corporate capitalism depended on a deep commitment to business methods rooted in a growing consumer culture that targeted agrarians.[70]

The typical League organizer followed in the long tradition of traveling salesmen. Despite the rise of catalog shopping offered by mail-order firms such as Sears & Roebuck and Montgomery Ward, commercial travelers still visited individual farms on a regular basis. They plied a wide range of wares. Distantly related to the drummers of old, early twentieth-century salesmen read manuals on methods, shared strategies, and carefully studied consumer psychology. Their success depended on intentionally cultivating their persuasive skills. Transitory and yet ever present, they remained outsiders wherever they went. Yet they transcended that status (and the associated suspicions of rural consumers) by providing farmers access to the latest trappings of modernity.[71]

Usually trained at business colleges and part of a carefully organized effort of a large company, traveling salesmen embodied an up-to-date corporate sensibility. Likewise, "the organization methods" of the League "were and are scientifically practical. They are studied and deliberate." Years later, an observer noted that "the key to the sudden growth of the Nonpartisan League lay in the large-scale application of intensive book agent or oil-stock selling technique."[72]

Like the best modern salesmen, League employees focused on the attitudes and culture of potential members. One early organizer wrote: "'Psychology' is a favored word in League circles. 'I don't think it would be good psychology,' is the most frequent objection to an argument or a method of approach." The NPL recognized that "the psychology of the farmer varies in different states and under different conditions." Much depended on the crop that year, party affiliation, and community standing. League employees familiarized themselves with local conditions whenever possible.[73]

For instance, once they started enlisting farmers outside North Dakota, organizers admitted that in some cases "there was no 'discontent' among the farmers." "Before you can organize them," one argued, "you first have to 'create' a discontent in their minds." Indeed, "figures were put on the blackboards by speakers and organizers, showing that the average wheat farmer could gain about $1000 a year through state ownerships of mills and terminal elevators." The best NPL employees always tried to cater to farmers' wants, needs, and desires—and sometimes worked to create new ones.[74]

Even among highly motivated farmers, organizers always followed the first rule of sales—build trust fast. This proved to be the biggest hurdle. In the earliest days of the League in early 1915, "several of the organizers were arrested on charges of obtaining money under false pretenses, so strong was the suspicion in some quarters." Alfred Knutson, a veteran League organizer, noted that "salesmen came around trying to get the farmer's money on some shady deal, so they were wary." One farmer in Alberta wrote the NPL office to make sure he gave his money to a genuine organizer. "I am a poore man and can not aford to fool away one sent. . . . I have been jiped a good many times in my life," which meant he hated "to have a good ta[l]ker walk of[f] with a few dollars of my money." But "the worst cases to tackle," wrote organizer George Cronyn, "are those who no longer believe any good of their fellow-men, or even of their neighbors."[75]

Three tactics warded off skepticism. First, NPL employees tamped down wariness by recruiting a local farmer to accompany them in the day's work. Upon arriving in an area not yet combed by League employees, the organizer often spent "a whole day sometimes just getting one member." Doing even that could involve cutting a deal. Henry Martinson remembered a farmer outside Carrington, North Dakota, who insisted on using his labor for two days before setting out together to enroll neighbors in the League. "It was a relief to climb into the Model T after a hectic two days of shocking the heavy bundles of grain cut by the farmer," he noted. A newly enrolled farmer in Minnesota "did not feel as though [he] could devote [his] time for no pay," so he charged the organizer $1 for the use of him and his team of horses to visit and convince neighbors.[76]

Second, NPL organizers showed prospective members the "names and address of well-known farmers in the state who had already joined the League" or "names of those in the neighborhood who had signed up as League members." Third, putting down traveling salesmen—Leaguers often referred to them as "prune peddlers" or "tin-horn drummers"—firmed up farmers' confidence in the organizer's integrity. In each case, NPL organizers used local community ties to quell initial fears.[77]

FIGURE 7. Nonpartisan League organizer E. A. Young, signing up a new member, circa 1918. (Minnesota Historical Society, J1.4 r37)

After winning a farmer's trust, the organizer pointed out the individual gains made possible by creating a strong agrarian presence in politics. From the earliest days of the movement, a contemporary reported that organizers told agrarians, "'We farmers can't get what we want, because we are not organized as farmers.' . . . The business men have their organizations; the union working men have theirs; the doctors, the lawyers, the editors, all have their organizations to which they pay dues." Prospective members then learned that "heads of the big corporations like the packing trust, sugar trust, machinery trust, and others, give thousands, if not millions, to election funds. That is a fine business investment for them. They get the men they want elected." Organizers then insisted that farmers should spend money in politics to get their way, just as others did. In fact, they claimed that "organization is good farming, good business, if you please." In Alberta, Leaguers went so far as to tell farmers that "you are paying more than $15.00 each year for your own enslavement. Pay $15.00 for your own emancipation." They sold the NPL as the vehicle for exactly that.[78]

✳

Finding the finances to fund the employees necessary for expansion became a primary concern of League leaders. Townley announced to one colleague that "there's plenty o' money—plenty of it. The farmers've got it, and all we gotta do is go out an' get it—get it—kick 'em in the goddamn guts an' get it." The colleague noted that "of course Townley meant no disrespect to the farmer. . . . That was just his rude way of expressing his belief that the supply was inexhaustible." To keep getting the money, "Townley frequently called in his organizers for consultations. At these sessions, methods and arguments were exhaustively discussed. The organizers' meetings became schools of salesmanship."[79]

In fact, "no sales organization ever worked more carefully and intensively with its men than Townley worked with his crew of organizers." James Manahan, the lawyer for the Equity Cooperative Exchange, attended one training session. There he heard Townley implore: "Make the rubes pay their goddamn money to join and they'll stick—stick 'til hell freezes over." Apparently, he offered "many picturesque and even more profane variations" on the same theme. Townley then "drilled his men how to 'organize' the farmer in his barn yard; how to 'surround the rube,' one man in front and one on each side, facing him, and all urging him to join the farmers' non-partisan league, at the same time agreeing with him, good naturedly, on everything else."[80]

Even as he used condescending language to describe agrarians, Townley insisted that League employees respect rural folk. In one instance, the NPL leader "took an organizer to task because he had berated a waitress serving him in a small-town restaurant when the food and service were not to his liking." In fact, "it was an organization rule that an organizer should take great care not to lose his composure, and to be sure that, in the event that he didn't want or refused to sign up a League member, to be friendly with the farmer and leave him in good humor." Townley told another organizer to "take off that diamond ring and that gold fob and those clothes." Instead, the employee needed to "dress like a farmer" and "talk their language." Respecting the social norms of lower-middle-class culture demanded great care, especially when discussing controversial political issues.[81]

Despite these efforts, hard-sell tactics sometimes backfired. On a "cold and windy day in the winter" outside Brentfield, South Dakota, one NPL organizer "talked and talked" to Herman Haven, a local farmer. Exasperated,

Haven finally gave the organizer "fifteen dollars and told him to get the hell out of here." Haven later remembered that he "joined . . . to get rid of the cuss." In another case in North Dakota, organizer Alfred Knutson "was the first to scurry to the security of the Model T . . . parked nearby" when "a cross-eyed farmer driving fence posts into the ground with a sledge hammer" vehemently berated the League "in Polish and broken English." In places where organizers felt unwelcome, an employee "was always on the alert and prepared to depart from where he was, with or without dignity."[82]

By early 1917, the push for expansion forced the NPL to create a new correspondence course "on farm economics and methods of putting our case before the farmers." Formerly North Dakota's best-known Socialist, Arthur LeSueur and his new wife, Marian LeSueur, designed the program. Before joining the NPL, Arthur briefly ran the Socialist Party–affiliated People's College in Fort Scott, Kansas. There he administered a correspondence-based law degree. Marian directed the publicity effort for the college and authored an English textbook for worker-students. They brought their correspondence-school experience and grounding in political theory to the League's educational work.[83]

Seeing the intellectual empowerment of farmers as crucial to NPL expansion, the LeSueurs believed in rooting organizing in ideology. One lesson trumpeted that "as an organizer it is of the greatest importance that you understand as thoroly [sic] as possible the history, development and meaning of the co-operative movement." Unlike "commercial salesmen" who upheld "the old system of competition," League organizers, "as salesmen for this organization, are building the new system of co-operation, to which the old system of competition must yield."[84]

The many lessons of the correspondence course instructed readers in topics such as markets, monopolies, bookkeeping, wages, mortgages, tenancy, consumption, and taxation. One pointed out that "the farmer is the only business man in existence who sells at wholesale and buys at retail." Another argued that "our modern form of society is called capitalism because it means the supremacy of capital" before dissecting corporate organization. Finally, the classes staked out the League's central belief: "We are not planning to take away any terminal elevators or packing houses or anything else from anyone who owns them now. What we want to do is establish institutions of our own where we can have a square deal. . . . We say to them, 'Keep your property but you cannot skin us anymore.' . . . That is not Socialism—that is plain ordinary common sense."[85]

One League employee reported that "inside of a few months more than one thousand" were taking the lessons. By early 1919, "more than two thousand" had taken the course. Successful students of the course usually went to work for the NPL, but their success depended greatly upon further training in the field. These less formal sessions emphasized sales tactics rather than ideology. Alfred Knutson remembered that "organizers would take turns on a subject (the farmer), one who was difficult to convince about the efficacy of the NPL, in a mock demonstration, showing how to breakdown his resistance, convince him, and sign him up as a League member." Years later, former employees remembered these experiences as more useful than the correspondence training. Ideology and deep knowledge of economics did little to address the difficulties facing organizers in the day-to-day enrollment of new members.[86]

Those difficulties punctuated the monotony of an organizer's life on the road. George Cronyn described his experience as "a matter of corduroys and gray flannel shirts, of tin washpans, lumpy beds, fried potatoes and pork, and the Ford, my constant companion; of endless fields of young wheat, endless miles of passable county roads, endless talks with keen-eyed tanned countrymen." An exhausted Albert Bowen—the League's founder—wrote the NPL office from Thief River Falls, Minnesota, after learning of a pending reassignment. He liked speaking in northwestern Minnesota because it allowed him to "get home to my two sweet girls every Saturday." Always "on the *go*," the homesick organizer was "getting somewhat tired of the continual living on hotel fare, sleeping in cold hotels and being unable to have any regularity in life." An organizer in Colorado put it best: "Its [*sic*] a strenuous life."[87]

With the relentless pace of expansion, finding organizers proved the biggest challenge. An early NPL employee noted that organizers "were preferably men who had farmed in North Dakota—and in a majority of cases they were farmers or the sons of farmers." That said, one employee remembered that "it wasn't considered necessary to instruct former Socialist Party organizers in regard to the techniques to be used when organizing farmers. . . . Past experiences in the socialist movement was considered sufficient." Besides their ideological commitment, former Socialists' experiences on the circuit and the stump made them veterans of movement and electoral politics.[88]

Many former Socialists found a home in the League as organizers in the earliest days. Close connections to Arthur Townley mattered most. This sometimes created problems. For instance, the outspoken atheism of J. Arthur Williams, a longtime Socialist, cost the League votes. One organizer remembered that canvassing in Pembina County, North Dakota, proved more difficult after Williams's speeches. "I was chased from four farm places in succession," he wrote, "because of the intense feeling created by his atheism."[89] Townley's trust in former Socialists to do the organizing work of the NPL sometimes proved misplaced. Furthermore, he and his colleagues soon exhausted their personal networks. The ongoing expansion required hiring from a much broader range of backgrounds.

*

Those concerned with the farmer's plight often sought employment with the League. George Cooper, a Methodist minister who preached the social gospel in Ellendale, North Dakota, worked for the NPL in 1916 and 1917. A. E. Baine, who organized for the League in 1916, worked at his local post office before joining up. Early boosters who accompanied organizers on their rounds sometimes became organizers themselves. Other organizers held no political commitments. Like Walter Quigley, they saw it as "strictly a bread and butter proposition."[90]

No wonder, since "men who were successful were making $100 to $200 a week." Becoming a League organizer could bring real wealth. Commissions determined one's salary. Thus the incentive to enroll members proved strong. But much depended on the individual. Failure soon weeded out the less effective men. In Alberta, for instance, the League struggled because "organizers have quit on account of not being able to make good."[91]

Those who stayed remained committed to the NPL for financial as well as ideological reasons. As early as April 1915, Leon Durocher admitted that he left the Socialist Party to become a League organizer because "the wives of all the organizers with the exception of mine have their summer clothes." "Most of the organizers are men who never have made nearly as much at anything else," one recalled. "Consequently they are very keen about taking and obeying orders since they do not care to lose their positions."[92]

When an annual membership in the Nonpartisan League cost a farmer six dollars, organizers got "paid one dollar per member" and cared "for their own cars." The rest covered organizational costs. As time passed, Townley refined the commission structure. When biennial memberships in the NPL went to sixteen dollars, the organizer's commission depended on whether or not the farmer paid with cash or a postdated check. The latter provided an opening to sign up a cash-poor farmer but put less in the organizer's pocket. Even so, once the employee signed up enough farmers, the League-provided Ford he drove from farm to farm became his.[93]

Some NPL organizers "signed more than 100 members in a week," and one claimed that "forty a week is very ordinary." In 1918, Ray McKaig tried to induce an old friend to become an organizer because of the money to be made. "For every member you get into this League," he could earn four dollars. McKaig avowed that "any fellow that has got the ability and has the right chance, and the weather is right, can do the same."[94]

By 1919, one's commission depended not only on payment type but also location. In places where the work proved more difficult, employees earned more. North of the border, a slightly different structure held sway. In Saskatchewan and Alberta, organizers earned $3.75 for every $15.00 annual membership sold—unless they used one of the League's cars, in which case they earned $3.00 per membership. As one organizer put it, "The work depends upon the ability of the man and the amount of hustle he can put into the work."[95]

Commission-based recruiting ensured that the corporate culture of scientific sales defined more than just the signing up of farmers. The movement depended entirely on the income raised from memberships. Indeed, Townley noted that "the success of this organization after everything else is said and done ... depends upon the number of men who have signed up and paid $16." Careful accounting for both members and monies became crucial. In later years, organizers even filed daily reports. Mailed to NPL state headquarters, they included information on where the organizer canvassed, who joined, and the amount of money collected. Finally, the League employee made a list of potential members missed for whatever reason.[96]

Thus the organization's basic structure mimicked the methods of efficiency-oriented companies. Resulting from "the driving energy of a few individuals applying to the political organization of farmers the methods of modern salesmanship," the Nonpartisan League's destiny depended entirely on selling the promise of empowerment. In fact, it produced what the *Nation* called "the unique spectacle in American national life" of a political organization "financed, and adequately financed, from the bottom up, by the rank and file, instead of from the top down."[97]

These methods also created challenges not unlike those found in business. Success in a new sphere sparked competition. Enrolled members provided the organization with hundreds of thousands of dollars. For League organizers and speakers, the opportunity to cash in always loomed. Large amounts of money made on commission led to temptation. Early on, rogue employees occasionally copied the model created by Bowen and Townley and started their own version of the organization.

With Townley occupied by expansion into Minnesota in 1916, staff members beat him to the punch in South Dakota. Early that summer, O. M. Thomason prematurely "attempted to start a league of his own" in South Dakota. Leon Durocher assisted in the effort. An angry Townley got wind of the plan "and sent a force of men to South Dakota" to take over the "Thomason-Durocher league, and absorbed it." Then he fired both men (though within days, both Thomason and Durocher again found jobs with the League). A year later, D. C. Coates, who "piloted the first North Dakota league legislature in 1917," split with Townley on tactics and briefly "quit the League to start one in Washington." Only Townley's personal plea "induced him to stop."[98]

All this made Townley wary of his subordinates. Strict rules emerged. As a result, "the policy of the association is not to allow a man to work too long in one state. After an organizer, speaker or state manager has acquired a certain degree of prominence, he is either dismissed or transferred to another state." Burned early, Townley took "no chances of having a revolt."[99]

Another difficulty stemmed from success. By mid-1917, NPL organizers signed up new members at the rate of twenty-five hundred a week. Not all of the farmers could pay cash. The League's policy of accepting postdated checks provided financial flexibility for potential members. The method proved well suited to creating a membership-based organization in a cash-strapped economy. Less reliable than cash, however, the checks often turned out to be worthless. "In the case of League membership checks in North Dakota," one organizer noted, "these defaults ran to approximately 15 per cent." In other cases, small-town bankers refused to honor them—especially after NPL success in local elections.[100]

Representing lost income, the checks became a priority for the organization. By the end of 1918, the NPL held over $500,000 in unpaid checks. At League headquarters, a growing staff coordinated a wide range of tasks, including promoting petitions and paying organizers. But postdated checks received special attention from "Miss Julia Lee, who sends these checks through the banks several times and collects quite a percentage on the second or third trial. The first time the check is sent it may come back 'payment refused' or 'no funds,' but succeeding attempts sometimes get the check past the banker's eye." At one point, "this department was taking in $12,000 a week." The extension of credit made it possible for cash-poor agrarians to join. But it caused its own difficulties, requiring a serious effort to account for every possible dime.[101]

★

Every dollar the League collected counted. After all, the entire structure existed to support not only political action and employee salaries but also the technology that made the NPL possible—automobiles. An oft-told joke described a banker and a farmer who saw a League organizer pass by in a Ford. The banker said: "You've got $16 in that car." The farmer responded: "'Yes, I know ... but it is not bothering me near as much as the $16 I have in that car,' pointing to the banker's automobile." Agrarians knew that League employees' cars represented a sizable investment. They also knew the NPL's success depended on them.[102]

FIGURE 8. NPL staff in Saint Paul headquarters holding signed petitions from farmers, circa 1917. Whether handling petitions or postdated checks, the work of the central office proved crucial. (Minnesota Historical Society, J1.4 p10)

Charles Edward Russell, for instance, reported in 1920 that soon after starting the League, "Townley had 10 automobiles rolling through the state, with canvassers inside of them taking subscriptions and enlisting members; then 20, then 40; and then 60. He bought machines, ran in debt for them, chartered them, borrowed them. Steadily the figures of membership went up." Recognizing the significance of cars, the NPL leader entrusted his brother with the supervision and maintenance of the organization's automobile fleet. Enlisting large numbers of dispersed agrarians demanded a technology that efficiently traversed the wide-open landscape. Automobiles powered by internal combustion engines and fueled by gasoline fit the bill.[103]

Born in western North Dakota, where vast distances between towns defined daily life, the Nonpartisan League embodied rural North America's preoccupation with cars. City-dwelling bicyclists looking to the country for recreation demanded good rural roads as early as the 1880s. Agrarians disliked urban cyclists and resisted their push for road improvements. As automobiles became more popular in the 1890s, farmers initially resisted their presence as well. Concerted reform efforts emanating from cities slowly transformed agrarian attitudes. Farmers began to envision roads as a technology serving a broader social good rather than as a local affair. Transforming rutted dirt paths into real thoroughfares demanded state aid and thoughtful engineering.[104]

The push for all-weather roads in rural America gained real momentum, however, when the US Post Office began offering rural free delivery in 1896. Just three years later the federal agency refused to deliver mail directly to farm homes without the assurance of year-round passability of roads. In one stroke, modernity and prosperity came to entirely depend on improving rural road systems. Rural free delivery offered not only convenience but also new connections between homesteads and the world. Farmers began supporting road improvements and buying cars with gusto.[105]

Across the northern plains, this national pattern replicated itself. By 1905, North Dakota legislators set urban and rural speed limits. Car dealers, auto clubs, and service stations sprang up in many towns. In 1909, farmers purchased one-fourth of all the cars sold west of the Mississippi. When Arthur Townley gave up hitchhiking and bicycling in 1914 so that he could organize in a car, North Dakota ranked fifth in the nation in per-capita automobile ownership. In the first seven months of 1915 alone, state residents purchased more than 6,000 cars. By 1921, the 650,000 residents of North Dakota owned more than 100,000 cars.[106]

Cars proved popular among farmers across the North American West because they reoriented experiences of rural space. They allowed more frequent social visits and trips to town. Automobiles collapsed space and time, just as railroads had years before. Cars even furthered existing conceptions of community by making sociability readily available.[107]

But not until the early 1910s did automobiles become a necessity. The emergence of mass-produced vehicles—especially Henry Ford's Model T—and agents hawking installment plans made all the difference. Indeed, that decade saw the price of the Model T dramatically decrease. By 1916, the cost fell to $360 per car. Mortgage-ridden farmers on both sides of the border found automobile ownership within reach. Branded as the "the Farmer's car," the Model T offered modernity and efficiency even as it appealed to rural thriftiness. It sported a high clearance, a powerful engine, and a lightweight but especially strong chassis. Readily available, easy to run, and simple to fix, Ford's flivver proved the perfect tool for NPL organizing. It helped League employees transcend the limits of distant farmsteads and a rugged climate.[108]

To be sure, rural isolation and the weather still mattered. In 1918, League employees in Idaho anxiously awaited "information concerning road and weather conditions" in the northern part of that state. Heavy storms sometimes made roads impassable. When they did, NPL organizers turned to section lines or even the service trail alongside railroad grades. When cars failed them altogether, organizers improvised. Alfred Knutson burnished his reputation as a dedicated Leaguer by spending his first winter as an NPL employee in North Dakota traveling from farm to farm on skis. Drawing on his Norwegian heritage, "he brought in new members all winter long while other organizers were stymied by blocked roads." Likewise, Gilbert Lee, a farmer in Griggs County remembered for being "100% behind A. C. Townley," worked "tirelessly, often traveling on snowshoes in the dead of winter to induce farmers to join the movement."[109]

Regardless, traveling from farm to farm on a commission basis demanded efficiency. Time was of the essence. In an energy-diffuse landscape, gasoline-powered cars like the Model T offered breathtaking mobility and freedom. The average Model T got about twenty miles per gallon. In small towns without filling stations, general stores, hardware stores, and even blacksmith's shops sold gasoline. Farmers grew dependent on the convenience autos offered. Though the need for fuel, as one organizer noted, occasionally demanded "an hour into the little city for gas, and another sixteen miles out home again," internal combustion engines that ran on concentrated liquid energy almost literally made time.[110]

For these reasons, the Model T became a centerpiece of the Nonpartisan League's culture. Ford's car proved to be the perfect vehicle for popular politics. The auto maker's obsessive application of scientific management brought the costs of cars in line with a mass market of middling Americans. The constant effort to standardize manufacturing and eliminate waste famously undermined the power of wage laborers in Detroit. But the very techniques that degraded the status of auto workers and concentrated wealth and power in a management class produced the vehicle that animated anticorporate insurgency across the rural North American West. In combination

FIGURE 9. Nonpartisan League organizer with his automobile and an issue of the *Nonpartisan Leader*, 1917. (Minnesota Historical Society, J1.4 r27)

with modern sales techniques as well as a strict organizational hierarchy and carefully managed organizer training, the Model T made for a literally Fordist form of populism.[111]

The shift from mobilizing an already agitated voting public in North Dakota to organizing new constituencies in neighboring states and provinces demanded innovation. Townley and other NPL leaders responded by systemizing employee training, adjusting the commission system for organizers, and buying automobiles by the dozen. Success depended on tactics and technologies. It also demanded a cohesive message that embodied the group's ideas and brought them into the everyday life of farmer members. Newspapers proved to be crucial to achieving that. The League's message reached individual agrarians via print. Indeed, the medium proved essential for making the personal political.

★

Newspapers delivered by the post office became the perfect tool for shaping opinions, reflecting reader sentiment, and delivering news. The rural free delivery of mail not only encouraged road improvement across the North American West (it came to Canada in 1908) but also made it possible to communicate directly with farmers on a regular basis. The Nonpartisan League's chief media outlet, the *Nonpartisan Leader*, offered lifestyle columns, serial stories, and editorials as well as official announcements and reporting.[112]

Funded by memberships as well as advertising, the League's official organ defined most individual farmers' experience with the organization. In fact, the *Leader* often served as the only formal contact with members in the months between elections. The publication claimed that "all newspapers owned by individuals or private companies are run for money-making purposes." Thus there remained "only one way to get pure news and honest support for the farmer." They needed to "own and run their own paper." Members rightly believed the *Leader* to be their own. It consistently voiced agrarians' concerns.[113]

The paper also served as a recruiting device. Farmers shared copies with neighbors, circulating back issues throughout their communities. Prospective members who read the *Leader* learned about NPL goals, critiques, efforts, victories, and defeats. Circulation grew alongside membership. By 1919, more than two hundred thousand subscribed to the *Leader*.

By mid-1917, League leaders also began encouraging farmers to work cooperatively to produce local papers. This initiative supplemented the offerings in the official Nonpartisan League publications with more focused content. It also became a tool for further empowerment. When farmers became stockholders in a newspaper, they created civically engaged community and found their own voice. To support this effort, the League's national office in Saint Paul established the Northwestern Service Bureau. It proposed to "help in establishing farmer-owned papers in every county in the territory in which the League is operating." Encouraging the extension of pro-NPL views into a wider variety of media outlets, it also offered syndicated content to allied newspapers.[114]

Renamed in late 1918, the Publishers' National Service Bureau published pamphlets showing farmers exactly how to found or purchase a local newspaper. Advice included tips on stock distribution, dividend rates, and ensuring sustainable circulation. In Long Prairie, Minnesota, League organizer W. C. Coates told farmers that the "time has come for the establishing of an independent newspaper . . . a paper that stands for justice and a square deal for all—a farmer's paper that is up and on the bit all the time." Hoping that control of a local press outlet might "expose the rottenness in county and state politics," he implored "live-wire workers" to participate.[115]

In some cases, agrarians purchased existing papers. In Walsh County, North Dakota, editor Frank J. Prochaska approached League organizers in 1917. Openly supporting the NPL, he offered to sell his newspaper to them. It took farmers a few months to raise the money. When they finally purchased the *Park River Gazette News* in December 1917, they changed the name to the *Walsh County Farmers' Press*. Prochaska stayed on as manager, noting that in the past "the advertisers and the bankers who loan money to the paper virtually" wrote the editorials. Under new ownership, "for the first time in many years," Prochaska "felt the full freedom of action. With the farmers back[ing] of the publication there was no more harkening to the dictates of any local interest." In fact, the editor claimed that "the 'freedom of the press' after languishing locally for many years was restored."[116]

In at least one case, NPL leaders assigned an editor to an existing paper. When Sheridan County, Montana, Socialists turned over their newssheet, the *Producers News*, to the League in 1918, NPL leaders sent longtime Socialist editor Charles Taylor to Plentywood, the county seat. Taylor had left International Falls, Minnesota, to volunteer his services to the NPL earlier that year. Arriving in Montana, the new editor quickly took charge. He organized a stock sale to local farmers and began vigorously touting the NPL platform.[117]

By 1919, League farmers had founded or purchased more than seventy county papers across the North American West. Used effectively, they helped agrarians imagine themselves as part of a much bigger community. Sustaining the vision of farmers as a distinct class with common problems and specific political needs depended on it. In turn, class cohesion hinged on capturing hearts as well as minds. Cartoons and slogans made newspapers a crucial tool in organizing.[118]

✶

The *Nonpartisan Leader* became more famous for its iconography than its ideology. Arthur Townley knew that cartoons offered an accessible way to frame a message and motivate members exhausted by farm work. In mid-1915, he demanded that the *Nonpartisan Leader* "have some cartoons—pictures— something the farmers can laugh at—something they can get without having to read." The paper soon hired John Miller Baer, the Democratic Party–appointed postmaster at Beach, North Dakota, as a cartoonist.[119]

Despite his private concerns about the League's potentially "socialistic" approach to government, Baer began creating images that enthralled members and offended League opponents. By early 1917, his fears dissipated. Baer wholeheartedly embraced the League.[120]

Baer's cartoons quickly became a beloved feature of the paper. One contemporary called "his drawings . . . the very heart of the Leader. " His careful use of symbols and analogies outlined League thinking more effectively than any writer, speaker, or organizer. In the September 30, 1915, issue of the *Nonpartisan Leader*—the first—Baer identified the problems facing farmers in his first official NPL image. "Getting Ready to Carry the State" featured fashionable fat men in top hats representing the Minneapolis Chamber of Commerce, bankers, and the railroads sitting down to deal with North Dakota's kingmaker—Alexander McKenzie—as well as the sitting governor, Louis Hanna. In this cartoon, Baer drew on long-standing tropes in populist, anti-monopoly cartooning. In this illustration of the corporate manipulation of politics as well as the complicit corruption of compliant politicians, a familiar representation was born.[121]

The engorged, suited, grinning businessman became a regular feature of Baer's cartoons. Within months, a well-dressed obese man, often wearing a fancy hat, came to stand for what the League's speakers regularly railed against as "big business." It replaced more intricate portrayals of the multiple interests arrayed against the farmer. This simplified symbolization of complex and entangled corporate structures—in railroads, flour milling, grain trading, and banking—made a mercurial and often invisible opponent easy for farmers to identify. In the mind of lower-middle-class agrarians, bigness connoted illegality, immorality, and ill-gotten gains. Soon known simply as "Big Biz," by mid-1916 this stock character showed up again and again in Baer's contributions to the *Leader*. His cartoons suggested that the ever-nefarious "Big Biz" lay behind most efforts to defeat and discredit the League.[122]

Even as he identified the enemy, Baer created another recurring figure. This one embodied the paper's readers. Hiram Rube—a deliberate contraction of "Hi, I'm a rube"—played on the frustration of League members. They believed that bankers, small-town merchants, politicians, and grain traders held little respect for agrarians. Farmers also imagined themselves as taken advantage of, just as a rube might be. Reclaiming a derogatory term, Baer crafted a symbol that represented League members' political awakening. Including farmers in his images and storylines, he furthered their transformation by showing Rube's growing empowerment. In turn, NPL members fully embraced the Hiram Rube character.[123]

The white-haired, goatee-sporting, and overalls-wearing Hiram Rube bore more than a passing resemblance to Uncle Sam. Adroitly melding farmers who "seldom got out of their overalls" with the very personification of the United States, Baer crafted an image that played on long-standing romantic characterizations of agrarians. The Jeffersonian notion that posited farmers as the backbone of republican values in America took on new meaning in this context. Evoking Uncle Sam allowed the NPL to symbolically conflate the specific class interests of agrarians with the broader interests of the nation. In this way, Hiram Rube encouraged readers to imagine themselves as the most thoroughly American segment of society. This justified their insurgency. In

FIGURE 10. Cartoon by John Miller Baer: "Hiram Rube Reflects on His Experiences in North Dakota,"
Nonpartisan Leader, May 11, 1916. (North Dakota Institute for Regional Studies, NPL00095)

Baer's cartoons Hiram Rube symbolically synthesized past populisms with the distinctly contemporary popular politics fostered by the League.[124]

Hiram Rube—like so many NPL members—bore the knowledge of years of toil and struggle. His physical strength seemed matched by hard-earned economic enlightenment. In May 1916, for instance, Rube appeared in a cartoon that tracked the descent of optimistic middling agrarians into a muddle of mortgages and debt. In the first image, a vigorous young farmer shows off his work ethic and takes on debt to purchase horses. By the last panel, the older, wiser Hiram Rube laments years of struggle and further financial loss. In identifying the existential threat that mortgages posed to middle-class lives, Baer hit the mark. Providing a backstory for the character that captured so many farmers' experiences, Baer made sure that readers saw their own experience reflected in Rube's. Empowered by experience, Hiram Rube became a signifier of strength, action, and engagement for farmer readers.

Hiram Rube represented farmers in action. Analogies drawn from agricultural work—whether domestic or in the fields—predominated. In early 1917, Baer portrayed Hiram Rube at the washtub, hanging out newly laundered bills (passed by the NPL-controlled legislature) to dry. By 1918, Hiram Rube, bag of "liberty seeds" in hand, sowed democracy for a future harvest against an American flag sky. Whether dangling legislative laundry on the line or spreading kernels of democracy, Rube came to symbolize the farmer-members of the League as no other character could.[125]

The cartoonist even gave the League its most useful slogan. In June 1916, he penned the phrase "We'll Stick, We'll Win." It merged the duties of members with the shared goal—political success. Acknowledging that previous agrarian movements fell short because of internal struggles, the slogan called for solidarity. It also suggested that determination in the face of powerful forces might deliver victory. Featured in countless cartoons in the

FIGURE 11. Cartoon by John Miller Baer: "By the Dawn's Early Light," *Nonpartisan Leader*, May 20, 1918. (North Dakota Institute for Regional Studies, NPL 00033)

Nonpartisan Leader and soon found on pennants, stick pins, and banners, the motto defined the defiance of farmers who joined the NPL.[126]

Well known among North Dakota's farmers for his weekly contributions to the *Nonpartisan Leader*, in 1917 Baer emerged as a leading candidate in a special election called upon the death of North Dakota's US House of Representatives member H. T. Helgeson. Never intending to become involved in national politics, the League nonetheless decided to take a careful first step into the new venue. NPL members immediately confirmed their suggestion of Baer as the best candidate. The League's first attempt to elevate a representative to national office proved successful. Voters ensconced Baer in Washington, DC, where he still found time to shape farmer sentiment in an occasional cartoon for the *Leader*. Many years later, he joked that he "caricatured [his] way into Congress."[127]

The League's deliberate and thoughtful approach to media proved foundational to its success. Newspapers delivered to a farmer's mailbox did more than examine current events. They became a way for the League to shape and reflect members' opinions. Appealing to emotion as well as intellect, slogans and cartoons supplemented more journalistic fare. Farmer-owned newspapers took League strategies a step further. Recognizing the significance of print media in a preradio age, NPL leaders and members created a dense communications network to unify middling farmers.

This effort transcended the international border. League organizers in Canada also appreciated the power of newspapers to deliver images and encapsulate ideas. From 1913 on, Arch Dale's cartoons for the Winnipeg-based *Grain Growers' Guide* pilloried the Liberal and Conservative parties as well as special interests. Cartoons published in NPL newspapers in Saskatchewan and Alberta followed the example of both Dale and Baer. Alberta League newspapers even occasionally printed versions of Baer's cartoons with modified captions.[128]

Unable to deploy Uncle Sam as a national symbol embodied as a farmer, Canadian cartoonists more often included women in their images. The powerful impulse for women's participation in western Canada's agrarian politics encouraged this trend. In the United States, women appeared in League cartoons only as subordinate farm wives or as the more traditional stylized representations of democracy or the nation. But women proved an especially powerful force in NPL activities in Saskatchewan and Alberta. A cartoon on the back cover of the Saskatchewan Nonpartisan League's pamphlet *The Why of the Farmers' Nonpartisan League of Canada*, printed in early 1917, displayed an agrarian woman chasing a greedy pig identified as "big biz" away from the profits meant for farmers represented as babies. It highlighted women's power at the expense of men's.

In the United States, Baer's cartoons consistently represented physically strong and morally upright adult men defending their livelihoods—and by extension, their families. Canadian cartoonists proved more willing to acknowledge a space for empowered women in the movement. Farm women there nonetheless found their political expression constricted by the assumption that their power stemmed from their domestic responsibilities. However constrained, women would mark one of their first postsuffrage successes by electing a woman to Alberta's parliament under the NPL banner. At least one farm woman wrote to the NPL office in Calgary noting that she found "in this League an anchorage for the women voters." Indeed, women proved crucial to the early success of the NPL in Canada.[129]

✶

Though the Alberta NPL and Saskatchewan NPL sported only a few thousand members, many were women. They tested the electoral waters—and whether or not the League idea stood a chance in a different country—early. The parliamentary system forced them to operate as a third party. In Alberta, events moved quickly. Elections came just four months after the meeting that brought the NPL into being. Despite having "only a skeleton of an organization," the Alberta NPL decided to enter the June 1917 fray in the ridings where most of their members lived. Affiliated with a candidate in Acadia and putting forward their own candidates in South Calgary, Gliechen, Nanton, and Claresholm, the League seized the latter two seats.[130]

The victory of James Weir in Nanton showed the power that organized farmers brought to politics when they worked together. Weir, a longtime United Farmers of Alberta (UFA) activist and vice president of the provincial organization, came from a community that shipped more than a million bushels of wheat east every fall. Already engaged in cooperative purchasing and UFA initiatives, Weir and his neighbors worked hard to bring equity to farm economies. The Nonpartisan League became another, more potent, vehicle for doing so.[131]

Louise McKinney's success in Claresholm made her the first woman elected to a parliamentary body in the history of the British Empire. It also made her an instant celebrity. Fellow League members crowed that the "Non-Partisan League platform is the only place where the high-minded women of this Province can stand with dignity and clean feet." Indeed, her election proved to Alberta's women that the NPL took their concerns seriously.[132]

McKinney embodied the central role of women in the Canadian iterations of the Nonpartisan League and inspired others to not just join but to work for the NPL. In the months to come, at least one farm woman, Jean Stevenson, actively organized for the League in eastern Alberta. She wanted "all our speakers" to "appeal especially to the women" the way McKinney did. Only when every Leaguer realized "fully that a political responsibility is a moral responsibility," she said, "and not as it now is a kind of grabbing match I believe you will find that women will tackle the job with enthusiasm and common sense."[133]

Empowered by unlikely victory, the Alberta NPL moved to expand both its membership and its message. William Irvine later remembered that though the NPL won only two seats, "the Liberal Party was astounded that an upstart new party should have met with such success." Furthermore, the two new MPs "voiced the grievances of the people with honesty and without fear." Irvine marveled at "the amazing fact that the peoples' political organization could . . . debate with Ministers in Parliament with credit." Finally, "the farmers took heart when they saw the possibilities opened up by the Non-Partisan League." So did laborers. Calling for a "democratic administration of the people's business," William Irvine soon busied himself with organizing Calgary's wage workers along the same lines.[134]

In Saskatchewan, the burgeoning NPL made an effort to challenge the ruling Liberal Party, which had long held sway there. Through the early 1900s, they consistently positioned themselves as the party for farmers. They also controlled the Saskatchewan Grain Growers' Association (SGGA). In February 1917, the League ejected SGGA director P. L. Craigen over his queries about financial impropriety in the NPL. That alienated the already wary leadership of the most powerful farmer organization in the province. Three months later, at the SGGA's annual convention, Leaguers in attendance tried to push the farm organization into nonpartisan electoral politics. Liberals on the SGGA executive committee skillfully quashed the effort. Seizing the moment, they then immediately called for provincial elections.[135]

With only a few weeks to select candidates and generate a campaign, Leaguers found themselves stretched thin. Further hamstrung by a shortage of organizers, the NPL struggled to make face-to-face connections with farmers. Making matters worse, what they described as "the Liberal machine" concentrated on the ridings where the NPL showed strength. Nonetheless, counting on their months of organizing in south central Saskatchewan, Nonpartisan Leaguers put forward eight contenders for the province's legislative assembly. They included Zoa Haight, vice president of the Saskatchewan Women Grain Growers' Association (and Saskatchewan NPL founder Silas Haight's daughter-in-law). Anxious to push for "many drastic reforms that will eliminate the woeful woman-waste and useless consumption of woman-power at present going on every farm in Saskatchewan," she became the first woman in Saskatchewan to accept a nomination as a candidate for the provincial legislature.[136]

Suffragists across Canada swelled with pride. The newly elected Louise McKinney, for instance, congratulated her friend. The Albertan hoped for Haight's success, since, she wrote, "I shall feel less lonely if I can clasp hands with some other woman across these prairies of ours." The NPL touted Haight's life as "a farmer's wife" as well as her commitment to the League's program. Despite great hopes, the elections proved disastrous. Haight went down to defeat. Only one NPL candidate won a riding—and the Liberals and Conservatives endorsed him as well.[137]

Across provincial lines, the Alberta NPL felt strong enough to advocate for more. Openly calling for the conscription of wealth alongside the conscription of men, the NPL adamantly attacked the policies formulated by the wartime coalition of Liberal and Conservative politicians that ran the federal government. Enduring three years of massive casualties, many farmers insisted that the war threatened democracy at home. This made the League an easy target for flag-wrapped critics. The surging NPL entered four candidates into ridings for federal elections in December 1917. All of them lost. "The Win-the-War and Conscription issues succeeded in diverting the attention of the electors," one Leaguer reported. The defeat sparked intense reflection about the future of the NPL in Alberta. Some began to envision a union of their beloved UFA and the NPL. The emerging plan to do so would dramatically shape the future of both agrarian organizations. Meanwhile, League members south of the border came up with new plans of their own.[138]

*

Whether they planted potatoes on irrigated land in Idaho, sowed sugar beets along Colorado's Front Range, or herded dairy cows in hilly Wisconsin, the NPL's success attracted agrarians' attention. Townley and his circle of advisors recognized that a program developed for wheat farmers exploited by Minneapolis millers might not work everywhere. In fact, in 1917 a League staffer noted that "our program varies somewhat for the different states." Expansion into other parts of the country demanded a great deal of leaders who only knew the Dakotas, eastern Montana, and Minnesota well. In those states, the League filled a clear void. But in other places, existing organizations run by farmers had already effectively confronted local problems and established networks for action. Establishing good relations with those organizations proved crucial to the NPL's successful expansion.[139]

To learn more about the possibilities for organizing elsewhere, in late December 1916 NPL leaders sent Ray McKaig—staunch Leaguer and president of North Dakota's Grange—on an investigative tour of Nebraska, Colorado, Idaho, and Washington. Under the cover of sharing the League's story with fellow farmers, they wanted McKaig to assess the potential for NPL success in different states. Decisions about where to commit the NPL's limited resources depended greatly on his findings.

More than any other prominent Nonpartisan Leaguer, McKaig knew—and was known to—farmer leaders across the country. He carried with him both credibility and connections. McKaig grew up in a family fueled by the social gospel. A 1900 graduate of Hamline University in Saint Paul, he followed his father into the Methodist pulpit. Resigning from ministry in 1913 for health reasons, he resettled his family in south central North Dakota. There he became deeply involved in the Grange and then the League. Spreading the gospel of the League appealed to McKaig. He took on his assignment with vigor.[140]

McKaig arrived in Nebraska in mid-January 1917 for that state's Farmers Union convention. He noted that "the Union wants us to come but . . . are afraid we might swallow up their organization." From the western part of the state, the Nebraska State Grange pledged its support to the NPL. Its leader, James D. Ream (himself a former Populist), reported that "our conditions are different from ND—because we have gotten the RR under fairly good control . . . and less unbearable unjust conditions." But Ream remained convinced that the NPL would "strengthen the Grange and Union" in Nebraska.[141]

Ray McKaig's tour next took him to Colorado, where he "really began to see genuine enthusiasm." The Congress of Colorado Farmers welcomed the League representative to the state agricultural college in Fort Collins. Before McKaig addressed the group, a speaker there declared the NPL "the only farm movement ever founded" that was "economically sound." He further asserted, "You can't beat it, you had better meet it." The next day, McKaig spoke to the assembled Colorado State Grange. Farmers there met him with acclaim.[142]

A general impulse for reform in Colorado made McKaig's way easier. A meeting with labor representatives in Denver went much like the farmers' meeting. The union men pledged to "absolutely co-operate . . . to every extent." Recognizing the potential for success, McKaig promised an audience in Fort Morgan that the League would come "to the state in July with a fleet of Ford auto submarines and will torpedo the stand pat gang of rank partisanship." Finally, the Colorado Farmers Union met McKaig with open arms. Agitated by the power sugar processers wielded when pricing Front Range sugar beets, agrarians there ached for change. Shouting out, "Do you want us to come?," the Leaguer received an affirmative answer.[143]

Moving on to Idaho, McKaig addressed the State Federation of Farmers in Boise. He flattered the crowd with his assertion that the "farmers' Congress" put together by the federation represented "an attempt at a Nonpartisan league." Indeed, frustrated by the false promises of private irrigation companies on the Snake River Plain as well as low prices for sugar beets from processers in Salt Lake City, Spokane, and Seattle, farmers demanded the protection fostered by state-owned enterprises. They also wanted state-run banks for easier access to credit. Finally, they hoped to undo party politics while avoiding the creation of a third party. The failure of previous efforts by Populists, Socialists, and Progressives in Idaho showed the futility of the more conventional strategy. Within minutes, the federation insisted that the League begin organizing in Idaho no later than May 1917. After McKaig spoke at the mostly Mormon Intermountain Equity Congress in Pocatello a few days later, farmers there overwhelmingly approved a similar resolution. Even a Socialist local in northern Idaho sought out the NPL to begin working in that area.[144]

In Washington State, McKaig used his Grange connections to organize meetings. From Spokane to Seattle, the Leaguer spread word of NPL efforts in North Dakota. Farmers in the Evergreen State—whether they grew grain or fruit—struggled with monopolistic water companies. Already inclined to make Washington "a real nonpartisan state," a small group of Socialists and labor leaders in Seattle had organized their own Nonpartisan League in March 1916. The effort to break the hold of political parties through "the passage of nonpartisan election laws" did not advocate for state-owned enterprises but did embrace the NPL's tactics.[145]

The twenty-thousand-strong Washington State Grange promised McKaig their "unanimous support." The leader of that organization, Casey Kegley, a former Populist with socialist leanings, jumped on the chance to bring the League to Washington. So did the leadership of the much smaller Washington Farmers Union. McKaig reported that the farmers, however, "demanded" that NPL planners include the state's labor movement. Kegley had worked to build relationships with labor for more than a decade in an effort to pass the recall, the direct election of US senators, and the direct primary. Having created a joint state legislative committee with unions in the 1910s, he and his fellow farmers refused to throw aside their hard-earned alliance.[146]

In his final report, McKaig told NPL leaders in Saint Paul that "Colorado and Idaho . . . must be organized immediately this summer as Farmers' Nonpartisan Leagues." Nebraska featured "too many propserour [sic] farmers . . . but it can be organized." In contrast, Idaho and Colorado "will be very easy to organize." While he considered Colorado "very ripe," McKaig noted that "Idaho has more grief than Colorady [sic] in its farm products." Yet in Washington, the NPL needed a different tack. There, McKaig noted, "We must take in others besides farmers because the most progressive people in the state are not only farmers but consumers." Acting on McKaig's assessment, the

League sent off its most trustworthy organizers to as many of these states as it could.[147]

∗

Townley and his advisors first targeted Colorado. Traveling to Denver with Casey Kegley, master of the Washington State Grange, NPL organizer Alfred Knutson began signing up members. Most of those who joined the League were also Grange members. Knutson set up a skeletal organization before moving on to Idaho and Washington State. R. W. Morser, a former Socialist from Bowman, North Dakota, took over. Colorado Grangers soon provided the League with a list of potential allies across the state. Drawn from Progressives, labor leaders, and ministers, the resulting coalition rendered farmers a minority. As a result, divisions permeated the Colorado NPL. It took months—and an intervention by the national office—to begin rural organizing in earnest.[148]

When Alfred Knutson arrived in Idaho, he joined with William Scholtz, president of the Idaho State Federation of Agriculture, to help local farmers organize an executive committee. Knutson also began enrolling members in the small farm towns just west of Boise. Soon after, he traveled north to call on agrarians on the Camas Prairie of northern Idaho. Finally, he also visited the Twin Falls area. The League message held a special appeal to farmers working land on big irrigation projects in the Snake River valley. From Nampa to Minidoka to Blackfoot, these agrarians became the core membership of the Idaho NPL. Soon, the Idaho Federation of Agriculture—insisting on local control of the state organization—turned itself into the Idaho branch of the Nonpartisan League.[149]

To help the Idaho NPL prepare for the 1918 election, Townley sent Ray McKaig to Boise. McKaig immediately went to work "taking out organizers, breaking them in, speaking, and standing by Idaho." By July 1917, McKaig reported to the League's Saint Paul office that they had already enlisted twelve hundred members. He hoped for more, but noted that "we need cartoons . . . make them send us forty or fifty . . . so we can use two, three, or four an issue to help our paper out." Luckily, local branches of the Farmers Union and the Grange continued to lend their support. By the end of August 1917, the NPL had signed-up more than four thousand agrarians in Idaho. McKaig soon eyed northern Utah as another potential bastion of League support. Irrigation-based farmers there shared the Mormon faith of southeastern Idaho's farmers and faced the same agricultural processing monopolies.[150]

Washington State saw the itinerant Alfred Knutson come into his own. Recognizing his trustworthiness as well as his organizing abilities, the Saint

Paul office quickly designated him state manager. Sending out organizers to rural precincts, Knutson focused his efforts on talking to county commissioners and township supervisors. If they joined the NPL, efforts for direct democracy might gain momentum. Yet they usually resisted League entreaties. Despite the backing of many in the state Grange as well as labor leaders, Knutson later remembered that "our best success was with socialists or former socialists." At least one League intimate offered a reason for the slow work in Washington State. He noted that membership drives there stalled because of "the very diverse character of agricultural production." As late as December 1917, Knutson reported that Leaguers in the state were only "plodding along."[151]

Finally, the NPL turned its attention south to Nebraska. In May 1917, O. S. Evans, another former Bowman, North Dakota–based socialist, began organizing farmers there. As a hotbed of Populist activity in the 1890s, Nebraska seemed a likely field for growth. Within three months, Evans employed fifteen organizers to canvass the eastern and central parts of the state. But the Nebraska Farmers Union ultimately rejected the NPL's call for action. This formal resistance to the NPL meant that things moved slowly. By August 1917, only a thousand Nebraskans had joined the League. Membership growth there remained incremental into 1918.[152]

Trying to enlist a wide range of agrarians, the NPL needed to convince other farm groups to endorse the effort. Where they succeeded in that, they prospered. When they did not, it became difficult to find new members. The necessity of convincing many different kinds of farmers that they held more in common than not also became clear. The shared identity that powered the movement in the spring wheat belt proved difficult to immediately replicate elsewhere. Nonetheless, the notion that farmers might control a state government continued to provoke interest among agrarians well beyond the reach of the League's capacity to organize them.

✦

Oregon offered the best example. In late January 1917, representatives from the state's Federation of Labor, Farmers Union, and Grange met in Portland to formulate a political program and create a joint organization. Charles E. Spence, the master of the Oregon State Grange, reported to his friend Ray McKaig that the assembled decided "to organize an 'Oregon Non-partizan league' . . . but also voted to cooperate with the N.P.L. if it comes into Oregon." That said, Spence told the NPL organizer that the Farmers Union "has instructions to keep the League out of Oregon."[153]

McKaig, already stretched thin, took the warning seriously. He refused to commit limited resources to Oregon. Portland's leading newspaper, the *Oregonian*, viciously attacked the NPL efforts in Idaho and Washington. Spence, anxious for the NPL to expand into Oregon, assured McKaig that "the pounding of the Oregonian . . . has bourne [*sic*] fruit" by convincing some farmers of the League's worthiness. But despite Spence's best efforts, Oregon's Grangers decided to remain free of political affiliations. Further attempts to convince Oregon's Farmers Union that agrarians needed to join the NPL failed.[154]

Farmers from the winter wheat belt also reached out to the NPL. In January 1917, Ellsworth, Kansas, newspaper editor Milton L. Amos offered to "entertain and route . . . organizers." He touted the strength of local cooperatives as well as the local Farmers Union chapters in central Kansas. Agrarians there hoped to organize as a class to exert more political power in an agriculturally oriented state.[155]

But A. P. Reardon, the elderly head of the twelve-thousand-strong Kansas State Grange, did not trust the NPL. Fellow Granger Ray McKaig privately characterized him as a "bonehead." After Reardon stepped down in early 1917, younger leaders responding to their membership proved more interested in the NPL. They wrote to McKaig, asking him to come speak in Kansas. Just weeks later, however, Grange leaders again became suspicious of the NPL. This back-and-forth created confusion amongst rank-and-file Grangers in Kansas. McKaig responded by assuring his colleagues that the League could be trusted.[156]

The national NPL office gave up on the Grange and turned to the Farmers Union. After sending representatives to meet with "leaders of the various farm and co-operative organizations," they targeted Saline County, which the *Kansas City Star* referred to as a "hotbed" of Farmers Union organizing. Guided by prominent Farmers Union members, one correspondent suggested that "the Kansas men take to the League idea of political action." By December 1917, enough Kansans became Leaguers that farm leaders across the state took notice. Governor Arthur Capper, a Republican, even provided references for the League's rental application for office space in Topeka. In January 1918, however, leaders of the state Farmers Union—almost one hundred thousand strong—changed their minds. They publicly labeled the NPL's views as "purely socialistic" and urged members to avoid the League altogether.[157]

Finally, League leaders watched Wisconsin closely. Milwaukee stood out as a Socialist stronghold. Reformers rallied behind US Senator Robert "Fighting Bob" La Follette, who played a central role in state politics. Internal struggles

over taking an active role in electoral politics divided the otherwise strong
Wisconsin Society of Equity, sponsor of a sympathetic cooperative movement. In late February 1917, a farmer-owned packing company in Wausau
took matters into its own hands. It called a meeting "for the purpose of organizing the farmers of Wisconsin into a non-partisan league." League organizers told the assembled that they needed to "have the aid of all the rest, or you
will all fail." An unofficial group of equity members from across the state took
their advice and joined together to insist on affiliating the state organization
with the NPL.[158]

Beecher Moore (a former Socialist Party gubernatorial candidate in Minnesota and a former employee of Townley's Organization Department) immediately began organizing in north central Wisconsin. Simultaneously,
Joseph Gilbert (the former editor of Seattle's *Socialist Herald* and Townley's
newest close associate) led tricky negotiations to bring the rest of the equity
leadership into line. But their persistent squabbling sank League efforts to
build a strong movement. So did the suspicion and skepticism of Milwaukee's Socialists. While La Follette openly championed the NPL, many of his
supporters feared that a local branch of the NPL might threaten his power.
They stayed out of the fight. As a result, the Wisconsin Nonpartisan League
struggled.[159]

All told, the potential for League success beyond North Dakota depended
on creating good relationships with existing farm groups and incumbent
politicians as well as deploying tried and true technologies such as automobiles and newspapers. It also rested on the ability of organizers to foster a
sense of shared experience and identity that might motivate agrarians. Those
conditions—already present in North Dakota in 1915—needed to be carefully
constructed elsewhere.

Unable to resist entreaties for organizing farmers beyond North Dakota,
League leaders—who initially hoped only to take over North Dakota's state
government—spread themselves thin. Townley's organizing model depended
on animating already agitated agrarians. Persuading weary and wary farmers
to join the League proved more difficult than signing up already organized
ones. Convincing agrarians to reconsider their self-identity was harder than
tapping into existing sentiment. Outside North Dakota, Townley's tactics often faltered in the face of more complicated circumstances.

One trenchant observer of the League's expansion suggested that the
movement depended entirely "for its efficiency upon its solid and compact
organization," which needed "to be slowly builded." Because conditions in
other states and provinces differed greatly from those in North Dakota, "in
making the entire Northwest safe for democracy the League confronts a task

that will certainly occupy it energetically for some time to come." Even so, the movement spread rapidly. The NPL push for equity and nonpartisanship even appealed to farmers in western Canada, who used it to advance their interests in a different electoral context.[160]

Clearly, many farmers across the North American West hoped to make the NPL a vehicle for achieving their hopes and dreams. Whether it would prove to be the instrument for doing that remained to be seen. When the United States entered World War I, the political calculus shifted. League opponents found new and more effective tools for attacking the NPL. Faced with claims that they opposed the war effort, farmers in the NPL soon came up against stiff opposition that made organizing more difficult than ever.

Opposition

The most prominent leaders of the League were playing the game of sedition and disloyalty . . . seeking to acquire power by pandering to and influencing the base spirit of
greed and envy and ignorance and class hatred. They are trying to do what Lenine and
Trotzky have done in Russia.
THEODORE ROOSEVELT, quoted in *New York Times*, October 6, 1918

Reverend W. W. Deal—a Methodist minister, master of the Idaho State Grange,
and national Grange chaplain—looked out at the crowd of his fellow farmers with satisfaction. Taking advantage of the good feelings engendered by
the warm summer evening, he addressed them with heartfelt emotion and
praised their patriotism. As one of the featured speakers at the annual meeting of the Washington State Grange, the beloved Deal held a special place
in their hearts. Agrarians had come to Walla Walla in June 1918 to hear him
and other speakers discuss politics, to raise money for the Red Cross, and to
telegram President Woodrow Wilson their support of his war policy. Proud
parents wore buttons proclaiming their sons' service in the armed forces.
Women listened attentively while knitting "for our boys at the front."[1]

Suddenly, a man ran to the podium. Deal stopped in midsentence. Told
that the group had fifteen minutes to vacate the premises, the Idahoan shared
the shocking news. Rumors spread through the hushed crowed. Panic ensued.
But soon the truth came out. Already refused at the major meeting halls in
town, the farmers learned that the Walla Walla school board, questioning the
organization's loyalty, decided to eject the Grange from the one space available for their assembly—the local high school. Fury replaced fear.[2]

"We are 100 per cent American," one woman shouted. She demanded that
the assembled "fight for [their] rights." Noting "a body of about seventy-five
huskies, held in readiness in the shadow of the building," Grange leaders decided otherwise. The six hundred farmers, farm women, and children hast-
ily departed the school. Defiantly, they sang "The Star-Spangled Banner" and
"My Country, 'Tis of Thee" as they filed out. Days later, one woman proudly
wrote that despite the "insults and taunts of the Walla Walla citizens," the

Grangers "kept themselves under control. It was *worth while* to see what kind of folks our Grange is made of."[3]

These trumped-up charges of disloyalty stemmed from the Washington State Grange's public embrace of the Nonpartisan League. Not only did the assembly elect William Bouck—an NPL booster—as the new state Grange master, but the Grange also cheered on NPL leader Ray McKaig during his speech to the assembly the day before Deal's address. For most residents of Walla Walla, support for the League equaled "I.W.W.ism"—a reference to the Industrial Workers of the World (IWW), a much-feared radical labor union that took an antiwar stance. Indeed, just days before the Grange met in Walla Walla, locals threatened to tar and feather an NPL organizer. McKaig and Bouck quietly left town before meeting the same fate.[4]

Seeking redress, the Grange immediately called on Walla Walla's mayor and sheriff. The mayor refused protection. The sheriff told the farmers that "a 'Home Guard' armed with loaded Winchester rifles added greatly to the danger." Vigilante violence loomed. Telegrammed pleas to Washington's governor Ernest Lister brought no response. Assembling for a worship service the next morning, the Grangers found every church closed to them on account of their being "disloyal and pro-German in sympathy." Shut out and facing an angry mob, the agrarians decided to retreat. Officers retired to their hotel rooms to finish the Grange's business. Groups of shaken members sang patriotic songs as they boarded trains to return to their farms. Meanwhile, at a local rally organized by Walla Walla's businessmen, pro-NPL speakers "were hissed and forced to quit."[5]

Amazingly, the dramatic incident in Walla Walla received little coverage. The Associated Press wire service, fearing a free-speech controversy, attempted to squelch news of the event. When word finally did reach Washington, DC, President Woodrow Wilson ordered an investigation. Federal agents—busy surveilling NPL members all over the country—were already on the case. Before the episode, none could "discover any actions" that suggested the "Nonpartisan League has been disloyal to the government." Wilson then learned that, in late 1917, an anti-NPL group of small businessmen formed an organization called the Employers Association of Washington. Knowing that the Grange planned to meet in Walla Walla later that year, the group actively spread vicious rumors about the League. Walla Walla's leading citizens took matters into their own hands from there.[6]

Investigators also drew on information collected by the Washington State Secret Service—an agency formed by Governor Lister to investigate revolutionary organizations across the state. Lister's detectives secretly kept leading members of the Nonpartisan League under close observation during 1918. Federal agents also interviewed dozens of participants in the Walla Walla incident. But state-sanctioned spying turned up little proof of disloyalty. Deciding that the situation "was a very complicated one," federal investigators declined to press charges against either the Grangers or local citizens. Though the people of Walla Walla violated the constitutional rights of the farmers, it was almost impossible, in the words of one government attorney, to "indict a community." Nothing came of the inquiry.[7]

Authorities and newspaper editors continued to harass the Washington State Grange's leader, William Bouck. In the wake of the Walla Walla incident, the *Spokane Spokesman-Review* wondered if Grange members knew that "the Townley nonpartisan league is an un-american organization." Bouck bravely responded by repeating his pro–Nonpartisan League convention speech at other venues across the state. By August, the unrepentant Grange master faced a grand jury indictment for violating the Espionage Act. His refusal to wilt in the face of intense pressure put him in legal jeopardy. William Kaufman, another prominent pro-League Granger in Washington State, received a five-year prison sentence for similar charges. But in December 1918, federal prosecutors finally dropped the case against Bouck for lack of evidence.[8]

When the smoke cleared, Bouck wrote to Nonpartisan League leader Arthur Townley. "For the whole year past," Bouck noted, "the reactionaries have raised Heaven and Earth to ruin the grange and all progressive organizations." Desperate for help, he pleaded with Townley for aid. Bouck worried that many in the Washington State Grange suspected the League might "sit idly bye and see us crucified." Already under attack in multiple states, the Nonpartisan League could offer little succor. In fact, back in Minnesota, Townley himself faced wartime indictments by zealous county prosecutors who argued that "the League is the Kaiser's hope." League opponents, who slowly gathered strength through 1916 and early 1917, took advantage of wartime sentiment. Small-town businessmen, in particular, affronted by the NPL's ire, became the first bastion of resistance. Corporate leaders in Minneapolis and Saint Paul emerged as another constituency that feared the NPL. World War I crystallized their collective efforts. Charges of disloyalty became a political bludgeon.[9]

Anti-German and anti-Red hysteria during World War I—which fueled more famous incidents such as the deportation of union activists at Bisbee, Arizona, in 1917 and the jailing of noted Socialist Eugene Debs in 1918— shaped the trajectory of the Nonpartisan League. Targeted by local and state officials, League members struggled to articulate their stance on the war.

Pegged as antiwar and un-American, the NPL elicited a violent response. Motivated by misunderstandings as well as the desire to translate public fear into political power, the League's opponents used every tool at hand to challenge the NPL's influence.[10]

From the start, those who fought the Nonpartisan League inadvertently paid tribute to its potential to remake American politics. Indeed, the ferocity of the opposition depended entirely on the fear that League members might actually foster the change they hoped for. Covert investigations by federal and state agents as well as private detectives matched mob violence and innumerable indictments for disloyalty. In a few cases, adversaries even renewed the medieval practice of tarring and feathering. Civil liberties questions sparked by controversies over the Nonpartisan League reached the US Supreme Court.

Even so, the NPL's complicated position on the war made the organization vulnerable. US involvement in World War I bore few connections to the League's political goals. Noting the imperial origins of the conflict, Leaguers held strong to the notion that the war needed to create a better world. They coupled clear-eyed support for the country's efforts to a critique of war profiteering. In their minds, the conscription of wealth needed to replace the conscription of everyday Americans. The war could be used to strengthen participatory democracy instead of putting more power in the hands of corporations. Yet as a wave of less thoughtful patriotism swept over the nation, it became difficult for the League to simultaneously maintain both its loyalty and its criticisms. For many Americans, the latter put the former in question.

In some states, officially sanctioned councils of defense and loosely organized local militias aggressively pursued violent means to stop the League. Ethnicity and religion also played a powerful role in opponents' efforts to frame the NPL as anti-American. Slowly, a multistate network of anti-NPL activists emerged. Even as they blunted the NPL's spread, efforts to coordinate resistance to the League often backfired. Such coordination provided the NPL with clear evidence of the forces arrayed against farmers. This proved the League's point that entrenched economic and political interests would stop at nothing to keep farmers from their rightful place in politics. Indeed, League members often clung more tightly to the organization in the wake of vicious anti-NPL campaigns. But at the end of 1918, when disgruntled former Leaguers in North Dakota began channeling their energies into stopping the organization's spread, a new and more effective opposition emerged. It would play a major role in the decline of the NPL in the years to come.

*

G. W. Reed, a ninety-three year-old Nebraska farmer, believed "that the Non-partisan league is the best thing that has happened for the farmers and work-ingmen since . . . the Civil war." Many farmers across the Great Plains and Pacific Northwest felt the same way. In fact, in October 1918, the *Nation* sug-gested that the Nonpartisan League had gained "national prominence, having emerged from the stage of mere sectional existence and importance." Another commentator believed that "the League has already provided the progressive forces of the nation with the inspiring and illuminating spectacle of a great social revolt." But newly empowered farmers undid the delicate balance be-tween small-town merchants and agrarians who together made up the rural middle class. The local businessmen who depended on farmers as custom-ers and as cocreators of community often felt alienated by the rise of the agrarian-oriented Nonpartisan League.[11]

Even before World War I—or NPL electoral success in North Dakota in 1916—rural merchants, bankers, and insurance salesmen became the earli-est opponents of the League. In fact, the farmer-only organization initially targeted these fellow members of their communities. As early as December 1915, the *Nonpartisan Leader* argued that "it is useless to deny that there is a growing hostility between 'local' business men and the farmer." Yet "the little business man cannot exist, as such without the farmer." The newspaper sug-gested that "if the little business man does not change his attitude toward the farmer he will lose him and lose him for good."[12]

This antagonistic attitude caused trouble in small towns. Politics divided churches and community clubs. Things got worse when traveling League lecturers followed with more pointed attacks. At least one NPL member re-membered that "imported speakers" told North Dakotans that local business-men created problems for farmers. One orator referred to local merchants when he suggested that League success meant that "we will have these Skunks driven out of here." In a few cases, NPL employees even organized local boy-cotts when they met resistance from small-town businesses.[13]

Businesspeople across North Dakota pushed back against the NPL. Wor-ried about the League's insistence on state hail insurance, in March 1916 the insurance agents throughout the state joined together to form the Insurance Federation of North Dakota. One man told an NPL representative that "we have built up a great business which is valuable to us, and we are going to defend it as anyone will defend what is his own." The new group envisioned the League's insurance plan as "the entering wedge" of a broader campaign to shift the industry to state control. The National Council of Insurance Federa-tions grew so worried that it sent its field secretary, Harry Curran Wilbur, to North Dakota to lead the effort.[14]

Bankers, too, responded quickly. Small-town cashiers refused to cash the checks NPL organizers received from new members. Leaguers protested, wondering "what banker has ever held up a check of a farmer that was made out to pay for a mortgage, elevator, mill, railroad . . . or other service?" As the NPL grew in strength, more organized opposition emerged. In a mid-1916 letter to its members, the North Dakota Bankers Association described the ascension of the Nonpartisan League as a "crisis which faces the government of the state." The association believed that "the good name" of North Dakota hung in the balance.[15]

Pressured by local advertisers, small-town newspaper editors called organizers "professional agitators," openly suspected the sales tactics used to enroll members, and occasionally critiqued the commission-based structure of the organization. Once enough local farmers—to newspapermen, subscribers—became NPL members, small-town newspapers adjusted their stance. But *Normanden*, a long-standing Norwegian-language daily and one of the biggest newspapers in the state, consistently attacked the League. In March 1916, the paper claimed the NPL would undo statewide prohibition and suggested faltering allegiance to the movement among some leading members. Calling the charges an "untrue and vicious slander," Leaguers responded by referring to *Normanden* and other anti-NPL papers as the "kept press" of big business.[16]

Even in North Dakota—where the League proved popular—this resistance to the NPL occasionally sparked open conflict. Dorothy St. Arnold, who grew up on a farm outside Ellendale, North Dakota, remembered a telling incident. In order to see William Langer, the League candidate for state attorney general, one afternoon, "every farmer in the neighborhood" got up at "3:00 in the morning" to get their chores done early. When the speaker finally arrived, "the Sheriff wasn't going to let Langer speak, and he had a few men with him." The farmers responded by getting "auto cranks and monkey wrenches from the cars, and that sheriff let Langer speak."[17]

Recognizing the burgeoning resistance—as well as the political power of these businessmen—the NPL tried to make amends. In February 1916, the *Nonpartisan Leader* claimed that the League had "no fight to make on the business man—as long as that business man does not fight the farmers' organization." The paper suggested that agrarians and businessmen, sharing middling aspirations, "ought to make their interests identical, as far as it is possible."[18]

After selecting Lynn Frazier as its candidate for governor, the NPL sent him on a May 1916 tour "to meet personally and talk with businessmen." A number of them "welcomed the farmers' candidate very warmly." Right before the June 1916 elections, the NPL publicly pleaded with the state's businessmen to remember that "if each North Dakota farmer could retain his share of the wealth that now goes to make millionaires outside the state he would have at least $1500 more to spend each year." That windfall would give "the farmers and their wives . . . money enough to buy these things at home" rather than in mail-order catalogs. League leaders even created "social" memberships, which allowed businessmen to affiliate with (but not vote in) the farmer-only NPL.[19]

But the gulf could not be fully repaired. Jerry Bacon, editor and president of the *Grand Forks Herald*, became the most vociferous and visible critic of the League. As "a capitalist of Grand Forks," Bacon owned one of the state's largest newspapers, a dairy farm, and a hotel. His reputation as "a director of many of the most important industries and leading corporations not only of Grand Forks but of the entire state" made his opinion especially important. A staunch Republican, Bacon served in the state legislature at the turn of the century as part of Alexander McKenzie's corrupt political machine. Carefully avowing that he did not want to "discourage or knock the Nonpartisan League," Bacon focused his energies on assailing "the organizers of the League, who appear to us . . . to be simply agitators and boomers." Other prominent North Dakota businessmen joined in.[20]

Together, they created a new organization. Born in Fargo before the June 1916 primaries, the Good Government League attempted to steal the NPL's thunder. Led by Morton Page—the wealthiest man in North Dakota—and the Insurance Federation of North Dakota's secretary, H. G. Carpenter, the organization looked to enroll voters in a nonpartisan movement against "well-meaning but misguided citizens against other classes." Backed by the North Dakota Bankers Association, the Good Government League paid tribute to the NPL's success by copying its tactics. But, unable to defeat the NPL in the primary election, the organization announced that "the farmer and the business man . . . must work together" and promptly folded.[21]

Twin Cities businessmen used a similar tack after the ascension of the League in North Dakota politics and the rapid spread of the NPL into Minnesota. In March 1917, they started the "Minnesota Non-Partisan League." Advocating "women's rights, temperance, eight hour day, government ownership, and exemptions for personal property of less than $200," the organization offered to "save the farmers of this state considerable money" by charging only three dollars for a membership instead of the NPL's sixteen dollars. They even secured offices next to the real Nonpartisan League offices in downtown Saint Paul in order to have a similar address. The *St. Paul Daily News* dubbed the entire effort "a despicable scheme." Ever vigilant, the *Nonpartisan Leader* ensured that farmers knew the difference between the two Leagues. Within months, the phony effort failed.[22]

Finally, the emergence of the NPL in North Dakota attracted the attention of a little-known traveling salesman from Lindsborg, Kansas. Philip Zimmerman spent his days selling spark plugs to hardware stores and automotive repair shops across the northern plains. After returning home from a trip to North Dakota in June 1916, Zimmerman shared the news of the emergent NPL with esteemed journalist William Allen White. The Spanish American War veteran warned his fellow Kansan about the potential evils of the League, but to no avail. So Zimmerman turned to Henry J. Allen, editor of the *Topeka State Journal* (and soon-to-be governor of Kansas), pleading with him to run a story on "Townley's intention to immediately put his organizers in the other west-central states." Believing that Arthur Townley was "a crook of the worst type," Zimmerman hoped to "lend a somewhat mediocre hand in crabbing his act." Allan paid little heed. Over the next year, the drummer decried the NPL to anyone in Kansas who might listen. But few in the Jayhawk State took him seriously.[23]

All told, early efforts to stunt the NPL's growth ended in failure. Under direct attack, insurance men, bankers, and corporate leaders tried to respond. Unable to convince farmers that League leaders looked to embezzle their money, businessmen then questioned the legal status of the organization. Frustrated again, NPL opponents paid tribute to the movement by attempting to copy it. Farmers refused to rise to the bait. Outside of North Dakota and Minnesota, few worried about the League. Those who did found themselves ignored.

*

World War I changed everything. On April 2, 1917, with German submarines sinking American merchant ships, President Wilson told Congress that "the world must be made safe for democracy." Though he built his 1916 reelection campaign around avoiding it, the president reversed his public stance on the European conflict. Securing the support of leading social and political figures, he asked for a declaration of war against Germany and its allies. Wilson got his wish four days later, when the United States entered the war.[24]

NPL-dominated North Dakota proved resistant to America's entry in World War I. The rural middle class retained deep-seated concerns about the uneven concentration of capital made possible by war profiteering. In March 1917, a newspaper in Bowman, North Dakota, told fellow citizens to "beware of Predatory 'Patriotism.' . . . It is murder and makes money for some." The day before, a newspaper editor in LaMoure, North Dakota, assured readers that war meant Americans would be "farmed by a capitalist class." Fearful that jingoistic fervor might blind the nation to the possibilities for corporate profit, farmers found themselves in a tight spot. That said, a number of young men from North Dakota surreptitiously crossed the border before 1917 to join the Canadian army in its fight against Germany—including a son of one of the NPL's 1916 electoral candidates.[25]

The Nonpartisan League held a similar bifurcated stance, claiming loyalty and rejecting American entry. North Dakota governor Lynn Frazier reflected his constituents' sentiments by declaring flatly: "North Dakota is not in favor of war." In March 1917, Arthur Townley told a Saint Paul audience: "Let capital throw its resources into the war game unselfishly and the 100,000 farmers of the Northwest will throw their resources and their blood if necessary into the game, just as enthusiastically." The NPL even launched a "Peace until War Is Necessary Campaign"—endorsed by North Dakota's state legislature—in multiple states. To that end, Arthur LeSueur addressed a crowd in New York City. He assured them that "in case of war with Germany" farmers would "serve the nation as loyal patriots." But the League hoped to "avoid war if possible" and "take the profits out of war if hostilities are inevitable."[26]

Though Congress made the war—and, soon thereafter, the draft—official, the NPL's stance changed little. In fact, the organization insisted that the war's origins lay in "rival groups of monopolists . . . playing a deadly game for commercial supremacy." Yet simultaneously, the League hoped that Wilson might transform the war into a quest for improvement. In its formal stance on the war, drafted in June 1917, the NPL "unreservedly pledged to safeguard, defend, and preserve our country." Leaguers nonetheless held specific ideas about the war's goals. "Private monopolies" needed to be "supplanted by public administration of credit, finance and natural resources." Justified only by ending "the rule of jobbers and speculators," American entry into the war needed to produce "a real democracy." Charting its own path, the NPL firmly stood behind the nation, even as it offered a powerful critique.[27]

On a speaking tour in North Dakota in June 1917, Arthur Townley justified the organization's opinion. He claimed that, in wartime, the United States was "working, not to beat the enemy, but to make more multi-millionaires." "In the heat and haste and confusion of war," Townley argued, "they multiply their millions many times at your expense." Such charges invited fierce opposition. As a result, Townley remarked, "they have charged us with treason." But "this nation of farmers are so patriotic that even though the government today may be in the hands and the absolute control of the steel trust, and the sugar trust, and the machine trust . . . we are going to do our best by producing all we can."[28]

The League's complex take on America's role in the conflict empowered a wide range of detractors. Caught up in prowar sentiment, NPL opponents refused to allow for any nuance. Adversaries immediately accused the organization of treason and sedition. Jerry Bacon's *Grand Forks Herald* suggested that Arthur Townley "either be shot or hung." Quoting from Townley's recent speeches, the *Minneapolis Tribune* argued that "where the influence of these meetings has been most felt, there . . . is to be found the most marked lukewarmness" toward the war. Indeed, the paper openly wondered if League members could tolerate "the sort of poison he has been spewing at a time when the Nation needs the whole-souled support of every citizen." From Boston, the *Transcript* suggested that the League was nothing more than "a sorry combination of old-time grangerism, socialism, I.W.W.-ism, and a form of pacifism with a distinctly German basis." The NPL responded to these slanders by arguing that "the anti-farmer press thinks it can use the League's resolutions against the League, without regard to whether the resolutions are right or wrong in principle." To some extent, it could.[29]

For instance, Theodore Wold—chief administrator of the brand-new Federal Reserve Bank in Minneapolis—charged the League with "hampering the campaign" for Liberty Loans with farmers across the northern Great Plains. As a fund-raising effort, the federal government hoped to sell thousands of these war bonds to help pay for the national mobilization. In response, the NPL announced that crop losses left farmers with little money to purchase war bonds. Governor Lynn Frazier told agrarians they might better support the war effort by instead investing their meager funds "into supplies and seeds and labor for the crop." Leaguers were "not opposed to the loan . . . but . . . to this method of raising the money." Rather than putting "a heavy burden of debt on future generations of workers," Leaguers wanted "the government to use the surplus wealth which the nation has created and is creating to bear the expenses of the war." Even though NPL-controlled North Dakota eventually oversubscribed to the Liberty Loan program, for the League's enemies, this position simply confirmed what they already believed.[30]

*

In this context, loyalty became a political weapon. Used by NPL opponents, it defined wartime responses to the Nonpartisan League. Charges of un-American behavior or disloyalty carried real weight. At this crucial moment, the NPL made two serious mistakes.[31]

With rural manpower stretched thin by the draft, many farmers worried about having enough labor to bring in the upcoming harvest. R. J. J. Montgomery of Tappen, North Dakota, president of the state's Farmers Union and an NPL member, proposed that the League cut a deal with the Agricultural Workers Organization (AWO)—a branch of the IWW. Most in America feared and derided the controversial IWW. Calling for "One Big Union," Wobblies, as IWW members were called, embraced revolution, decried capitalism, and took a decidedly public stand against the nation's entry into the war. Federal agents and businessmen alike specifically targeted the IWW as a disloyal organization.[32]

Using his considerable clout, Montgomery convinced first Townley and then NPL members attending their state convention in May 1917 that negotiating hours and pay directly with organized migrant laborers in the AWO would prevent potential disruptions in the fall. After all, "probably over three quarters of the farm laborers who come to North Dakota each year . . . are members of the Union." Suggesting that "it would be 'a case of the organized farmer making an agreement with the organized farm laborer for the benefit of both,'" the League used the *Nonpartisan Leader* to inform members of the initiative.[33]

In June 1917, representatives of the AWO and NPL produced a draft agreement. Leaguers promised to hire only members of the AWO, and in turn, AWO laborers agreed to work only for NPL members. Migrant workers would receive forty cents an hour for a ten-hour day. NPL leaders looked for approval at meetings held across North Dakota. In Minot and Bismarck, a majority of Leaguers endorsed the agreement. At gatherings in Valley City and Devils Lake, however, the membership firmly rejected the plan. In the latter, suspicion of the IWW ran so deep that the meeting almost turned violent. Concerned about creating divisions in the NPL, Townley decided to drop the initiative altogether.[34]

Living in a lower-middle-class culture already apprehensive of migrant laborers, many farmers saw the attempt to negotiate with the IWW as distasteful. Some disliked the rootless lifestyle of single men who lived on the road. Others, envisioning AWO members as untrustworthy, worried about whether the agreement would hold. Indeed, the reputation for violence associated with the IWW challenged middle-class conventions and threatened to undercut rural community. As news of the failed agreement spread, League opponents used the NPL's attempt to reach out to the AWO as proof that the NPL not only was wrong on the war issue but also was socialist. Because the Socialist Party adopted a firmly antiwar stance at its convention in Saint Louis in April 1917, each charge fed off the other. That the League worked to establish state intervention in the economy did little to discourage the accusation—even as the Socialist Party envisioned the NPL, made up of property-owning farmers, as inappropriately bourgeois. Furthermore, NPL staff such as Henry Teigan and Henry Martinson quietly kept up their Socialist Party membership for many years.[35]

By 1918, at least one NPL antagonist asked readers if they noticed that all Socialists and radicals "have the same facial cast." He claimed that both "Gene Debs and A. C. Townley . . . have high cheek bones and a pointed chin, a kind of coffin shaped face a la Russian." Phrenology—a now discredited evolutionary view linking bodily features and racial types—suggested that such men "cannot help it, they are born that way." Claiming that one could recognize subversives by their physical features, the adversary literally lumped together Socialists and Leaguers.[36]

The second blunder took place just a few weeks later. In September 1917, the League hosted a Producers and Consumers Convention in Saint Paul. Facing federally fixed prices for their wheat, agrarians hoped that the government might also curb the prices of consumer goods. The growing cost of living seemed to offer an opening for a closer relationship with organized labor. Speakers included League leader Arthur Townley, US Senator William Borah of Idaho, and Jeannette Rankin, the US representative from Montana (and first woman to serve in Congress). Minnesota State Federation of Labor secretary George Lawson also addressed the crowds. The most important speech, however, came from Wisconsin's US senator, Robert La Follette. La Follette, a former presidential candidate, voted against the declaration of war on Germany in April 1917 as well as the draft and the Espionage Act. This made him the most prominent antiwar figure in America. His presence brought national attention to the conference.[37]

Opening the meeting, Townley assured the assembled that an "unalloyed atmosphere of loyalty" defined the proceedings. Then he critiqued the efforts of wheat traders in Minneapolis, Kansas City, Duluth, and Chicago to influence the federal government's decisions on wartime grain prices. Townley concluded by characterizing corporations as pro-German. "They are not patriots, because they possess billions and billions of dollars of war profits wrung from the agony, sweat and toil of starving men and women," he noted. In fact, one might "get a German helmet, place it upon THEIR head, and YOU SEE THE KAISER HIMSELF."[38]

Skeptical anti-NPL reporters accurately described Townley's opening speech as "patriotism tempered with protest" and "loyalty tinged with discontent." Indeed, in a series of resolutions by the League members in attendance, the organization reiterated its notion that "the moving causes of this world war was and is Political Autocracy used to perpetuate and extend Industrial Autocracy." It called on Woodrow Wilson to "extend the political democracy which we, in the United States, enjoy in order that political democracy be safe

in our own land and that it may be used to accomplish its historic purpose—
Industrial Democracy." That said, the gathering pledged "our lives, our fortunes and our sacred honor to our country and our flag in this OUR WAR."[39]

At the end of the conference, more than eight thousand people gathered
to hear Robert La Follette. He hoped to make a strong statement about free-
speech rights during wartime. Learning of his intentions, NPL representatives pleaded with him to tone down his remarks. They knew that charges
of disloyalty lurked around every corner. Furthermore, Leaguers saw police
detectives lurking in the crowd, waiting to report on any misstep. Encouraging the Wisconsin senator to instead improvise a speech on war financing,
they hoped to avoid potential disaster. The men received more than they bar-
gained for.[40]

La Follette resorted to a standard stump speech about the power of representative government. Then he "lauded the Nonpartisan league to the skies."
Finally, he turned his attention to paying for the war. But then La Follette
briefly digressed to discuss his vote against American entry. He vigorously
defended his right to dissent. The senator pointed out that in the months
leading up to the nation's declaration of war, German submarines had sunk
ships carrying not only American citizens but also American-made munitions bound for Germany. When someone in the audience yelled that this
stance made him "yellow," La Follette calmed the crowd. Then he replied directly to this charge: "I don't mean to say that we hadn't suffered grievances;
we had—at the hands of Germany. Serious grievances!"[41]

The rest of the address went off without a hitch. The assembly cheered
him wildly. Federal agents in the crowd reported that nothing in the speech
"could be considered treasonable." But the Associated Press misquoted that
section of La Follette's oration. Its transcript suggested that he had said,
"I wasn't in favor of beginning the war. We had no grievances." The next
day, thousands around the country learned of the Wisconsin senator's supposed disloyalty. Aroused Americans ignored the rest of his speech. Over-
shadowing the conference's declarations of loyalty, the flawed transcript
provided another opening for League opponents.[42]

With the faulty wire service transcript in hand, newspapers across the
country smeared both the Wisconsin senator and the Nonpartisan League.
Scathing anti-NPL editorials in the *New York Times* forced Townley to respond directly. He called out the paper for making "the purposes of the
league seem unpatriotic" in "several statements utterly without foundation
in fact." He pointed out that public officials in Minnesota did not find "any
sedition" in the official conference transcripts. Townley then assured readers
that "if sedition existed" the NPL "could be trusted to ferret it out."[43]

Few believed him. The damage had been done. Under pressure to expel La Follette from the senate, his colleagues launched a full-fledged investigation. Across the country, opponents of the NPL pointed to the League's acclimation of La Follette's speech as evidence of the group's un-American stance. Not until May 1918 did the Associated Press retract the error. In December 1918, the US Senate finally cleared La Follette's name. Until then, many around the country accepted the inaccurate version of La Follette's speech as fact. Even after, they held firm to this proof of the Nonpartisan League's disloyalty.[44]

*

The war also created a new institution that helped League critics take advantage of these openings. As early as January 1917, state legislatures began responding to the growing call for preparedness. Focused on surveying manpower or increasing food production, war-readiness efforts swept across the United States. In May 1917, nearly every state sent representatives to a meeting in Washington, DC, organized by Newton D. Baker, Wilson's secretary of war. They left with a charge to create a council of defense in each state.[45]

Administered locally and made up of gubernatorial appointees, these quasi-official bodies ranged widely in power, composition, and scope. All existed to implement federal plans and to make wartime mobilization appear democratic and locally based. Some state councils of defense received wide-ranging legal powers to curb potential sedition and espionage. A few even immediately replaced the National Guard with a newly organized "Home Guard" for emergencies. Others held little sway over the civilian population, largely serving as figureheads. Much depended on local politics. But in nearly every case, the leading men and women of the particular state—largely businessmen and party politicians—staffed these councils. Partisanship permeated them.[46]

Regardless of their reach, most state councils of defense made loyalty a watchword. The call for Americanism in a time of national crisis penetrated every aspect of daily life. At the federal level, President Wilson created the Committee on Public Information (CPI). Led by journalist George Creel, this agency deployed posters, films, and pamphlets to promote the draft, define the enemy, and create a unified citizenry. By creating a prowar atmosphere, the CPI ensured that public patriotism reigned supreme. Taking their cue from the CPI, most state-run councils of defense focused their attention on ensuring loyalty not only to the nation but also to America's participation in the war. Selling war bonds, meeting draft quotas, raising funds for the Red Cross, and unquestioningly supporting government initiatives all became indicators of one's stance on the war.[47]

In some states, the councils of defense became especially powerful. Even before America's entry, Minnesota's legislature began to prepare for the Great War. It proposed a Commission of Public Safety (CPS) to coordinate preparedness efforts. Approved soon after Congress's formal declaration of war, the legislation created an agency with sweeping powers. Members appointed by the governor worked with a $1 million appropriation to exercise "supreme authority in matters related to defense." One backer privately admitted that "there are provisions in it that are unconstitutional and palpably so." In fact, the commission held the power to review and remove elected officials, raise its own militia, condemn property, hold investigative hearings, and enforce the law. Essentially, the legislature handed dictatorial authority over the state to an unelected body.[48]

From the start, the commission—acting as the state's council of defense—questioned the loyalty of the Nonpartisan League. The League's rapid ascension struck fear into Minneapolis and Saint Paul businessmen. As a result, one federal agent noted that "a coalition of privileged interests" helped write the legislation for the CPS and then manned it. Using connections with local businessmen, the CPS hired private detectives to investigate known Socialists, members of the Industrial Workers of the World, and the Nonpartisan League. To further check the power of the latter, the commission organized county-level public safety bodies and empowered local sheriffs with the authority to break up League meetings in rural districts. Eliminating freedom of assembly made it difficult to for the NPL to organize. By February 1918, twenty Minnesota counties had banned League meetings.[49]

The Producers and Consumers Convention debacle, however, gave the commission its long-awaited chance to directly attack the NPL. Immediately after Senator La Follette's speech in Saint Paul, the CPS appointed member Charles Ames—an executive at Saint Paul's West Publishing Company—to investigate the NPL's "method of operation, their financial methods, their purposes, their various activities, and the effect thereof." Ames, who believed the League to be "about the most dangerous organization in America," threw himself into the work with gusto. Ames issued a subpoena for Arthur Townley and briskly interrogated the League leader.[50]

A lack of evidence forced Ames to relent. He found no evidence of outright disloyalty on the part of the NPL. Nonetheless, the League remained under official investigation for the rest of the war. With self-admitted "secrecy" as "the essence of their methods"—and with the assent of the CPS—local businessmen actively supplemented ongoing inquiries with surveillance of their own. They also met with CPS members in Saint Paul to coordinate an anti-NPL campaign in the state's newspapers. Centered on the promotion of loyalty, the campaign they crafted built patriotism to a fever pitch. This left little room for fine distinctions on the war question.[51]

Fully unleashed, the Commission of Public Safety tacitly suppressed the NPL. It refused to offer members or leaders any protection from violence or illegal repression. In late February 1918, the League officially queried Governor J. A. A. Burnquist, anxious to determine its legal status. Burnquist obliquely suggested that as long the organization obeyed the law, it would not face persecution. Then, just days later, he rejected the NPL's formal invitation to address a League meeting in Saint Paul. Noting that "the National Nonpartisan league is a party of discontent," the governor claimed the NPL had "drawn to it the pro-German element of our state." Furthermore, "its leaders have been closely connected with the I.W.W. and the Red Socialists." Soon thereafter, another member of the CPS, John F. McGee, told the US Senate that "a Non-Partisan League lecturer is a traitor every time. . . . Where we made our mistake is in not establishing a firing squad in the first days of the war." Inflaming local passions, the commission empowered small-town businessmen to take matters into their own hands.[52]

In other states, councils of defense proved equally aggressive. Nebraska first organized its state Council of Defense in late April 1917. Vested with investigative power, the council immediately turned its attention to ferreting out antiwar and pro-German sentiment. A secret service, made up of sixty-five appointees, reported any whiff of un-Americanism. The council then banned the German language, prosecuted professors at the University of Nebraska for presumed disloyalty, and targeted the Nonpartisan League. Soon after, J. L. Albert, a member of the state legislature, wrote a friend confidentially. He noted the unconstitutional powers of the state council but kept quiet publicly because he feared "that tide has set too strongly against sanity and safety in this state for any word . . . to stem it."[53]

Recognizing the threat, the NPL offered its services to the state Council of Defense. With seventy-five organizers and fifteen thousand members across the state, it proposed to "do a great work in educating the farmers of Nebraska as to what the Council of Defense wants." Leaguers argued that their "workers" might "promote the sale of thrift stamps, impress the necessity of food conservation, the necessity of increased production" and "the advantages of the sale of Liberty Bonds." The council scoffed at the offer. Within weeks, local sheriffs and councils were jailing Leaguers without charges or warrants or "the slightest semblance of authority."[54]

Frustrated, NPL members responded with anger. Raymond Beach, a farmer from Pleasant Dale, Nebraska, wrote an open letter to the state council. In it, he laid claim to the "duty of American citizenship to stand up for that which I believe to be right." The League was "a movement of democracy." Beach wondered why "we farmers have been denied these constitutional rights and privileges to hold meetings." The farmer demanded clear evidence of League disloyalty and wondered why the NPL, alone among political organizations, faced such scorn. Beach answered himself, knowing that "we are stepping on somebody's toes hence this squeal and opposition against us." Finally, the agrarian laid bare the council's use of "the increasing number of outbreaks against the League as evidence why the League should be suppressed." In fact, he said, "The blundering stupidity or devilment of your council is largely if not solely responsible for this." The letter had little effect.[55]

In June 1918, the Nebraska NPL officially filed suit against the state Council of Defense to stop the extralegal persecution endorsed by that body. The state council and League leaders eventually reached an out-of-court settlement. Members and organizers would no longer be deemed disloyal. Farmer district and state conventions would be permitted. In return, Oscar S. Evans, the state NPL manager, would be replaced "with a Nebraska farmer." Furthermore, the organization must cease distributing the literature "which the council had branded as disloyal." The League also agreed to "withdraw all other paid organizers."[56]

In Montana, Governor Sam Stewart established the state Council of Defense in April 1917. Staffed by leading businessmen, it remained mostly powerless. Most of its attention focused on the Industrial Workers of the World and food production. For the rest of the year, members of the council clamored for more authority. In February 1918, the governor pushed through legislation giving the council formal and wide-ranging powers. With its official charge to "perform all acts and things necessary or proper so that the military, civil and industrial resources of the State may be most efficiently applied toward maintenance of the defense of the State"—the exact language used in Minnesota to establish its Commission of Public Safety—the council began organizing at the local level.[57]

Newly empowered county-level councils used their unchecked authority to settle local scores. More than 130 Montanans—mostly from the agrarian southeastern part of the state—landed in jail on sedition charges. Meanwhile, Will Campbell, the rabidly antiunion editor of the *Helena Independent*, joined the state council. Campbell directed the Montana Loyalty League and published the *Montana Loyalist*, an anti-NPL newssheet. Both viciously attacked the anti-American "Townley and his gang of socialist soapboxers." Friends on the council joined him. Soon the Montana Loyalty League and the state Council of Defense worked together to squash NPL organizing. When the pro-NPL Butte newspaper editor William Dunn challenged the state council's authority to crack down on laborers and farmers, he landed in jail.[58]

Formed in June 1917, the Washington State Council of Defense—unlike those in Minnesota, Nebraska, and Montana—earned accolades from the national office for its efficiency and good order. Headed by Henry Suzallo, president of the University of Washington, the fifteen-member council directed mobilization efforts with little legal authority and a tiny budget. It provided prowar propaganda for local newspapers and Liberty Loan drives. When organizing county-level activities, the council devised formal procedures to protect those being investigated for disloyal behavior from violence. They also worked hard to negotiate deals during IWW-inspired labor strife.[59]

Even so, the council reserved special spite for the Nonpartisan League. Noting the presence of "short-sighted and theoretical" organizations in "opposition to our form of government," in December 1917 the state council called on the counties to "organize a volunteer secret service." Though "numerous complaints ... concerning the unjust acts or alleged acts of chairmen or members of County Councils of Defense" came in front of the state council, the latter chose to "rely very largely on the judgment of the local body." Effectively, this meant that League organizers who faced mob violence or antagonistic local sheriffs got little sympathy or protection from the council.[60]

John G. Kelley, head of publicity for the council, also owned the *Walla Walla Bulletin*, the newspaper whose virulent anti-NPL slant fueled the antagonism toward farmers in Walla Walla in June 1918. Indeed, the council admitted that "representatives of the Council of Defense were said to have participated" in the incident. Once the Department of Justice opened a formal investigation, however, the state council retreated from any direct investigation of its own. This move protected Kelley from any criminal prosecution and gave the whole affair an indirect stamp of official approval.[61]

In South Dakota, the state Council of Defense focused largely on selling war bonds, fuel conservation, and agricultural production. Governor Peter Norbeck, who recognized the League as a clear challenge to his administration, publicly tried to keep South Dakota's council from egregiously overreaching its powers. Yet he appreciated the political power that came with suppressing dissent. Privately, he believed that "these Non partisan fellows do so much lying" that they needed to be stopped.[62] He confidentially directed W. Harry King, his campaign manager, to give "some publicity ... to the disloyalty of League members." Moderating the views of his boss to prevent voter backlash, King strategically told Norbeck's allies that "it is well now to show up the disloyalty of the League *leaders*, and all their socialistic connections." Norbeck's aide used loyalty to quietly insist to the South Dakota Farmers

Union that the League threatened the war effort. He also secretly contacted South Dakota's US senator Thomas Sterling, asking him to investigate the NPL-operated *South Dakota Leader*'s "right to the use of the mails." Finally, King began directing a campaign of anti-NPL editorials in leading newspapers across the state.[63]

Norbeck made sure that the state Council of Defense did not engage in illegal activity. Combined with a publicity campaign that questioned League loyalties, he avoided a backlash. When county officials prevented an NPL meeting in Madison, South Dakota, in February 1918, Governor Norbeck demanded that "no political organization should be interfered with . . . except such as are held for the purpose of embarrassing the government in time of war." Weeks later, when Arthur Townley tried to speak in Britton, locals forcibly deported him to nearby North Dakota. Soon thereafter, Norbeck directed the state Council of Defense to ban Townley from speaking in the state. The next day, it rescinded the order—as long as Townley promised to no longer encourage disloyalty.[64]

Through 1918, Leaguers continued to face harassment. In October 1918, Norbeck publicly reiterated that he "made an honest effort to protect the League organizers in their meetings" but later "came to see . . . that the League leaders did not want protection. They live and grow on trouble." If needed, he argued, NPL organizers "proceed to manufacture it." Simultaneously, Norbeck carefully avoided alienating agrarian voters. "The farmers are not socialists," he declared. But in order to ensure victory in the upcoming November 1918 elections, his campaign hammered the League as disloyal.[65]

*

In Kansas, Colorado, and North Dakota, governors pointedly refused to allow the persecution of the NPL. In the former, Governor Arthur Capper refused to turn the state Council of Defense into an anti-League weapon. For a few Kansans, fund-raising drives, Red Cross work, and food production started to seem less important than disloyalty. League opponent Phil Zimmerman told the chair of the state Council of Defense that "the entire leadership of this League" were "Red Socialists, IWWs or Anarchists . . . of the Pro-German Pacifist type." The Kansas Farmers Union—whose president, Maurice McAuliffe, sat on the state council—also believed that "the organizers for the Non-Partisan League are selecting communities where they hope to find German sympathy." He accused the NPL of distributing "disloyal German literature."[66]

In February 1918, the manager of Kansas City's Federal Reserve Bank sent a confidential letter to Governor Capper, sharing reports that "workers for this league" busied themselves "spreading the wildest sort of anti-government propaganda." Capper penned a public response, asking county councils of defense to forward reports of NPL disloyalty. League supporter Milton Amos wrote from Ellsworth—a center of NPL activity in Kansas—to defend the organization.[67]

The state's US district attorney agreed with Amos. After reviewing NPL publications, he assured the governor that "this literature is not of a nature that would justify the government" prosecuting the League "for disloyal utterances or the circulation of anti-war propaganda." Governor Capper declared the matter settled. Nonetheless, League opponent Philip Zimmerman insisted that the governor and the state Council of Defense continue the investigation. When the governor politely refused, Zimmerman accused him of cozying up to the NPL for his upcoming run for the US Senate. "Any Kansas politician who flirts with, or accepts a Townley endorsement for any job whatsoever," the League opponent told the governor, "well, he'll wish that man Zimmerman had never been born." Frustrated by the council's inaction on the NPL, Zimmerman even began to suspect that George Creel—the federal official charged with prowar propaganda—might himself be an IWW supporter or anarchist.[68]

Capper noted the emerging anti-NPL sentiment in Kansas among small-town businessmen like Zimmerman. Though League members throughout Kansas agreed with his policies and voted for him in the November elections, the governor carefully distanced himself from the NPL throughout his campaign. In fact, Capper discreetly ignored occasional eruptions of anti-League sentiment. For instance, at war's end the county council in Ellsworth—where a strong NPL presence deeply divided the community—proudly trumpeted that "every Non-partisan League organizer or pro-German was driven out of the county." Capper deftly headed to Washington, DC, with his reputation intact.[69]

In Idaho and Colorado, the state councils of defense never suppressed the NPL. Idaho's governor, Moses Alexander, deliberately ignored the business orientation of that state's council. Displaying prudence, Alexander neutered the council's effort to squash the League—though the council, in turn, charged the German Jewish governor with disloyalty. Julius Gunter, governor of Colorado, appointed the master of the Colorado State Grange (which strongly supported the NPL) to the state Council of Defense alongside labor leaders and businessmen. The council noted the "absurd reports" that flooded in from local officials throughout 1917 and called gossip and rumor mongering "the German way of doing things."[70]

In North Dakota—where the Nonpartisan League controlled the judiciary and the executive branches—the state council appointed by Governor

FIGURE 12. North Dakota State Council of Defense parade float, which included Governor Lynn Frazier (*center, without hat*) and Attorney General William Langer (*center right, with hat*), emphasizing the state's commitment to winning World War I, circa 1918. (State Historical Society of North Dakota, Col 44-14)

Lynn Frazier surpassed their Liberty Loan quotas and sent many young men off to war. Nonetheless, establishment Republicans and NPL members on the council fell into deadlock. As a result, Frazier formally reorganized the state Council of Defense in January 1918. The new version featured Frazier as president and included NPL state attorney general William Langer. All disloyalty reports now funneled from the county level up to the new council. In most cases, it protected people from illegal persecution. Federal judge Charles Amidon—in contrast to local Bureau of Investigation (BOI) agents—further shielded the League from the most outrageous attacks. Amidon's children both worked for the *Nonpartisan Leader*.[71]

The other three large newspapers in the state—the *Fargo Forum*, *Bismarck Tribune*, and *Grand Forks Herald*—accused the NPL of disloyalty. The anti-NPL frustrations that episodically erupted in public—such as painting a building yellow, threatening those who did not buy war bonds with tarring and feathering, and forcing the purchase of Liberty Loans—all met with strict rejoinders instead of encouragement. NPL-affiliated officials insisted that North Dakota's state Council of Defense stand out nationally for its levelheadedness.[72]

No less an authority than George Creel—head of the federal government's Committee on Public Information and the man responsible for manufacturing intense public sentiments around the question of loyalty—believed that the NPL consistently suffered at the hands of state authorities across the United States. Creel called disloyalty a "campaign weapon" used by League opponents to great effect. Indeed, in many places, "the disloyalty issue" became the "means by which" establishment politicians "hoped to crush and destroy the . . . organization." Yet "even campaigns of terrorism could not drive its membership . . . into disloyalty." In fact, Creel believed that "North Dakota, where the League elected every state officer, had a war record of which any state might be proud." In 1919, the CPI even awarded the *Nonpartisan Leader* a citation in recognition of its "patriotic services" during the war.[73]

*

Patriotism aside, political persecution animated by state councils of defense, businessmen, and state politicians forged new relationships between League opponents. They looked across state boundaries and began coordinating their efforts. The emergence of official structures for opposing the NPL made it possible for their adversaries to find each other and quietly coordinate their work. Insisting on federal investigations of the NPL, they even drew President Wilson into the fray.

In North Dakota, *Grand Forks Herald* editor Jerry Bacon reprinted Minnesota governor J. A. A. Burnquist's anti-NPL letter from 1918 as one of a series of pamphlets for wide distribution. In South Dakota, Governor Peter Norbeck's friends consistently disseminated Bacon's materials to obliquely attack the Nonpartisan League's legitimacy. In one case, Norbeck's campaign manager sent an ally in the South Dakota Farmers Union one thousand copies of a Bacon-penned anti-League screed. Philip Zimmerman, the self-appointed NPL opponent in Kansas, also used Bacon's publications to spread rumors about League disloyalty. Bacon proudly touted his role in hurting the League. He rightly insisted that his "pamphlets had something to do with assisting in the defeat of Townleyism in all other states except North Dakota."[74]

Norbeck and his advisors stayed in direct touch with Bacon. Sometimes they even shared information. In one instance, W. Harry King informed the North Dakotan that a federal raid on a German-language newspaper in Sioux Falls turned up correspondence "between the League leaders and these Pro-German editors." In another, King informed Bacon that "no Non-Partisan League organizers" had yet been "arrested in this state for disloyal utterances." He planned to keep the North Dakotan updated. After Norbeck's reelection in November 1918, Bacon reciprocated by personally congratulating the governor, noting his "gallant fight" against the NPL.[75]

The South Dakotans also corresponded with members of Minnesota's Commission of Public Safety, the Nebraska State Council of Defense, and the Montana Loyalty League. In August 1918, the governor insisted to one member of the Minnesota body that the "organization can't last." In another instance, Norbeck's adviser turned to a prominent anti-NPL Minnesotan for more detail on the editor of the *South Dakota Leader*. Just weeks before the 1918 elections, King contacted the state council in Lincoln for "information regarding one Charles Dean, Non-Partisan League organizer." The vice chairman of the Nebraska council responded that the "man was arrested" in Norfolk earlier that year. King even wrote to League opponents in Montana. There, the editor of the *Butte Miner* shared news on what he called the "Nonpatriotic League."[76]

This web of businessmen and politicians often deployed private agents to spy on the NPL. The Business Men's Protective Association in Omaha— which included members of the Nebraska State Council of Defense—hired detectives to infiltrate the state NPL organization. The businessmen insisted that "these damn fool farmers have to be stopped." They hoped to find something in the "personal habits, political and religious beliefs and past history" of League staffers that could be used against them. After months of work, the detectives failed to gather the hoped-for proof. One business leader then suggested that the group "get together and manufacture some incriminating evidence." The association even sent men to instigate mobs across rural Nebraska for the tarring and feathering of NPL organizers. The entire project collapsed when one detective who "felt the pinch of the capitalist claws" switched sides and started feeding League leaders information about the plot.[77]

The Minnesota Commission of Public Safety pursued a similar tactic. From 1917 on, the commission hired private detectives to supplement the citizen surveillance of potentially disloyal groups. In early 1919, the commission lost its legal authority. Without the outright support of law enforcement or state government to gather information on the NPL, members (who continued to meet well into 1920) became even more dependent on a private detective agency based in Minneapolis. The head of the agency bragged that "one of our men has ben an active member of the Nonpartisan League for more than three years and is well-known to all the leaders in Minnesota and North Dakota." A constant stream of reports from an operator inside NPL headquarters in Saint Paul helped local businessmen keep track of League thinking and planning through 1919 and 1920.[78]

Businessmen in Colorado Springs pursued a similar strategy. While monitoring the NPL in August 1918, Denver-based Bureau of Information agents learned that local bond traders "placed an operative into the League." He had earned "the confidence of the League organizer" and made "daily reports . . . regarding their activities." Though the mole procured an NPL membership list, he consistently proved "unable to get any direct evidence that they have violated United States law." Nonetheless, the BOI agent passed along material gathered by the informant to the US district attorney.[79]

Colluding businessmen and politicians used their clout to insist on the NPL's disloyalty and demand federal surveillance of the NPL. BOI agents actively monitored League organizers throughout the region. From Minnesota and Iowa to Oklahoma and Texas—and west to Washington State—they kept close watch on the NPL. Agents listened to League speakers, surreptitiously interviewed NPL employees, collected League literature, and gathered information harvested by hardworking opponents. Most federal investigators remained evenhanded. So did most US district attorneys.[80]

Their reports landed on desks in Washington, DC. Unsure about the NPL, Woodrow Wilson's administration nonetheless initially seemed to support it. Soon after the war began, the US Post Office issued an order to remove the *Nonpartisan Leader* from the nation's mail. Wilson countermanded the directive. In November 1917, Wilson's advisers refused to send a high-ranking official to Minneapolis for a loyalty rally unless the organizers also permitted Arthur Townley to speak. When the Commission of Public Safety refused to allow the NPL leader on the platform, Wilson sent Assistant Secretary of Agriculture Carl Vrooman as his representative. Vrooman—an NPL sympathizer—lambasted war profiteers and stuck up for the nation's farmers.[81]

That same month, George Creel invited Townley to Washington, DC. He hoped to firm up the NPL's support of Wilson's prowar stance and agricultural policies. Townley met with Wilson and Herbert Hoover (the nation's wartime food administrator). Received warmly, Townley then instructed NPL congressman John Miller Baer to formally offer the League's services to Wilson. They would prove the NPL's loyalty once and for all by offering the organization as a vehicle for wartime propaganda to western farmers.[82]

Appreciative of the League's inroads in the otherwise Republican strongholds of Minnesota, North Dakota, Nebraska, and Montana, Wilson responded to Baer with vague approval of the NPL's efforts. The president appreciated "the work of the sort you outline" because it contributed to "the universal cause." Knowing that "evidence from many quarters" suggested the

NPL had "self-seeking and untrustworthy" leaders, Wilson did not explicitly take Townley up on his offer. Later, he noted that he didn't take a stand against the NPL because "I don't like to send a message which would undoubtedly be used as a slap in the face to the whole organization."[83]

NPL leaders saw the friendly welcome for Townley as an endorsement. Ray McKaig excitedly wrote a friend to tell him that "Wilson, Carl Vrooman . . . and all that represent the Federal government, are with us." The League rushed to publicly support Wilson at every turn. NPL organizer Walter Quigley came up with the idea to distribute Woodrow Wilson's *The New Freedom*—a 1912 campaign booklet that outlined the president's plans for the nation—alongside other tracts. NPL employees everywhere sold over ten thousand copies within a year. Using the pamphlet to defend the NPL against charges of socialism, McKaig called it "our text book." Speakers even cited it in public debates with League opponents. Quigley, for one, found it "amusing . . . to watch the expression on the faces of various County Defense officers or Public Safety Commissioners when they heard these so-called radical utterances read from a book by the then commander-in-chief of the United States Army and Navy."[84]

In early 1918, George Creel further committed the administration to supporting the NPL. Overstepping his authority, he publicly suggested that the League sponsor Committee on Public Information speakers. Critics—including Minnesota US senator Knute Nelson—suggested that the partnership between the NPL and the Wilson administration proved that the Democratic Party would undermine the Republican Party by any means necessary. The angry president instructed his subordinate to backpedal. "We are getting into deep water in that part of the country," he noted, and "we had better pull away from them." Creel complied. He noted that "while I deeply resent" suggestions of NPL disloyalty, "we cannot afford an open break with the State authorities."[85]

Further influenced by an anti-NPL memo prepared by David Houston, Wilson's secretary of agriculture–and source of the constant stream of BOI reports—Wilson avoided full-blown support of the NPL. Soon thereafter, the president told a subordinate that he did not "feel at all safe in the hands of Mr. Townley or the Non-Partisan League and would like to proceed rather carefully in dealing with him." Even so, Wilson did not retract his tacit backing of the NPL, telling one colleague that "the League has rendered consistent assistance and very effective assistance where it could to the cause of the war."[86]

Nonetheless, Wilson did little to protect Arthur Townley when persistent charges of disloyalty brought him to Washington, DC, in May 1918. Debating the pros and cons of a bill permitting the court-martial of any American citizen accused of sedition, the US Senate hoped to learn more about supposed dissent. Townley testified to the NPL's loyalty in front of the Senate Military Affairs Committee. "We do not want to stand before the country and the world branded as traitors," he proclaimed. He then denied the many charges made against the NPL and pointed out civil liberties violations in state after state. The bill did not pass.[87]

Soon thereafter, the Republican Party threw former president Theodore Roosevelt into the fray. The bombastic reformer admitted to one newspaper man "that when it was first organized he had real sympathy with" the NPL's "aims and 'some of' its methods." But Robert La Follette's September 1917 speech made it clear that the NPL stood for the antiwar "sentiments of La Follette and that its methods seemed to be working towards the erection of a 'machine.'" Despite his desire to "remedy every injustice or wrong or mere failure to give ample opportunity to the farmer," Roosevelt, "with sincere regret," asked "every American severely to condemn" the NPL.[88]

Staunchly Republican members of the Minneapolis Chamber of Commerce reached out to the former president soon thereafter. They tried to convince him that the League was the "most dangerous movement in America." The Commission of Public Safety engaged in a similar campaign. After a series of missed meetings, in May 1918, Donald Cotton, director of the Ramsey County Public Safety Commission, forwarded Roosevelt a long article from the *Nonpartisan Leader*. The piece took issue with the New Yorker's criticisms of Woodrow Wilson's war policies. Roosevelt responded calmly, noting only that "later on I may wish to do something about the Non-partisan League."[89]

Two months later, he did. In a *Kansas City Star* editorial, Roosevelt assured readers of his affection for "honest and loyal farmers of high character" who joined the NPL to remedy economic imbalances. But he opined that they were duped by dishonest leaders. "Covert disloyalty," Roosevelt claimed, characterized the League. Concerned about Democratic Party inroads in firmly Republican states, party leaders urged him to directly campaign against the NPL in the coming election season. Believing that "Americans should organize as Americans and not as bankers, or lawyers, or farmers, or wage workers"—which would be "thoroughly anti-American and unpatriotic"—Roosevelt agreed. He soon launched a speaking tour to boost the Liberty Loan program.[90]

At stops in Wichita and Kansas City, he encouraged crowds to buy war bonds. But in Alliance, Nebraska, and Billings, Montana, Roosevelt's tone changed. He accused the NPL of offering German "spies, agents, and propagandists" much "aid and cooperation." Admitting that corporations mistreated farmers, Roosevelt called for federal intervention in the grain trade. Yet he decried the League's "state socialism" and categorized membership in the NPL as a form of "domestic treason."[91]

Leaving Montana, Roosevelt returned east. At a stop in Fargo, he again poked at the NPL. In Minneapolis, he broke with his schedule and stayed for more than a week, giving anti-League speeches and leading prowar rallies. He implored crowds to buy war bonds and suggested that NPL leaders spouted "anti-American" attitudes. Returning home, Roosevelt received a cautionary letter from the noted conservationist Gifford Pinchot. Pinchot pointed to the NPL's "thoroughly sound" approach to reform as well as the "business interests" behind charges of League disloyalty. Worrying that Roosevelt hurt rather than helped the Republican Party with his direct attacks on the NPL, Pinchot expressed disappointment. Roosevelt ignored him.[92]

League members noted that high-ranking businessmen in the Republican Party sponsored Roosevelt's tour. The NPL accused the former president of backing special privilege under the guise of patriotism. Though illness would soon take his life, Roosevelt stood firm. He remained proud of his attempt to defeat "Townleyism in agricultural districts."[93]

Frustrated by Roosevelt and convinced of the underhanded methods used to attack the NPL, President Wilson nonetheless remained silent. Privately, he admitted that though "it has given me a great deal of concern . . . there is nothing that can be done by federal authority and nothing that can be done in restraint of the state authorities." Wilson decided that "it is a situation which will have to work itself out." In August 1918, the president finally told administration officials that "under no circumstances" should they be pulled into the "very savage partisan fight in the Northwest." Ongoing federal inaction meant that state and local authorities working alongside concerned businessmen and politicians continued their campaigns against the League unabated.[94]

★

The combination of institutionalized opposition and no federal protection proved potent. The National Civil Liberties Bureau—a precursor of the American Civil Liberties Union—noted that most politically motivated violations of constitutional rights during wartime involved the Nonpartisan League. But the wisest enemies of the NPL recognized that the persecution of rank-and-file farmers might backfire. Thinking ahead to future elections, they instead targeted NPL leaders and organizers.[95]

Opponents tried intimidation. When local officials kept Ray McKaig from speaking in Ashton, Idaho, in March 1918, "500 farmers rolled into town and began to walk up and down the streets." The agrarians stimulated "the liberality of . . . these pin-headed business men," and "the mayor changed his mind and let me speak," McKaig said. In September 1918, Boise's businessmen plotted to deny Arthur Townley the right to speak at a huge NPL rally. In response, the local sheriff—son of a farmer—deputized "husky sons of toil" from across the Boise valley and waited for the mob to strike. When almost 5,000 Leaguers and their leader walked to the steps of the state capital, 150 men began roughing up the NPL crowd. Then "the farmers mobbed the mobbers." Wearing NPL stickpins, the deputies quickly rounded up the vigilantes and then forced them to listen to Townley's rousing address.[96]

When intimidation failed, incarceration loomed. The loyalty committee that grabbed NPL organizer J. A. "Mickey" McGlynn in Miles City in April 1918 locked him in the basement of the Elks Club before putting him on a train out of town. On the same trip, businessmen refused McGlynn the use of a hall in Terry, even after he drafted affidavits attesting to his Red Cross donations and war bond purchases. In O'Neill, Nebraska, a banker (and member of his county council of defense) took the law into his own hands. He threw League organizer J. W. Bissel "in jail for four days without a warrant being issued." A sheriff arrested Walter Quigley at the rural home of a leading Leaguer outside Wahoo, Nebraska, in June 1918. Charged with sedition and selling stock without a permit, Quigley eventually got the charges dropped.[97]

Others faced worse tribulations. In Kenyon, Minnesota, one night a gang of men dragged an NPL organizer "from a moving picture show" and put him "on an outgoing train." In 1918, a mob in Bonesteel, South Dakota, captured two NPL employees. They turned them over to the sheriff, who in turn entrusted the men to "some guys on horseback" who "drove them a foot just like you drive a couple of old cows" toward the Nebraska state line. They eventually put the men on a train and told them never to come back.[98]

Threats of lynching became commonplace. In Holly, Colorado, a crowd threatened to hang one League organizer, but he fled. A group of men "carrying a big rope" roused two NPL organizers from sleep in their hotel in Winner, South Dakota, and "told them they better get dress[ed] and leave town." Outside Mizpah, Montana, ranchers seized the long-suffering McGlynn, intending to use a rope to drag him down a nearby creek bed. He narrowly escaped. In Nebraska, an armed mob seized a League employee in the middle of the night. They took him to an island in the Platte River and put one end of a rope around his neck. The other end went around a tree limb. A journalist

FIGURE 13. NPL gubernatorial candidate Charles A. Lindbergh Sr. hung in effigy in Stanton, Minnesota, 1918. (Minnesota Historical Society, J2 1918r1)

reported that "he was given his life only on his promise to sell his League automobile, give the money to the Red Cross, and himself enlist in the United States army." Unable to find an NPL organizer to threaten, opponents in Stanton, Minnesota, instead hung in effigy NPL gubernatorial candidate Charles Lindbergh (father of the soon-to-be-famous aviator).[99]

Physical—as well as psychological—harm was not out of the question. A vigilante group in Montana "calling themselves the Musselshell Hundred" seized two League organizers and beat them badly. On April 29, 1918, young thugs abducted NPL organizer Joseph Golden in Sultan, Washington. They threw him in their car, beat him with a revolver, and then tarred and feathered him. Emil Sudan, working to organize farmers southwest of Mitchell, South

Dakota, woke one morning to a tar-and-feathering. Toughs then hooked the otherwise naked Sudan to a wagon and dragged him for miles. The medieval practice of tar-and-feathering was meant to both harm and humiliate. The application of hot tar often burned the skin. Applying feathers to the tar made it harder to remove the sticky pitch. Symbolically, the practice proved especially powerful. Those so publicly ridiculed never forgot the experience.[100]

In 1918, an "armed and drunken mob" in Hinckley, Minnesota, seized Nels Hokstad, a local NPL booster. After severely beating him, they applied hot tar and chicken feathers and told him never to return. Assuring the men that "if you want me to stop organizing, you better swing me," Hokstad boldly returned to Pine County two days later and gave a public address. In the meantime, his wife had a nervous breakdown. Members of the Commission of Public Safety–sponsored Home Guards soon captured another local organizer. Hooded men seized Rupert Kennedy from his home. When local farmers listening in on a party line learned of the incident, they gathered their guns and set off to the Kennedy farm. But they got there too late. Kennedy returned home tarred and feathered. Another NPL organizer, tarred and feathered outside neighboring Turpville, suffered worse indignities. There, men tied him up, brought him to the town hall, and rang the fire bell to draw attention to their work.[101]

In Winlock, Washington, wary businessmen surreptitiously installed a dictograph to capture the private conversations of NPL organizers Alfred Knutson and W. R. "Roy" Edwards. The evidence gathered convinced them of the men's disloyalty. After dragging them out of their hotel rooms, "a crowd of 50 citizens" tarred Knutson—head of the League effort in the state—and then covered him in cotton. Edwards escaped a similar fate "because he had been considered only a tool in the hands of the leaders of the league and had promised to sever his connections" with the NPL. Just a few days later, when citizens in nearby Toledo learned that Edwards had "pulled down $24 in commissions by signing up six farmers in an outlying district," they took action. Marching to the farmhouse where he slept, they tarred and feathered the recalcitrant organizer. They also secured his promise never to return. When Edwards tried to press charges, the local prosecutor knew that at least one "well-known politician" headed up the gang. He refused to investigate the matter.[102]

One League worker met a worse fate. Daniel McCorkle, a Presbyterian minister in Montana, organized for the NPL. Like others, he himself faced physical assault. But in 1922, he wrote a friend about a fellow NPL booster who disappeared during an earlier campaign. Learning that his colleague landed in the Montana State Hospital for the Insane, McCorkle held suspicions about how he got there. He remembered that some neighbors threatened to put his friend "in the asylum as crazy." Though he thought nothing of it at the time, McCorkle believed that the booster "was put in the asylum to counteract his influence in favor of the League and to get him out of the way."[103]

When they failed to stop organizers, League enemies intimidated NPL farmers. League-sponsored rallies and parades became a primary target. At an NPL parade in Anoka, Minnesota, in June 1918, three hundred men deputized by local authorities attacked over fifteen hundred Leaguers and their families. Assaulting women and children as well as men, they broke car windows, tore NPL banners from vehicles, and even ripped away American flags. Slapping deputies who assaulted their families, farm women hoped to shame the men into calm. The tactic failed. At a similar event in Austin, Minnesota, later that year, an unruly mob "tore buttons, badges and other NPL insignia from the person of the paraders." One organizer drolly noted that "there was no discrimination against women."[104]

Local law enforcement banned League gatherings of any kind. In February 1918, a sheriff forbade a gathering of NPL farmers in Madison, South Dakota. He "alleged disloyalty and creating a disturbance," though the group "had not even assembled." In the mind of local law enforcement, Governor Norbeck's guarantee of free speech "made no difference." Even after cutting a deal with the state Council of Defense in Nebraska, NPL members there still faced a rocky road. In Wayne County, for instance, local officials prohibited League meetings for the remainder of the war.[105]

Meetings that did occur were often cut short. In October 1917, the mayor of Lake City, Minnesota—egged on by local businessmen—ordered policemen to turn fire hoses on the men, women, and children gathered for a League rally. At an NPL meeting in nearby Lewiston, Minnesota, someone poured formaldehyde "into the heating pipes of a furnace, causing gas in the hall and a hasty exit by the crowd." Denied a meeting hall in Waseca, Minnesota, League speakers moved the gathering to the city park. City fathers turned off the lights. Then a mob "threw tomatoes and vegetables at them while they spoke in the darkness." When Leaguers gathered in Wahoo, Nebraska, to form a cooperative county newspaper, representatives of the state Council of Defense—backed by Home Guards—broke up the meeting. "Two of the men" run out of the hall, one Leaguer reported, "were old Civil War veterans."[106]

In the worst cases, opponents tried to force League members to renounce their politics. For some, mere "membership in the Nonpartisan league was taken as prima facie evidence for disloyalty." A couple in Dayton, Washington, joined the NPL, earning themselves an interrogation from their local council

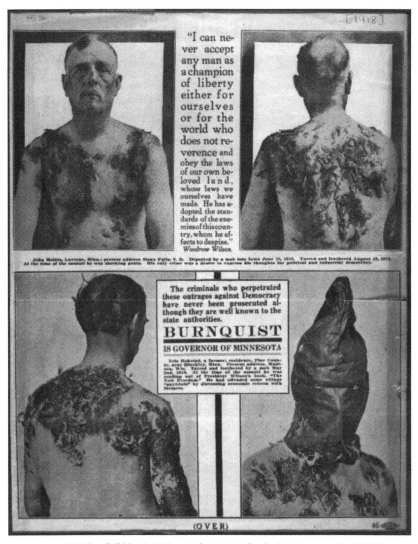

FIGURE 14. NPL handbill blaming Minnesota's governor for the intense persecution in Minnesota. John Meints, a farmer in Rock County, Minnesota, is featured on top, and Nels Hokstad, NPL organizer in Pine County, Minnesota, is featured on the bottom. (Minnesota Historical Society, J1.4 p26)

of defense chairman. John Schmidt, a member of the Nebraska NPL's state board, faced down a mob that "visited his home in an effort to indimidate [sic] and frighten his family" so that "he might sever his connection to the League."[107]

On June 22, 1918, in Luverne, Minnesota, "the Mayor and Marshall organized a 'mob' of loyalists" that "ordered members of the League in Rock

County to register and renounce their League membership." Three hundred and seventy farmers complied. Thirty-one refused. The mob deported two of the latter to Iowa. They told them never to return. One of them, a farmer named John Meints, came back a few weeks later. He hoped to help his sons with the harvest. As promised, a mob tarred and feathered Meints and then deported him again, this time to South Dakota. The farmer later identified a Methodist minister "who beat him with a rope after he was tarred and feathered." When another League supporter tried to go to Saint Paul to report on these injustices, a mob dragged him off his train. Meints himself did make it to Saint Paul, where he showed the US attorney general the mob's handiwork. Federal officials did nothing. Meints took matters into his own hands, suing his abductors in federal court. A jury acquitted them. Residents of Luverne welcomed the accused home with dignitaries and a brass band.[108]

Agrarians realized that this repression stemmed not only from fear but also from recognition of the League's growing power. In Nebraska, NPL state director Christian Sorensen noted that the charge of disloyalty "is a damnable lie born of fear and prejudice." A local committee of farmers in Lisbon, North Dakota, denied "most emphatically all charges of disloyalty made against the Non-Partisan League." They maintained "that the loyalty question was not an issue . . . but it was injected into" public discourse "by the enemies of the organized farmers for political purposes." When the NPL made that point in a published appeal to Congress, it went largely unheeded.[109]

In states with a sizable League presence, the vicious campaigns strengthened ardor for the NPL. Minnesota saw a thousand farmers join the League every week through spring 1918. Writing from Idaho, Ray McKaig compared events in the North Star State to revolutionary tumult in Latin America. "This Mexicanizing of the government of Minnesota, arresting anybody they want to" was counterproductive. Stories of League persecution in Minnesota made farmers in other states "bitter . . . and indignant." A League member writing from North Dakota assured McKaig that "we are with you to the last ditch in this fight for the common people." Ever defiant, NPL organizers and members nonetheless faced constant danger.[110]

*

The intense campaigns against the NPL drew from ethnic as well as economic and political antagonism. Across the country, anxieties over loyalty during World War I focused on immigrant communities. Anti-German hysteria, fed by provocative films, posters, and pamphlets produced by George Creel's CPI, swept the land. "Hyphenism" of any sort—the identification of one's cultural background in combination with American citizenship—proved problematic. As a result, NPL-related controversies sometimes played out along ethnic lines.[111]

Immigrants of every stripe voted for the NPL in North Dakota in 1916 and subsequent years. Elsewhere, it proved difficult to enroll German Americans. But the NPL's decision to both critique and support America's entry into the war attracted the attention of German speakers across the rural West. In 1917, League leaders began actively courting German Americans with a German-language supplement to the *Nonpartisan Leader*. German-language lectures soon followed. The NPL hired E. O. Meitzen, a radical agrarian from Texas, to coordinate these German-language activities. Eventually, the League published *Voksregierung*, a German-language paper, to serve this constituency.[112]

In Minnesota and Nebraska, the effort paid off. In response to the Commission of Public Safety's intense anti-German orientation—including outlawing the use of the German language—German American corn farmers in south central Minnesota began joining the League in large numbers in mid-1917. Because Germans made up the largest immigrant group in the state, this political shift mattered. The association of German-speaking citizens with the NPL in turn intensified the persecution of League members in the North Star State. The same thing happened in Nebraska. The state Council of Defense in Nebraska banned foreign languages, even as German American farmers began backing the League. One League employee who tried to sell German-language NPL newspapers in the state barely avoided a lynching. Targeting the Lutherans in the German-origin Missouri Synod, the council insisted that they were disloyal. This egregious violation of civil liberties further drove German-speaking citizens into the NPL.[113]

In South Dakota, German Americans faced the brunt of anti-NPL repression. The recently drafted John Warns, who lived on a farm outside Wentworth, faced disloyalty charges before leaving for the front lines. In February 1918, local officials outlawed a League meeting that Warns and his father—NPL members—hoped to attend. The *Wentworth Enterprise* charged those wanting to meet with "us[ing] the German tongue exclusively in their homes, in their religious services, and insist[ing] on being educated in German, and speak[ing] a broken English." As in Minnesota and Nebraska, many German Americans threw their support behind the NPL in the face of such campaigns.[114]

Scandinavian immigrants also became targets. Because the NPL backed Charles Lindbergh—an antiwar Swedish American—in Minnesota's 1918 gubernatorial race, Swedes remained under suspicion. That same year, Minnesota's Commission of Public Safety claimed that "the disloyal element in Minnesota is largely among German-Swedish people." Leading Swedish

Americans immediately demanded a retraction. Under intense pressure, the commission backpedaled from its claim. Swedes persisted in their NPL membership.[115]

Swedish Americans on farms across Nebraska also supported the League and earned themselves opprobrium. In Saunders County, for instance, Swedish Leaguers ran into opposition from their Czech neighbors. The Czechs hoped that their homeland would be liberated in the wake of an Allied victory and wholly supported American entry into the war. With their support, the county council of defense persisted in a campaign of harassment against Swedish American farmers in the NPL. The Swedes redoubled their commitment to the League. Long-standing ethnic tensions in the county persisted.[116]

Swedish agrarians as far south as Kansas found the NPL program attractive. Lindsborg, home of Bethany College, offered space to defy conformity. In 1919, the president of the college not only welcomed an NPL rally on the Swedish Lutheran campus but also presided over speeches made by League lecturers. Hundreds of mostly Swedish American farmers roared in support. Longtime NPL opponent Phil Zimmerman, who lived in Lindsborg, cried out for action, to no avail. In 1921, the president again threw open the doors of the college to League speakers, including Arthur Townley. His commitment showed the depth of Swedish American support for the NPL in Kansas.[117]

Norwegian Americans offered a different story. While many sided with the NPL, others firmly rejected it. Nicolay Grevstad, a close advisor of Minnesota US senator Knute Nelson, claimed that "the Norwegians, once so staunch and true, have become unstable and demoralized—drifting helplessly, like a rudderless ship, before any wave of popular craze. They are being roped in by League organizers in droves." Yet for all the Norwegian-speaking wheat farmers in western Minnesota who backed the NPL, Norwegian American enclaves in southeastern Minnesota offered only tepid support. *Normanden*, the leading Norwegian-language newspaper of North Dakota, continued its nasty attacks on the NPL.[118]

In 1919, the NPL launched its own Norwegian-language newspaper, *Nord Dakota Tidende*. Senator Nelson himself confidentially admitted that his fellow "Norwegians are at the bottom of our political troubles in the N.W." In fact, he worried that "the bitterness and malignancy of the Norwegian nonpartisan farmers, exceeds that of any other class of Nonpartisans," even as "the older class of Norwegian Imigrants [sic] & their descendants" stood "steadfast in the Republican faith." Class and generation drove deep divisions in the Norwegian American community.[119]

Religion also played a role in NPL affiliation during wartime. Closely correlated with ethnicity, faith became a tool for League opponents and proponents alike. Fearful of socialism in the NPL's platform, many clergy declared the League to be a worldly evil. Nearly as many sprang to its defense. Few denominations came to consensus on the NPL. This ensured that religious arguments remained useful to all involved.

North Dakota's two Catholic bishops, for instance, took opposite stances. Vincent Wehrle derided the League as socialist. James O'Reilly, on the other hand, believed "that there is no religious issue involved and therefore no special reason" to "take a public stand on the question." According to one Catholic newspaper, "The farmers of North Dakota, including Catholics, are almost solidly behind the League." But as opposition mounted, more and more Catholics fell away from the controversial NPL.[120]

The League also divided Lutheran clergy. The registrar of Luther Theological Seminary in Saint Paul informed South Dakota governor (and fellow Norwegian Lutheran) Peter Norbeck that "it has grieved me very much that so many of our Norwegian Lutherans are so wrapped up in Townleyism." But as John Flint, pastor at Trinity Evangelical Lutheran Church in Bismarck, put it, many Norwegian Lutherans believed that "the League program is based upon and in harmony with the Golden Rule and other fundamental laws in Sacred Scripture." Swedish Lutherans followed the same tack. Investigating Augustana Synod members of the NPL in North Dakota, a committee from Kansas noted that "they attend services, make use of the means of grace," and "give financial aid not only to the local church but also to the institutions of synod and conference." This earned the Kansans' approval.[121]

The Norwegian Lutheran Church in America did not take an official stand on the NPL until 1920. The shift came only after League opponents charged the controversial NPL with supporting "free love." Even then, the church proved a tepid opponent. The Swedish Augustana Synod refused to join the fray. After all, the exploitation of farmers meant "that the people do not possess their own soul." Nevertheless, it admitted that the League seemed to be "a socialistic scheme." Noting that the NPL's "fundamental principles should be well known to our pastors," one frustrated Swedish Lutheran called on the church to decide whether or not the League was "right or wrong." The Augustana Synod remained officially silent.[122]

Clergy from other mainline Protestant denominations proved equally split. In 1920, pastors in Nebraska responded to an anti-NPL editorial. Because "many members of the Nonpartisan League" worshipped in "our respective churches," they argued that the NPL was "one of the strongest forces in America for real Christianity." Rejoinders soon followed. Responding to fellow Methodist clergy in Iowa with a clear anti-NPL sentiment, Ray McKaig's father, a Civil War veteran and prominent pastor, suggested that

"Jesus Christ came not only as a spiritual guide but as an economic defender of the rights of the common people." He envisioned the NPL as "'applied Christianity.'"[123]

Farther west, McKaig himself fought charges of "unpatriotic" behavior made in Salt Lake City newspapers. He knew that tight-knit communities of Mormon farmers in southeastern Idaho took their "cue from Salt Lake City." Latter-Day Saints (LDS) agrarians who stood against the League drew on negative reports deliberately spread by a group of Montana businessmen anxious to prevent the organization from taking hold in neighboring Idaho. Church leaders and sugar beet plant owners in Salt Lake City were "rather close for comfort," and thus the NPL hoped to tread carefully so as not to antagonize the LDS Church. The strategy worked. In March 1918, McKaig noted that the Idaho NPL had successfully appealed to "about two thirds of the Mormon church district."[124]

With such divisions in mind, both the NPL and its adversaries deliberately deployed clergy to make a case for or against the organization. Leaguers published a pamphlet loaded with endorsements by social gospel–oriented pastors and priests from the East Coast. Opponents, like one Presbyterian minister in Grand Forks, responded by calling NPL members a "class of undesirable, low down, indolent, non-productives." Longtime League opponent Jerry Bacon published this sermon to counter NPL attempts to align religious leaders with the organization. Just a few years later, this pastor's taste for invective vaulted him into the leadership of North Dakota's Ku Klux Klan.[125]

The struggle for the soul of western farmers left plenty of room for congregants to make up their own minds. Choices made locally by pastors, priests, and congregants defined the fractured religious response to the Nonpartisan League. During the first two years of Nonpartisan League organizing, pressing economic and political issues meant that ethnicity and religion made little difference. But loyalty debates during World War I highlighted the significance of immigrant background and religious faith. It often intensified persecution, as League opponents found it easy to prove disloyalty through ethnic difference or church traditions. Just as often, immigrants found the League more attractive by virtue of that same persecution. Ethnicity and religion became a crucial venue and vehicle for NPL-related controversies.

✳

Recognizing that controversies needed constant care and feeding, NPL opponents used the war as a pretext and the courts as a tool. Rivals seized on any chance they could get to allege the League's leaders were both Socialists and disloyal. Ongoing indictments clouded the legality of League actions and provided "proof" of disloyalty that opponents everywhere found useful.[126]

In Idaho, NPL enemies accused Ray McKaig of openly supporting noted Socialist Kate Richards-O'Hare. McKaig had, in fact, attended her December 1917 federal trial for sedition in Bismarck, North Dakota. During a recess he briefly spoke to Richards-O'Hare. League opponents luridly amplified this brief encounter into a full-blown conspiracy. Newspapers across Idaho smeared McKaig by asserting his intimate association with a known and convicted Socialist. The Leaguer responded by bringing libel charges against former governor Frank Gooding. Gooding coordinated the press campaign against the Idaho League leader after the NPL refused to support his US Senate campaign.

When a court considered the case in November 1919, Gooding and his allies used testimony from League opponents in Minnesota and North Dakota to conclusively link McKaig and the NPL with Socialists like Richards-O'Hare. The courtroom became a venue for destroying the NPL's reputation in Idaho. Not only did McKaig lose the libel case, but he also inadvertently aided the effort to use disloyalty as a weapon.[127]

McKaig was not alone. Outsiders noted that in Minnesota, opponents unleashed "the forces of hell" against the NPL and its leaders. Through 1917 and 1918, both Arthur Townley and Joseph Gilbert, a former socialist and the NPL organizing chief, faced numerous indictments brought by county attorneys in Minnesota. Encouraged by the Commission of Public Safety as well as Minnesota state attorney general James E. Markham, prosecutors proved especially aggressive.[128]

In February 1918, Martin County Attorney Albert Allen attempted to arrest Townley and Gilbert in Saint Paul on charges of interfering with wartime enlistments through distribution of NPL pamphlets. With Townley out of town, the local sheriff returned with only Gilbert. League attorneys immediately secured Gilbert's return to Saint Paul on grounds of illegitimate authority. Gilbert also filed suit against Martin County for kidnapping him. After sorting out the jurisdictional question, the Martin County court ruled against Gilbert and Townley. Months later, the Minnesota Supreme Court struck down the ruling.[129]

In neighboring Jackson County, Allen's colleague took up the case. Though a ban on NPL organizing and meetings made it nearly impossible for the League to have any presence there, the county attorney used charges of conspiracy to discourage enlistments to haul Townley and Gilbert into court. The prosecutor pointed to NPL pamphlets as evidence. With a disgruntled former NPL organizer as the star witness, the prosecutor found favor with the court. The judge—who, a few months before the trial, publicly avowed that the NPL was pro-German—refused to allow affirmations of League loyalty from federal officials. The jury explicitly excluded anyone with an NPL affiliation. Townley and Gilbert were denied the right to speak in their own defense. Despite the best efforts of League attorneys, the ruling against the NPL leaders seemed foreordained. Only ongoing appeals kept them from serving their sentence. Cited as evidence of Townley's disloyalty around the country, this case produced powerful national repercussions. Adversaries pointed to the conviction as overwhelming proof of League disloyalty, even as others, such as George Creel, asserted that the outcome of the poorly handled case put "the machinery of American justice on trial."[130]

In Goodhue County, the perpetually unlucky Joseph Gilbert again found himself indicted, this time for a speech he had given nine months earlier. Accused of discouraging enlistments at a public gathering, Gilbert denied the charges. At trial, seven witnesses, clearly coached by the prosecution, described Gilbert's disloyal address. The judge had ruled that no member of the NPL could sit on the jury. One frustrated farmer said: "Politics! Politics! . . . It's all politics and it's a-goin' to cost this county a mint o' money. They're going to convict 'em, but s'pose they do—it'll only make the League stronger." He was right. Convicted of breaking the state's new sedition law, Gilbert faced a year in jail and a sizable fine. When Gilbert was granted a stay, the NPL appealed his case. In December 1918, the state supreme court admitted to the many irregularities but upheld the lower court's conviction.[131]

In January 1919, League lawyers appealed the case again. They hoped to convince the US Supreme Court that Minnesota's sedition statute—passed as one of the many pieces of legislation that granted the Commission of Public Safety wide-ranging powers—violated Gilbert's constitutional right to freedom of speech. War, they argued, did not give the state the right to revoke a citizen's ability to speak freely. Only the federal government could do that. The court agreed to take the case.[132]

The Supreme Court's ruling in *Gilbert vs. Minnesota* represented a low point in the history of American civil liberties. Justice Joseph McKenna wrote the 7–2 majority decision upholding the lower court's decision. The majority believed that the Minnesota statute was an appropriate wartime measure. The emergency allowed the state to curb particular constitutional rights.

Only the newest justice—Louis Brandeis—provided a thoughtful dissent. Brandeis pointed out that the deeply flawed Minnesota law clearly applied not only to wartime but also to peacetime. This meant that "it abridges freedom of speech and of the press, not in a particular emergency . . . but under all circumstances." Any such law seemed clearly unconstitutional and "invalid because it interferes with federal functions." In particular, the Fourteenth Amendment's guarantee that "no State shall make or enforce any law which shall abridge the privileges or immunities of any citizen of the United States" suggested to Brandeis that federal rights trumped state regulation of those rights. Not until 1925 would a majority of the court see fit to rule in the same fashion. But by then it was too late for Gilbert and the NPL. Calling his imprisonment "the height of nonsense," Gilbert spent a year in jail.[133]

✳

Fueled by fear, politics, ethnic prejudice, and controversy, the intensity of the anti-NPL effort hindered the organization's wartime attempts to expand to other states. A bad reputation, earned unfairly, hampered relations with other farm organizations—a crucial factor in League success beyond North Dakota. Even the Grange, whose state-by-state support of the NPL followed from Ray McKaig's strident pro-League advocacy in 1916 and 1917, became divided over the NPL. By 1918, "some Grange leaders in the Northwest" believed that "the new movement" was not "helpful." In the face of disloyalty charges, McKaig could not even convince the editor of the *National Grange Monthly* to publish a single story about the League.[134]

The League immediately felt the effects of this shift. In 1917, McKaig characterized the resistant leader of Iowa's Grange as a "damn fool." The NPL hoped to organize there despite this setback. Farmers in north central Iowa had expressed serious interest in the Nonpartisan League as early as late 1916. The first wave of organizers entered Iowa in response to their pleas. Unfortunately, "the man sent there to supervise the work announced his intentions through the newspapers before he had signed any members." "This publicity," reported one NPL employee, "created such opposition that a real foothold never was gained."[135]

Once the war came, Iowa's branch of the newly created Farm Bureau proved especially antagonistic. At the direction of Governor William Harding and with the support of county extension agents, they actively prepared a campaign to counter the entry of League organizers. James R. Howard, who became head of the organization just after NPL employees reentered the state in late 1917, churned out editorials asserting the NPL's disloyalty. Even *Wallace's Farmer*—a powerful agrarian paper run by the prominent agrarian Henry C. Wallace and his son (a future vice president of the United States)—took a stand against the NPL. Made up of small-town businessmen and boosters from around the state, the Greater Iowa Association "held community meetings in sixty counties" and "within forty days—turned the spot-light of truth upon the misrepresentations of the Non-partisan League organizers." It also fashioned "a corps of anti-League lecturers" and "a anti-League press service." The NPL organizing effort went nowhere.[136]

Six months later, at the behest of James Pierce, member of the Iowa State Council of Defense and editor of the *Iowa Homestead*—a farmer's newspaper—the NPL again sent organizers to Iowa. This time, the League opened a state office in Des Moines and purchased one hundred cars for organizers. Responding quickly, the Greater Iowa Association and the Farm Bureau both "appealed to the State Council of National Defense to consider the league as interfering with the prosecution of essential war work." The council agreed and directed NPL organizers to stop working and holding meetings. Self-described as "the only one out of twenty members of our state council of defense that is not bitterly opposed to the Nonpartisan League," the insistent Pierce soon found himself expelled him from the council.[137]

Despite fierce opposition, organizers enrolled fifteen thousand members in Iowa by October 1918. In his newspaper, Pierce lashed out at the Greater Iowa Association, noting that much of the antagonism stemmed from "the organized farmers of Iowa" threatening the "personal supremacy" of the leaders of the association in state politics. This brought on a hailstorm of Council of Defense–endorsed criticism. In Kanawha, Iowa, vigilantes snatched an NPL-boosting farmer and drove him forty-five miles to Mason City. There, the chief of police and mayor told him, "We are going to run the Nonpartisan League out of Iowa." After being forced to sign a pledge never to support the League again, the man learned that if he told the *Iowa Homestead* of the incident, "he would be put where he could not do anything more until after the war is over." In Oskaloosa, authorities arrested an NPL organizer. Such illegalities cowed Leaguers in Iowa.[138]

Other agricultural states attracted League attention. Oklahoma's well-known penchant for agrarian socialism suggested fertile ground for organizing. Arthur Townley sent L. N. Sheldon, a former colleague from North Dakota's Socialist Party, to Oklahoma in March 1917. Feeling confident, Sheldon established a state headquarters in Oklahoma City a few months later. Sheldon imported the League platform from North Dakota and hired organizers from the ranks of the Oklahoma Farmers Union.[139]

But wartime Oklahoma left little room for dissent. In August 1917, a full-scale draft revolt among farmers in the south central part of the state—which became known as the "Green Corn Rebellion"—ensured a brutal crackdown on all who seemed to oppose the war. It also made anything that smacked of socialism a clear indication of disloyalty. Empowered county councils of defense targeted anyone who spoke out against the war. Charges of NPL disloyalty emanating from other states made Leaguers obvious targets. In April 1918, the Washita County Council of Defense canceled an NPL speaker's appearance in Bessie. One NPL farmer protested this undemocratic act. Soon thereafter, the agrarian found himself seized in the middle of the night by unknown men who then tarred him.[140]

Existing farm organizations in Oklahoma, including the Farmers Union, pulled back from endorsing the League. So did many former Socialists—one of whom called the Leaguers "Rabbit-footed scab Socialists." Even so, three thousand farmers signed up. Disorganized and discouraged, they failed to produce a gubernatorial candidate for the 1918 elections. Sheldon shut down the NPL office in August 1918 because "poverty and the failure of crops" made it almost impossible for farmers to afford the sixteen-dollar membership fee.[141]

Similar problems emerged in Texas. There, longtime agrarian radicals E. O. Meitzen, E. R. Meitzen, and Thomas Hickey became interested in the NPL after a wave of repression against Socialists undercut their ongoing efforts to organize tenant farmers. Shaken by the intensity of the anti-Red effort in Texas, they drifted north to Minnesota, offering their services to NPL leaders. Believing that the League offered "the best plan yet" for disempowered farmers, they appreciated that the model "could bring together the irreconcilable" views of agrarians in politics. By November 1917, Hickey returned to Texas to start an NPL branch, establishing a headquarters in Waco.[142]

Because the League in Texas had emerged from the remnants of the Socialist Party, the platform for the Texas NPL deviated in crucial ways. It called for a tax reduction on farm improvements and equipment; the reduction of interest rates; state-run weighing and grading of agricultural commodities; the referendum, recall, and abolition of the poll tax; state-run terminal mills, warehouses, and an insurance system for storms and fires. Only the latter two stances stemmed directly from existing iterations of the League. Adapting the NPL strategy to the Lone Star State's particular issues—cotton, tenancy, and race—longtime radicals hoped to animate a new version of Texas populism.[143]

Internal issues and ongoing harassment dogged the organization. In 1918, state leaders fired Hickey, who drifted off into the business world. County councils of defense across Texas pronounced the League disloyal. Local law enforcement responded accordingly. That April, a sheriff in Quitman accused two NPL organizers of being pro-German. Fearing violence, the men fled to nearby Mineola to meet with M. M. Offutt, a farmer, Methodist lay preacher, and head of the state's NPL. There, police charged the two men with vagrancy and threw them in jail. A group of men then dragged Offutt away, shaved his head and whiskers with sheep shears, beat him, and put him on the next train

out of town. Meanwhile, a mob broke into the city jail, took the NPL organizers outside of town, stripped them, horsewhipped them, poured salt water
on their wounds, and finally told them "to run, while shots were fired after
them." Leaguers appealed to federal agents in San Antonio for protection. Already watching the NPL closely, they offered little aid.[144]

Questions about Nonpartisan League loyalty even stunted attempts to link
the Canadian manifestations of the movement with those south of the border. In 1918, Jack Ford—publisher of the *Alberta Non-Partisan*—hoped to get
Townley to address Alberta's 1918 NPL convention. But after learning about
the controversies surrounding the NPL in the United States, Ford noted that
"we are not exactly sure whether a speech from Mr. Townley could be re-
garded as an asset to the cause here." Indeed, "the Calgary Herald in particular has knocked us hard regarding our connection with the League across the
line." Louise McKinney, one of two NPL members of the Alberta parliament,
agreed. She made it clear she thought "that it would be a mistake to invite
Mr. Townley at this time."[145]

Wary of summoning League leaders—and the associated disloyalty—from
the United States, Ford settled on sending a simple request to the League's
headquarters in Saint Paul for "pamphlets, etc., to push the cause." Even so,
as news of NPL disloyalty drifted north, some farmers angrily canceled their
membership in the Alberta NPL. Ford went to great pains to assure one man
that "we have no connection whatsoever" with the "the Non Partisan League
in the States." But the damage had been done. Indeed, Ford wrote to C. W.
McDonnell, an NPL legislator in North Dakota who often visited his sister
and parents in Lougheed, pleading for an article that would discuss "Townley's arrest, showing up the incident in its true light" in order to "show the real
truth." McDonnell contributed such a piece to the *Alberta Non-Partisan*, but
it was too little, too late.[146]

*

The most important manifestation of organized opposition to the League
emerged in North Dakota at war's end. It did not grow out of a campaign
trumpeting NPL disloyalty. Nor did it depend on corporations, councils of
defense, or the courts. Instead, a small group of farmers became dissatisfied
with the NPL. Together, they created an organization—the Independent Voters Association (IVA)—that quickly garnered support from small businessmen and party politicians alike as the perfect vehicle to counter the League's
well-established power in the Flickertail State.

From its earliest days, NPL leaders promised members that the organization belonged to farmers. In November 1915, the *Nonpartisan Leader*

trumpeted the notion that "you farmers are the League." Local meetings to
select primary election candidates as well as annual conventions of members
seemed to back up these claims. But when the 1917 legislative session began,
Townley and his advisers carefully tutored NPL politicians in both methods
and policy. Insisting that their lack of experience made it necessary for Leaguers to meet in closed sessions, Townley invited charges of manipulation from
opponents. A disgruntled NPL-endorsed state legislator named E. W. Everson
challenged Townley.[147]

Everson complained that he "found a ready-made slate instead of the cus-
tomary free-for-all discussion of who was to do what." League leaders had
already drawn up committee assignments. "Believing that if the Non-Partisan
League was to be a farmers' organization it should be controlled by farmers,"
Everson started circulating a petition among fellow legislators to that effect.
When Townley got wind of the effort, he forced the Leaguers to retract their
signatures. Everson's effort went nowhere. So he quit the NPL and began organizing what he called an Anti-Socialist Conference in Grand Forks.[148]

Days before the conference, Theodore Nelson, whose experience as one
of the founders of the equity movement in North Dakota made him a prominent League member, wrote to the *Nonpartisan Leader* to defend himself
against charges that he spread anti-NPL sentiment. He argued that farmers
held an "undercurrent" of hard feelings toward the NPL. Nelson called for a
meeting to "prevent this undercurrent" from "doing the movement as a whole
an injury." He did "not suppose that it is to be considered treason if when one
member of the League in writing to another calls attention to all of the things
that are abroad and hurting the league movement." In response, the editor
suggested that Nelson's desire to "get a little prominence for himself" made
him "consciously or unconsciously . . . THE TOOL OF OPEN ENEMIES OF
THE LEAGUE."[149]

Nelson went to the anti-NPL meeting in Grand Forks. A wide range of
League opponents—from anti-NPL legislators to avowed League hater Jerry
Bacon—joined him. All agreed that the League's founders, mostly Socialists,
deliberately deviated from the stated democratic aims of the organization.
The small businessmen, establishment politicians, and addled agrarians who
attended, however, held little else in common. The former worried about
state-run enterprises ruining their businesses. Officeholders and party opera-
tives desperately wanted to restore their longtime hold on the state. The farmers in attendance "stood boldly for all of the economic reforms embraced in
the original Non-partisan League Program" but disagreed with the decisions
made by League leaders.[150]

Because such a coalition had little staying power, not much came of the

conference. But "the appearance of House Bill 44 confirmed" Nelson's "suspicion that the Nonpartisan League was not designed primarily to take up the
farmers problems where the A[merican] S[ociety] of E[quity] had left off."
As a result, Nelson gradually moved to the center of a small group of unhappy
farmers engaged in "a strong, steady, but quiet campaign" to challenge the
League.[151]

Soon charged with standing by "Big Biz" instead of North Dakota's farmers, Nelson insisted on his allegiance to the League's goals. He reminded one
friend that "my record . . . since boyhood shows that I have always been lined
up on the side of progressive politics and the common people." In a February 1918 letter to NPL counsel William Lemke, Nelson noted that "as you
know I have felt all along that there was room for improvement in the form
of organization of the League but it is equally true that I believe the League
form is a decided improvement over the old A.S. of E. form." If "the League
government is responsive to the will of the majority of its membership and is
accomplishing results," he promised to return to the NPL ranks.[152]

Meanwhile, Everson began organizing "the progressive element among the
farmers that had been active in the Farmers Union and Equity movements"
who disagreed with the NPL's approach into the State Legislative Campaign
Organization. They prepared a formal petition to the NPL's leaders. It urged
"the officers of the League . . . to frame a constitution and by-laws to embody
therein the initiative, the referendum and the recall in order to insure that
the wishes of the majority of the members shall at all times guide and determine the policies of this organization." The NPL members also called for "the
election of officers whose duties are definitely prescribed." Firmly focused on
fostering farmer unity, League leaders ignored their plea.[153]

Around the same time, a fed-up Theodore Nelson gave up on reform and
renounced his League membership. He joined Everson and a handful of other
former NPL members at a meeting in Minot on May 1, 1918. Those in attendance drafted "a platform of principles . . . having for its object electing
progressive men to the legislature who should be independent of Townley's
influence."[154]

Alongside angry leaders of the state's Democratic and Republican parties, the organization supported Stephen Joe Doyle, a fusion candidate, for
governor, to run against NPL incumbent Lynn Frazier in the 1918 election. It
desperately dubbed the NPL as disloyal. Yet North Dakota's farmers generally
ignored the organization and its candidate. One staunch Leaguer in Lisbon
noted that the "old corrpupt [sic] machine politicians . . . are not sincere in
their loyalty campaign." In the elections, Frazier successfully held on to his

post. The NPL also took charge of the state senate. It emerged with complete control of North Dakota's government.[155]

Disgusted by the election results, on December 5, 1918, Theodore Nelson, E. W. Everson, and six other disgruntled farmers met in Cooperstown, North Dakota, and made four crucial decisions. First, at Nelson's insistence, they became the Independent Voters Association. Second, they agreed to support the formation of a "State Federation" of agrarian-owned elevators and other forms of economic cooperation—which established their bona-fides as reform-seeking farmers. Third, they made Nelson the new organization's secretary. Fourth, they established their intent to open a headquarters in Bismarck. Then they called for a mass meeting in the same city in late January 1919.[156]

Bringing his many years of experience in the American Society of Equity to the group, Nelson infused vigor and political acuity into the anti-League organization. At the meeting in Bismarck, the assembled carefully separated the alleged "socialism" of the NPL from the original goals of the movement:

> Since the people of our State have twice voted in favor of the establishment of a State-owned Terminal Elevator and since there is some logic in the argument that a Terminal Elevator would be more likely to prove a success if a flour Mill was built and operated in connection with it, we are in favor of having established as speedily as possible, a Terminal Elevator with a flour Mill if it can be legally done.[157]

Strategically positioning themselves as allied with the League's aims even as they opposed the League's methods, the IVA carved out a platform that appealed to agrarians' economic concerns and concurrently pointed out the moral dangers of NPL political power.

Unlike businessmen in small towns or financiers and millers in far-off Minneapolis, Nelson and the other farmers who made up the IVA's leadership possessed credibility. Their reputations in profarmer politics were a matter of public record. So was their discord. Drawing on this strength—which distinguished the IVA from every other opposition group that emerged to take on the League—Nelson ensured that, within just a few months, the IVA would be well positioned to mount a challenge to the League in its home state.[158]

✳

The emergence of internal debates about the NPL in North Dakota created a space for a more effective opposition. The IVA proved more credible than its predecessors. During the League's first two years, widely scattered small-town businessmen failed to mount an effective defense. Corporate leaders in Minneapolis and Saint Paul fashioned feeble front organizations in a failed effort to sink the farmers. As the NPL established itself in multiple states, established politicians and businessmen struggled to cope with the emergent power.

World War I—and the League's awkward stance on US involvement in the conflict—offered them an opening. Vicious wartime smears, disloyalty charges, and illegal harassment characterized attacks on NPL leaders, organizers, and members. But with the exception of states the NPL attempted to organize after April 1917, solidarity accrued wherever persecution thrived. From the start, the League insisted that established politicians and "big biz" powered the opposition to farmers working together as a class. The many attacks that rained down on the League after America's entry into World War I seemed to bear out that claim. Organized opponents used not only newspaper attacks and sham movements but also state-sanctioned efforts to undercut the NPL's power. The blatant manipulation of government agencies and institutions simply confirmed what League leaders suggested—and what many farmers suspected—all along.

Despite its enemies' best efforts, at war's end the NPL proved stronger than ever. In the 1918 elections, the League seized the reins of state government in its most powerful stronghold—North Dakota. Rank and file joined legislators and leaders in celebration. The victory offered the NPL an opportunity to show the entire nation the organization's potential. Leaguers jumped at the chance to do exactly that.

4

Power

In my estimation, that which we have started in North Dakota is the one hope of putting the government of the various states and of the nation into the hands of the people.... This change can be brought about in a true American manner by the use of the Non-Partisan ballot.

GOVERNOR LYNN J. FRAZIER, *New York Times Magazine*, May 16, 1920

The men gathered in a one-room schoolhouse on a hot summer evening. Twilight approached. Weary from the day's many chores, the "Neighbors and friends Residents and voters of Dover Township" in rural Griggs County, North Dakota, drafted a declaration of their hopes for the primary election at month's end. Addressing the "voters of Griggs and Steele Countys" they declared themselves "believers in thrue [*sic*] democracy That is a government by the people of the people and for the people." Then they endorsed one of their neighbors as the Nonpartisan League candidate for state senate in 1918.[1]

Noting the intensity of opposition to the NPL, the men implored fellow voters not to "let big biz fool you this time they have fooled us enough in the past." Realizing that "our candidates may not be as polished as some of theirs," the agrarians argued that "the hearts in our ca[n]didates is [*sic*] in the right place." Furthermore, "the towns will stand united behind their man." If the League candidates failed to triumph in the primary, the farmers warned their neighbors that "they will give you the laugh afterwards . . . and say I told you they could not stick."[2]

Finally, the men responded to charges leveled by the emergent Independent Voters Association. "Big biz is trying to lead you to believe that our leaders might go crooked and run away with the money," the agrarians noted. This proved especially true in Griggs County. Four months earlier, the same group of farmers assembled to condemn E. W. Everson—who along with Theodore Nelson served as the leader of the burgeoning anti-League effort in North Dakota—"for forgetting that he was the peoples servent [*sic*] after he was elected as our representative." They publicly branded Everson "a traitor to the Farmers Interest." Indeed, "every farm movement that has ever sprung up and later gone to pieces" fell apart because "the opposition has always led us to believe that our leaders were crooked." Would farmers "let them fool us that way this time?" "The opposition seems to think they can stuff any thing down our th[r]oat and make us believe it," they asserted. "Friends, let us show them that we are not as foolish as they t[h]ink we are." Instead, they urged, "let us have the laugh ... on them this time."[3]

The farmers did have the last laugh. In the Republican Party primary, Dover Township went almost entirely for Nonpartisan League candidates. So did Griggs and Steele Counties. In fact, the NPL swept all but one of the statewide races, cementing its hold on North Dakota's most powerful political party. In the November 1918 general election, the League swept away the remaining obstacles to enacting its platform. As an example to agrarians across the West, North Dakota embodied the possibilities inherent in farmers sticking together.[4]

Gaining control of all three branches of government in North Dakota showed the NPL's growing power. In Minnesota, Montana, South Dakota, and Idaho, League gubernatorial candidates lost, but NPL minorities in those state legislatures began advocating for change. Pointing to the establishment of a state-owned flour mill and grain elevator in the Flickertail State, they touted the benefits of government competition in the marketplace. North Dakota's legislature also passed laws establishing a state bank, state insurance, and state-sponsored housing. Final approval for these initiatives came in referenda in 1919. North Dakota's voters—finally satisfied—overwhelmingly supported the NPL platform.

Jubilant farmers created poems, songs, and an iconography extolling the virtues of the League as an innovative vehicle for reclaiming American democracy. They shared their sentiments at picnics, rallies, and gatherings that illustrated the change in their own imaginations. Envisioning themselves as actively engaged in deliberative politics, they created a culture to match their newfound passion for political action.

Farm women, too, became especially crucial. As the backbone of rural communities, women already served as the primary connection across rural neighborhoods. But with the full extension of suffrage, they became direct players in electoral politics. The League's initial focus on agrarian manhood gave way. Desperate to curry favor, the League made women full members of the organization and supported the creation of women's NPL clubs. These auxiliaries provided spaces for rural women to expand their vision of self. Though the organization replicated a gender division often prized in both rural families and American society at large—and showed the limits of League efforts to empower—farm women themselves turned meetings into vehicles for citizen education and action.

The seizure of an entire state government and the emergence of a genuine movement culture in the League guaranteed more national attention for the NPL. It also ensured that unions began taking the League more seriously. This led to NPL-brokered agreements with organized labor not only in North Dakota but also in Minnesota, Montana, and Washington State. Furthermore, success extended the League's power beyond electoral politics and into American culture. Swept up in the excitement, leading figures such as Thorstein Veblen, Waldo Frank, Upton Sinclair, and Sinclair Lewis touted the democratic possibilities inherent in what many referred to as the "farmers' experiment." With more than two hundred thousand members in thirteen states, the NPL's power peaked in early 1919. For many Leaguers, a multitude of hopes and dreams seemed within reach.[5]

*

Marshaling resources for the 1918 campaigns, NPL leaders decided to focus their monies and energies on Minnesota, Idaho, and North Dakota. League membership in those states numbered a combined one hundred thousand farmers–half of the League's total strength—and they hoped that reassigning organizers and speakers from every corner of the country to those states would assure victory. Adding two states to the NPL column—including one outside of the wheat-growing belt—would make a bold statement about League power and propel the expansion effort elsewhere.[6]

Leaguers knew that repression and weak organizations in Iowa, Oklahoma, Washington, and Texas rendered any electoral success in those states unlikely. Farmers in Kansas and Wisconsin, worried that premature failure might hurt the organization's future chances in those states, also decided against nominating candidates for office (though NPL members in the former helped elect Governor Arthur Capper to the US Senate). South Dakota's NPL decided not to take on the powerful Republican governor Peter Norbeck in the primary. Instead, they looked to the November elections for victory. Forgoing the NPL's strategic innovation and running candidates as independents in the general election, they failed miserably. In Montana, Colorado, and Nebraska, Leaguers decided to show their growing strength by nominating candidates for the state legislature in counties with sizable NPL membership numbers.

The influence of twenty thousand Leaguers in Montana created cause for celebration in the August 1918 primaries. They ran legislative candidates in both parties. The League endorsement carried fourteen into the general election for state senate, and twenty-eight nominees for the house. The effort to be thoroughly nonpartisan, however, meant that the NPL did not capture a single party's machinery. That said, one observer noted that "in questions of State politics, the Republicans can be hardly differentiated from the Democrats." Both parties "declared their readiness to fight . . . the Non-Partisan League." With the help of "the worst blizzard in the history of the state," the entrenched politicians of every ilk succeeded. In November, only three Leaguers found their way to the state senate. Sixteen made it into the house. One NPL-endorsed judge made it onto the Montana Supreme Court.[7]

Nebraska and Colorado saw a handful of League legislators elected for the first time. During the summer of 1918, Nebraska NPL members met in district-wide conventions and endorsed men for the statehouse. They generally focused their attention on the Democratic Party. Many nominees were not Leaguers. Those who were often ran reluctantly. Reticence stemmed not only from the need to keep their farms and families going but also from fear of reprisals from League opponents. The state NPL recommended Democrat Charles Bryan (brother to William Jennings Bryan) for governor. He lost in the primaries. Then the League joined other farm organizations in endorsing Republican George Norris for the US Senate. Norris triumphed in the general election, alongside one NPL state senator and seven representatives. Farther west, in Colorado, a lack of institutionalized repression allowed the small but growing NPL membership there to elect two state senators and two members of the house.[8]

Much, then, hinged on victory in the League's signature states. Preparing for the 1918 elections in Minnesota meant overcoming the oppression fomented by the state's Commission of Public Safety. The NPL ably fielded a number of candidates in the Republican Party primaries. The most prominent, Charles A. Lindbergh—a longtime Republican member of the US House of Representatives from central Minnesota—vigorously campaigned for the party's nomination for governor. As the most prominent Minnesota politician willing to cast his future with the League, Lindbergh offered the best chance for success in 1918.[9]

Directly challenging the incumbent, J. A. A. Burnquist, Lindbergh fearlessly stood his ground despite numerous efforts to effectively prevent him from campaigning. At one rally, he finished his speech and returned to his car to find a mob beating the driver. Lindbergh calmed the crowd, put the Leaguer back in the car, and began driving away. Seconds later, bullets whistled by. Lindbergh insisted that the pair "must not drive so fast" so that the men would not "think we are afraid of them."[10]

However coolly Lindbergh handled opponents, his last months in Congress before becoming an NPL candidate proved especially controversial. A vote against American entry into the war in Congress in combination with unsavory comments about Catholics provided Burnquist with powerful talking points. Then Democrats called on their loyalists to cast ballots against Lindbergh in the open Republican primary in June 1918. Their participation made the difference. The League candidate lost to the incumbent by more than forty-five thousand votes. Even so, most of the men nominated by the NPL for the state legislature defeated their opponents. One of them, a pro-farmer lumberyard owner in Hawley named Knud Wefald, reported that his victory turned "the whole town against [him]." He proudly hoped to be "expelled from the co[m]mercial club" over the matter.[11]

Crushed by Lindbergh's loss, but enthused by local victories, union members and Leaguers met separately in Saint Paul in August 1918 to put together a joint farmer-labor ticket to challenge Burnquist in the general election. Though the farmer-labor nominees created little buzz—and lost resoundingly in November—Minnesota NPLers began exploring electoral options beyond seizing a party through the primary. Furthermore, the League sent twelve men to the state senate and thirty-four representatives to the house.[12]

Yet these legislative gains proved limited. Walter Day, a newly elected NPL member of the Minnesota house from Bagley, remembered that other legislators treated the League caucus just as white southerners did African Americans, by "step[ping] on them every chance" they got. Indeed, William Nolan, speaker of the house, intentionally organized the body "along Loyalty lines." Encouraged by colleagues not to let "the non-partisan league get one single solitary thing," Nolan ensured that Leaguers failed to receive choice committee assignments and got little traction in the assembly.[13]

Out west, the Idaho NPL readied itself to take over the long-dominant Democratic Party. At its convention in early July 1918, the League called not only for state-owned mills, packing plants, and sugar factories but also for state ownership and distribution of water power. The latter proved especially attractive to farmers, especially those on reclamation tracts in southern Idaho. Finding candidates for statewide offices as well as the state legislature, Leaguers put forward union men as well as farmers. Among them were progressive Republicans ready to run under the opposition party's banner. Reaching out to the state's most famous progressive Republican, the NPL also endorsed US Senator William Borah. Borah's push for the conscription of wealth alongside young men made him especially attractive to Leaguers. This arrangement also proved the organization's nonpartisan orientation.[14]

In the September 1918 primary, the NPL candidates swept the field. Leaguers on the Democratic Party ticket began campaigning across southern Idaho, where farmers—including Mormons in tiny hamlets such as Mink Creek, Grace, Oxford, and Thatcher—backed them in large numbers. In the face of the defeat in the Republican primaries in Minnesota just two months earlier, it looked to many as though the NPL would add Idaho to its column as a firmly League state.[15]

But League leaders did not successfully take over the party's state committee in the wake of their primary victory (as the NPL had in North Dakota in 1916). This left a resilient party structure that despised the League. Democratic Party functionaries withheld campaign monies, organized a non-NPL party caucus, and even charged the League with illegalities. The Idaho Supreme Court eventually refuted those charges, but the damage had been done. At war with itself, the state Democratic Party collapsed into chaos and consternation. Unaffiliated voters remained unsure about whom to back against the Republicans.[16]

Republicans and stand-pat Democrats alike pounced on the League. Pointing out the disloyalty charges facing the NPL in other states, they viciously and vigorously connected the League to socialism and the Industrial Workers of the World. Through coordinated attacks in the press, they called on voters to repudiate the attempt by the NPL to take over state politics. The scheme worked. Statewide candidates endorsed by the League—with the exception of the US Senate races—fell well short of victory. Of the many NPL legislators nominated in the primaries, only eight found their way to the state senate, alongside a paltry fourteen in the house.[17]

This left North Dakota. With somber news rolling in from across Minnesota and Idaho, League leaders and farmers alike anxiously awaited results from the NPL's birthplace. There, the NPL triumphed. Despite the best efforts of opponents, the League held on to nearly every office. It added the two remaining seats in the US House of Representatives. The NPL finally seized the state senate—one of only two remaining barriers to completely installing the League's program. The other barrier fell as well. Voters approved the NPL-sponsored initiative to alter North Dakota's constitution. Embodying House Bill 44 from the 1917 legislative session, the new amendments gave the state legal authority to own and operate state-run industries.[18]

Flush with victory, NPL politicians stridently proclaimed a "New Day" for the state. In celebration, the League legislators published a pamphlet outlining their responsibilities and touting their vision of the state's future. Addressing "the farmers and other workers of America," the men noted that "the people of this state have always been dependent for their existence on industries, banks, markets, storage, and transportation facilities either existing altogether outside of the state or controlled by great private interests outside the state." This "utterly unendurable" situation called for change. Unjust taxation on farm improvements had to end. "Farmers and working people," they contended, deserved credit at cost, supplied by a state-run bank. Wheat needed to flow to a state-run elevator and mill. Expanding on the NPL's original platform, the assembly even called for state-organized coal mining and natural gas production to bring light, heat, and power to a frigid state without a sizable electric grid. Finally, the legislators asked "citizens everywhere to be interested in our proceedings, to be informed as to the measures we shall adopt and . . . to aid in their adoption elsewhere." In the weeks that followed, the League lawmakers crafted and approved legislation unlike any ever seen in America, before or since.[19]

✳

Sending historian William MacDonald to Bismarck in early 1919, the *Nation* hoped to find out more about the exact nature of "North Dakota's Experiment." Upon his arrival, the evenhanded reporter learned that "even the hotels at Bismarck . . . were pro-League or anti-League, and the traveller could not escape 'atmosphere' even if he would." Even so, the new majority "was so obviously sensible, moderate, and intelligent." Indeed, the NPL representatives were "neither visionary theorists nor wild-eyed radicals." In fact, "outside the chambers there was the least possible formality—the executive officers, from the Governor down, were easily accessible." When MacDonald "asked for copies of certain bills" he was "shown the file rooms and invited to rummage for [himself]." Such access clearly marked their government as one of and for the people.[20]

MacDonald also reported that the governor and legislature "understand perfectly well that they are putting into effect a social revolution, and that it is none the less a revolution because it proceeds by due process of law." Firmly rejecting socialism, "the League seeks to increase the number of landholders and to enhance their profits." The governor, state attorney general, and state agricultural commissioner together made up an "industrial commission" to supervise the new state-run programs. Of the three, the governor retained a veto and thus held the most power. The state bank, "in which all State funds will be deposited, and which, besides doing the usual banking business," would "make first-mortgage farm loans at a low rate." Additionally, the League-controlled government began a house-building initiative "to erect dwellings for farmers and industrial workers." State-run workmen's compensation programs—along with now-compulsory hail insurance—provided a safety net for workers as well as agrarians. Finally, they not only exempted farm improvements from taxation to cut down on land speculation but also shifted the tax burden away from farmers burdened by mortgages to those with money.[21]

All these programs sought to encourage and expand the opportunity for citizens to own private property. League journalist Herbert Gaston noted that these plans would make it "easier for farmers to acquire land," enable "renters and farm hands to become owners on small capital," and provide "mortgage loans on a simpler and easier plan." Far from promoting socialism, the League hoped to use state government to put private ownership of land and homes within the reach of more citizens.[22]

In less than two months, the lawmakers remade North Dakota in the League image. Focusing more on their duties than the rush of press attention that attended the 1919 legislature, Leaguers made quick work of the session. In fact, it was the shortest ever recorded in state history. A journalist characterized the tenor of their deliberations as "grave, decorous, business-like, swift." Little of the legislation reflected personal patronage or pork. The legislature's efficiency saved the state $50,000.[23]

Elected to provide "a marketing service, a credit service, and an insurance service . . . free from all private monopoly" for farmers, who made up more than 80 percent of the state's population, the League legislature did exactly that. Using, in its own words, "methods entirely consistent with our constitutions and in harmony with American traditions," the NPL-dominated state government established new agencies. To build an elevator and mill, the legislature also crafted legislation giving the state authority to issue bonds. Legislators shifted the state's tax burden so that "privilege would bear the greater part of the burden, and that industry should be as little as possible hampered or discouraged." They also instituted a personal income tax and a new tax on corporations.[24]

Anticipating charges of socialism, the NPL expanded the initiative and referendum rights of citizens. The law creating state-owned industries "was so drafted as to permit a politically elective body to determine its course." To ensure that "the chain of responsibility is complete and unbroken, from the people to every officer and employee in their service," League-endorsed politicians drafted, passed, and signed a new recall law that applied not only to legislators and elected judges but also to members of the state's Industrial Commission—the governor, the attorney general, and the commissioner of agriculture and labor. Only then could the novel institutions stand "out in the open, subject to the scrutiny and criticism of every citizen." This simultaneously established the League's popular intent and countered the immediate criticism that the Industrial Commission, though made up of elected officials, constituted an autocratic structure influenced by petty politics. Faith in voters' abilities to rectify potential problems trumped concerns over political shenanigans. Because the recall legislation included magistrates as well as lawmakers, "it aroused a furious attack." The bill passed anyway.[25]

A similar effort to put more power in the hands of citizens became the most debated NPL proposal of the 1919 legislative session. The League enacted a law providing for the creation of a state printing commission. Acting on the recommendation of the NPL's Publishers' National Service Bureau, the proposed statute empowered voters to select an official newspaper in each county on election day. The chosen paper would publish all legal notices, bypassing the county commissioners, who usually selected the official source for notices in their locality. Since the next general election lay more than a year and a half in the future, a state printing commission would decide which papers earned official status until then. Leaguers believed this law provided the public with an easily identifiable source of the news they needed. Knowing that empowered citizens needed quality information to make informed decisions about government, the NPL argued for its implementation.[26]

The reaction proved swift and harsh. Opponents—coalescing around Theodore Nelson and the new Independent Voters Association (IVA)—countered that the law undermined the freedom of the press. The circulation of some small-town newspapers depended on sharing this information with readers. Rightly, the IVA believed that a state printing commission made up of NPL appointees would show favoritism to farmer-owned newspapers sponsored by the League. A few NPL legislators even joined with League opponents to challenge the bill. Within a year of its passage, sixty-one weekly papers across the state went out of business. Many had opposed the NPL. To Leaguers, their failure confirmed the nefarious nature of a press supported by county-level crony politics. To the IVA, the closure of so many newspapers proved League malfeasance. Such charges became the core of the IVA's argument to North Dakota's voters.[27]

✶

Fulfilling the wishes of a broad majority of those voters, League politicians began to create the long-awaited state institutions. The bank, formed in July 1919, became a public presence in capital markets. Lawmakers hoped it would keep money circulating in North Dakota's economy rather than sitting in banks in Minneapolis and Chicago. Governor Lynn Frazier declared it would "act in a similar capacity for our state as does the Federal Reserve." Pointing to the example of a federal central bank—advocated a decade before by financiers to create more stability and predictability in the nation's rapidly expanding economy—the League suggested it would relate to other banking institutions just as the regional Federal Reserve Banks did.[28]

F. W. Cathro, former president of the North Dakota Bankers Association, became the first bank director. He immediately invited his former colleagues to help establish bank policies. To make sure the state bank did not undercut local banks, lawmakers agreed it would neither establish branches nor accept independent deposits or accounts. To provide operating capital, the new legislation required that all local and state government monies be deposited there. It would not loan to corporations or individuals, except for serving as a source for credit at cost for North Dakota farmers interested in purchasing land. Meanwhile, banks across the state could use the bank as a clearinghouse for various financial transactions.[29]

The state bank, however, proved to be much more. Though tensions between small-town bankers and rural farmers lay at the heart of the economic conflict that spawned the NPL, the bank "strongly recommended" that borrowers seek mortgages by working through local institutions. They hoped this might restore harmony to rural middle-class life rent asunder by the rise of the NPL and the viciousness of its small-town opponents. In this way, the state bank could repair the economic and social relationship crucial to North Dakota's future prosperity.[30]

The bank also became the lynchpin for instituting the rest of the NPL program. As the vehicle for financing state-run enterprises, the success of all the other League-created legislation hinged on it. The tremendous amount of capital needed for the new mill and elevator placed great demands on the bank from the start. Desperate to raise finances, the state immediately began issuing bonds. Unfortunately, they sold slowly. Major banks in the Twin Cities—allied with virulently anti-NPL milling interests—consistently refused to come to the state bank's aid. Minneapolis bankers submarined attempts for succor from the city's branch of the Federal Reserve Bank. Ongoing economic hardship kept the state's farmers from buying enough bonds to support the bank on their own. Only a July 1920 loan from a major Chicago bank kept North Dakota's experiment solvent.[31]

Minnesota financiers even hired lawyers to challenge the constitutionality of the state bank. This effort offered the added bonus of undermining confidence in the bank among other customers. In April 1919, Minneapolis attorneys found one taxpayer in forty-two of North Dakota's fifty-three counties—including Jerry Bacon in Grand Forks County—willing to bring suit against the state. The lawyers—who regularly worked for Minneapolis millers as well as the Northern Pacific and Great Northern Railroads—suggested that

FIGURE 15. Counting money in the new State Bank of North Dakota, Bismarck, 1920. (State Historical Society of North Dakota, C3518)

the state's use of taxpayer monies to serve a private, industrial purpose was unconstitutional. In federal district court in Fargo, Judge Charles Amidon dismissed the case. He believed that farmers could only protect themselves from "the vast combinations of those terminal cities" by using "their state government." Soon thereafter, the NPL brought a friendly suit to the North Dakota Supreme Court, which upheld the state's right to created state-owned industries and banks.[32]

League opponents appealed to the US Supreme Court. The high court agreed to hear the cases. Ruling in June 1920, the justices found nothing untoward in the state bank or other state-owned institutions. In fact, they argued that since "the people, the legislature, and the highest court of the state" agreed that these actions fell within the "authority of the constitution" and the "authority of the state . . . we are not at liberty to interfere." Secure in its legality, the bank continued its quest to remain viable.[33]

Despite bond-selling campaigns, construction costs for the state grain elevator and mill continuously used up much of the institution's liquidity. Countering concerns, Leaguers argued that the processing facilities would not only keep more wealth in the state but would also encourage "diversified farming" and conserve the state's "soil fertility." Edwin Ladd, a scientist at North Dakota Agricultural College, insisted that "the amount of valuable fertilizers" taken each year by "the removal of the entire product from the farm and the state" numbered in the millions of pounds. Processing wheat in the state meant by-products such as fertility-inducing nitrogen, lime, and potash might be returned directly to farms.[34]

The state soon purchased an existing flour mill in the tiny town of Drake. Drake was chosen because NPL members in and around the town bought state bonds for the elevator with gusto. It began processing grain immediately. This small-scale experiment—a laboratory for government competition— paid farmers higher prices for their wheat and sold flour and feed back to them more cheaply than any private company. But within a year the operation slipped into the red.[35]

Undeterred, the Industrial Commission began casting about for a city in which to locate a much larger terminal mill and elevator. Looking for good railroad connections, cheap freight rates, and a community ready to back the publicly owned operation, the commission received proposals from James- town, Fargo, Valley City, Wahpeton, Ray, Devils Lake, Cassleton, New Rock- ford, and Bismarck. Grand Forks, home of staunch NPL opponents such as Jerry Bacon, remained divided over a bid. The commission decided that the city's location—with direct rail connections to Duluth, Minneapolis, and Winnipeg—best suited the state. A campaign to bring on the Grand Forks business community ensued. The promise of jobs, prosperity, and an influx of state spending soon won over the League's most ardent opponents. In November 1919, the leading men of Grand Forks pledged to buy more than $1 million in state bonds to construct the necessary buildings.[36]

Construction began soon thereafter. Part of the land needed for the site included a parcel owned by Jerry Bacon. Though he insisted he would "never allow any Socialistic experiment" on tracts he owned, Bacon quietly sold his land to the state in February 1921. But by then, construction crews had stopped working. More court cases against the state bank halted bond sales and choked off the capital for the project. Half-finished, the structures stood out against the prairie sky as a testament to the ongoing challenges faced by the NPL. Implementing the will of the newly empowered electorate meant taking on corporations, which proved more than willing to deploy their own power in the courts. Clouding claims to viability, businessmen from outside the state continued to interfere in North Dakota's affairs.[37]

*

Often overlooked in the uproar over these large state-run enterprises, smaller changes instituted by the Nonpartisan League in North Dakota in 1919 offer equally significant insight into the ways in which the state's farmers hoped to use government to strengthen small-scale profitability in the marketplace.

FIGURE 16. Construction crews worked on the State Mill and Elevator in Grand Forks in March 1920. Construction halted later that year for lack of funds. Both were finally completed in 1922, after the IVA took over state government. (State Historical Society of North Dakota, E0453)

Deploying novel ways to foster private economic success, the NPL remade government so that the economy might better serve the people.

The Home Building Association, for instance, looked to encourage the construction and purchase of new homes across the state. As a building and loan association directed by the Industrial Commission, the program built single-family houses as well as offering home mortgages. By cutting out real estate agents and construction companies, the association hoped to put home ownership within the reach of the working and lower-middle classes.[38]

The state offered mortgages to those who deposited at least 20 percent of the cost of the home. With the power to purchase and subdivide land, the Home Building Association then constructed the home for the mortgage holder. Upon completion of the house, the state transferred the title and deed to the participating citizen. In return, the association required monthly payments on either a ten- or twenty-year plan. Careful to set limits on the mortgage size—no more than $5,000 could be spent on an urban home, and no more than $10,000 could be spent on a farmhouse and outbuildings—the Home Building Association hoped to provide North Dakotans with "a complete home with all modern conveniences and comforts."[39]

In practice, however, the association did not live up to this promise. In fact, it violated the law that instituted it. Placing Robert B. Blakemore—a former Fargo city commissioner and longtime real estate agent—in charge, the Industrial Commission instructed him to sell $50,000 in bonds to get the project going. For months, little happened. Complaining that he could not secure enough depositors, Blakemore finally offered a loan to a government employee in September 1919. When the small, modern bungalow was completed in January 1920, all had gone according to plan. The future seemed bright.[40]

By May 1920, the association announced proposals to build an entire subdivision of homes in Bismarck. Soon thereafter, similar projects began in Fargo. Just as construction began, news broke that the Home Building Association had not only financed and constructed a $12,000 home in Fargo for William Lemke—a member of the Nonpartisan League executive committee—but also purchased an existing $10,000 home in the same city for long-time League allies George Totten and his son, George Totten Jr. The direct violation of statutes establishing the program destroyed public confidence in the Home Building Association. Subsequent investigations revealed misconduct and corruption. Instead of careful accounting and contracts, auditors found cost overruns and oral agreements. A number of depositors had no new home to show for their investments. To make matters worse, the association nearly ran out of money. Lawsuits ensued. The state lost thousands of dollars. In November 1923, the program ended, having completed only around fifty homes in Bismarck and Fargo.[41]

The implementation of a new tax structure proved more lasting. It shifted the tax burden, in the League's words, "from industry and enterprise to privilege." This meant raising taxes on land values and creating new exemptions for farm equipment and improvements. For the first time, North Dakota taxed personal income—on a progressive scale—as well as corporations. Individual monies earned through investments faced a higher rate than those earned through direct labor. The state also began collecting an inheritance tax. Effectively, the new code attempted to limit the concentration of wealth and push small, landholding agrarians toward prosperity. Those holding farm mortgages paid less. Those who held large amounts of property and capital paid more.[42]

These taxes opened the NPL-controlled government to charges that it raised taxes in order to pay for its many initiatives. In truth, only about 6 percent of the new tax revenues went to cover costs of the new programs. The state used most of the money to reinvigorate public education. Starved for years, North Dakota's schools required attention. Buildings were falling down, shrinking budgets led to continual teacher layoffs, and those who remained faced annual cuts in the budget for essential supplies and equipment. Firmly committed to the notion that the health of the state depended on well-educated citizens, the League legislature immediately began to rectify the situation.[43]

NPL opponents also claimed that the new hail insurance program raised taxes. In response, Melvin Holte, a farmer from Voltaire, pointed out the problematic propensity of League enemies to lump the hail insurance program assessment and county, road, and school taxes together with new state levies. Only then could they cite a huge tax increase for North Dakotans under the NPL plan. Holte admitted that under the League program, he paid $40 a year for $1,000 of crop insurance coverage for his land. But the year before, he had paid a private company $65 for the same protection. Holte angrily reminded League enemies who claimed the NPL duped agrarians that "we North Dakota farmers are not Bolsheviki, Socialists, reds, or I.W.W., but common, everyday farmers with level heads and grit and backbone enough to run our state for the common people."[44]

Most farmers agreed with Holte. In Hansboro, agrarians felt "greatly elated at the splendid achievements of [their] representatives at Bismarck—a bloodless educational economic victory for democracy." In their minds, the legislative success guaranteed "an even stronger League" in the months to come. Farmer Carl Olson agreed. As "one among those who have callouses on my hands instead of my seat," he said, he saw the session as a triumph for the people.[45]

*

For the NPL, this excitement ensured survival. The IVA successfully used the new initiative laws passed earlier in the year to circulate petitions and put seven referendums challenging the new legislation before voters in June 1919. Opponents hoped to strike down not only the state bank but also many of the other programs passed by the NPL legislature. But farmers across the state defended the League program with vigor.

Farmers cast their ballots in droves. Like A. H. Aaserude, many agrarians felt that the "old gang that at present is fighting the farmers have run this state since statehood & have had pretty fair chance to make good." He believed "it time to try someone else." George Schonberger, a farmer from outside Wheatland, North Dakota, put a finer point on it: "I remain a Nonpartisan at any price." The final tally made the will of the state's citizenry clear. Voters affirmed all of the League laws. In fact, the NPL earned more votes than it had in the 1918 elections.[46]

Voter turnout at the June 1919 referendum suggested a growing confidence among North Dakota's farmers. Emboldened to make the changes they hoped for, agrarians involved in League efforts took pride in their elected representatives and in their own newfound power. Destroying feelings of dependence—on bankers, railroads, millers, elevator men, commodity processors, and politicians—the NPL encouraged a renewed sense of self-worth and the sentiment that political participation made for true empowerment.

This shift proved to be as important as the innovative legislation passed by NPL politicians. In creating a membership-based organization, Leaguers stressed solidarity among farmers. Combined with exhortations for direct political action, this reoriented many agrarians. As one perceptive observer put it, the NPL attempted "the destruction of partisan politics" without the creation of "a new political idol." He further suggested that "on the ruins of political attachments [the NPL] hopes to build up group-cohesion for economic and social purposes." Working outside of the parties, the NPL tried to "break up old political ties and outworn traditions" and "help to reorganize American politics on a rational basis." This work produced "a lasting effect on the political psychology of the people . . . drawn into it." Working across ethnic and political differences, farmers began shedding a passive relationship with politics and embraced participatory democracy. This dynamic began to remake rural communities and even the League itself.[47]

On one level, this surprised many. NPL slogans, buttons, journalism, and cartoons came out of a sustained, intentional, and manufactured attempt to motivate farmers to engage a political system doing them a disservice. They trickled down from the top of a tightly organized effort that looked to efficiently affirm and apply the preexisting sentiments of aggrieved agrarians. The innovative membership-based model broke new ground—utilizing techniques that even today form the backbone of citizen mobilization most famously deployed by public interest research groups across the United States. But in doing so, the League initially articulated the needs of an agitated group without expanding the democratic capacity of participants.[48]

Yet NPL discourse deployed words and images that slowly transformed the way farmers thought about themselves. Though it hoped initially only to capitalize on and reinforce agrarian sensibilities, the NPL fostered a new capacity for civic agency in rural precincts by changing farmers' perceptions of what was possible. One agrarian in North Dakota put it this way: "Farmers had been negligent about being in office. . . . They woke up to the simple fact that they had to get out and take part in government." Solving the problems they faced in common demanded participation in cooperative public work.[49]

This shift depended entirely on the education proffered by NPL literature. Sent by her father to distribute League pamphlets on horseback among the neighbors, thirteen-year-old Nellie Allen, an African American farm girl in Pine County, Minnesota, "kept that horse walking so that [she] could read." She "just propped the material up on the saddlehorn to read it" between visits to neighboring farms. After devouring those tracts, she decided to write a school paper about the NPL and its enemy, "Big Biz." Her mother, fearing for Nellie's safety in the controversy that would likely ensue, forbade her to do it.

But the ideas Nellie encountered in League literature as a girl took hold. As an adult, Nellie Stone Johnson became a prominent civil rights and labor leader. Though she lived and worked in Minneapolis most of her life, Johnson always pointed to her childhood introduction to politics through the NPL and consistently referred to herself as a "radical farmer."[50]

Decades after the League disbanded, even Arthur Townley understood that the NPL's "educational value to the farmers" trumped its other legacies. In 1919, a Fargo businessman complained to a friend, "These Farmer Voters are like Newborn pups—they can't see anything." Yet the NPL "helped educate everybody" about the capacity for lower-middle-class agrarians to transform politics in the modern age. Henry Steinberger, a North Dakota farmer, recalled that "as time went on," he and his fellow agrarians "became more knowledgeable in political ways." He asserted farmer intelligence, pointing out, "Nobody has a corner on brains, you know." Many years later, Nellie Stone Johnson agreed, believing that her father's drive for educating himself and his family stemmed from the League's belief in "the rights of all people to get an education."[51]

Both of them were right. The NPL educated agrarians in the broadest sense. By providing spaces for farmers to work together across differences to solve common problems, the League turned members from occasional voters disillusioned by politics into everyday cocreators of community and governance. Agrarians began imagining themselves as able to actively shape public policy. In 1919, North Dakota farmer Ed Braun announced to League opponents that "we farmers in Slope County wont take no back seat. . . . WE believe in government by and for the people." A Leaguer in Colorado in 1920 proclaimed that "we do have a little pride . . . even if we are only hog-sloppers—and we don't mind telling the world (and the politicians) that we are out to win this fall."[52]

As a vehicle for political participation, the NPL did not just push more people to the polls. It also nurtured the recognition of common challenges and interests. A shared sense of purpose nudged farmers newly educated in their own empowerment toward a broader vision of citizenship, one rooted in local relationships. Facing powerful economic challenges demanded a struggle sustained not only by a League membership but also a collective discernment and patience. This tapped into preexisting networks of sociability and cooperation. Just as NPL organizers depended on neighborhood interactions for success, so too did farmers as they began to make the organization their own.

League organizers deliberately drew on the powerful past already present in rural communities. Careful to articulate the differences between earlier farmer movements and their own, NPL employees nonetheless stirred up long-dormant Populist embers of self-respect and independence. Nebraska state manager Jesse Johnson, for instance, cited the "great value" of "the experience and counsel of these grand old men." Indeed, the NPL drew many former members of that movement into its ranks. George Sprague of Emmitt, Idaho, referred to himself as "an old time populist of nearly 35 years standing" when he told readers of the *Idaho Leader* that he "JOINED THE LEAGUE FOR THE ENEMIES IT HAS." Others sounded warnings derived from deep experience. J. Y. Swigart, an elderly Farmers' Alliance organizer and People's Party politician in Nebraska, ominously predicted that "the leaders of both old parties will unite when necessary to defeat the people."[53]

Yet drawing on the past was not enough. In framing a common challenge and combining it with a new purpose, farmers unevenly but gradually forged a fresh identity for themselves in community. Working not just as individual political actors but instead in concert with other agrarians cultivating other crops in other states, the NPL's membership itself began to foster authentic empowerment. Cohesion stemming from a shared public perspective led to a shared public judgment that produced more than a voting bloc. It transformed rural neighbors into democratic citizens convinced that change might emanate from their collective work. The innovative attempt by the League to pull together landowning farmers of many sorts as a cohesive and nonpartisan political bloc depended entirely on the creation of intense individual attachments to that goal. After four years and some electoral success, a deeper devotion to the shared enterprise of the League's work began to emerge.[54]

A returning World War I veteran employed by the NPL in Idaho believed farmers needed to "get together and work for something better." He dedicated his time to the League because it had "the finest land and people in the world." "Let us have, not a sharing of wealth," he continued, "but a sharing of opportunities to make a home and a decent living." In southeastern Colorado, a farmer who agreed with that sentiment walked 110 miles to attend the closest county assembly of the League.[55]

The NPL engendered such sentiments because—at its best—it provided farmers with a language as well as the venues in which to articulate and act on an ideology they already possessed. In 1920, North Dakota insurance commissioner S. A. Olsness counseled a worried friend. "Why call us Leaguers Socialists, bolshevists, free lovers and all such rot," he asked, "when we simply ask for enough publically owned resources and enterprises to make . . . profiteering and the creation of private monopoly impossible." Identifying agrarians' latent vision of economic and social equity, the NPL combined that work with a realistic strategy for extending an agrarian vision of the cooperative society into politics. Fostering these feelings, the NPL grew more and more dependent on farmer-driven precinct meetings, picnics, and public expressions of loyalty to the League in poetry and song.[56]

∗

NPL-sponsored gatherings provided places for cooperation and comprehension during election season. League employees could neither run for office nor guide the proceedings. Organizers offered little more than procedural advice. At local precinct meetings, Leaguers selected delegates and candidates, spoke their minds, grappled with divergent opinions, and witnessed locally rooted democracy in action.

Nominating neighbors—always "members of the League" and always a "bona-fide farmer" according to the directive of one state manager—to represent them at conventions or in elections put politics within farmers' reach. Cooperative decision making trained participants to imagine themselves in the business of democratic work. One farmer, who came in second during a nominating process, nonetheless believed that "a better man won." Furthermore, he said, the "men of different religions and nationalities" in his township "seemed of one mind as we in turn discussed the issues nearest our hearts." The whole experience made him "think of those men when this nation was in the making . . . when they sat in convention drafting a Declaration of Independence." Rejecting the notion that politics should be left to politicians, Leaguers discerning a future together expanded their own sense of what it meant to be a citizen.[57]

The deliberation fostered in members-only meetings flourished because of preexisting relationships rooted in less formal affairs. Besides ethnic associations and churches—which only drew residents from specific communities—rural neighbors of every sort encountered each other in farmers' clubs, usually centered on individual townships. These clubs combined cooperative action with leisure. During temperate months, the clubs organized picnics. These festive occasions prompted participation from across the community, bringing men, women, and children together. Piling into cars filled with basket lunches, entire rural families drove up to thirty or forty miles to enjoy the community-sanctioned break from backbreaking work. Speeches, fund-raising, quiet conversation, and joyful play defined these daylong excursions.[58]

Picnics also served as sites for politics. In the 1880s and 1890s, popular politics in the rural West and South grew out of such gatherings. Despite this heritage, NPL employees and leaders did not immediately seize on them as an organizing venue. Instead, "farmers themselves" convinced League organizers

FIGURE 17. A typical township NPL caucus, Swift County, Minnesota, 1918. These farmers proudly displayed copies of the *Nonpartisan Leader* as well as the American flag. (Minnesota Historical Society, J1 .4 r52)

to use picnics as venues for creating farmer solidarity. During the 1916 campaign season in North Dakota, these NPL-sponsored events not only offered speakers—who, the *Nonpartisan Leader* carefully noted, would "not be allowed to interfere with the good times which all are expecting"—but also "brass bands . . . baseball games . . . athletic events, egg races for women" and "games for the children."[59]

Picnics did more than just supplement the work of organizers. As enjoyable gatherings that involved neighbors and friends, they transformed mobilized members and their kin into participants in an organized movement. In Montana, for instance, League employees encouraged NPL members to host "the biggest and best monster picnics" on their farms during the 1920 campaign. Called on to identify the best locations as well as boosters to spread the word, farmers essentially coorganized these gatherings.[60]

NPL picnics soon created such a demand for speakers that the League hired permanent lecturers to serve the busy circuit. Many of these rallies took place in off years. Loosed from upcoming elections, picnics became a full-blown component of the NPL's broader "educational campaign." With entire families participating, picnics served as a place for farm women and children as well as men to rethink their relationships with others in democratic community. One North Dakota farmer remembered attending a picnic at Willow Lake where "every man we saw," he said, "was wearing one of these League buttons on his overall bib or hat brim. . . . Because of this identification we knew that strangers and acquaintances alike were of the same political faith as we were."[61]

Purposely conflating pleasure and politics, these get-togethers gradually became more sophisticated. In July 1920, over two thousand people attended a League picnic in Manzanola, Colorado. Featuring a flag raising, a band, speeches, a "fancy airplane exhibition," "special movies," and a "grand ball" in the evening, the gathering became a celebration. Because the local organizer successfully reached out to the town's businessmen, Manzanola's mayor even presented a key to the city to the "Non-Partisan League farmers and their families who took the town by . . . a storm of good will." Days later, "about 1,500 plains farmers with their families and fried chicken held a Nonpartisan League picnic" in nearby Hugo, Colorado. A "Ford Auto Race" provided the main entertainment. The success of these massive meetings convinced local NPL leaders to turn all of Colorado's county League conventions into county picnics, where besides signing petitions, selecting candidates, and registering voters, the *Colorado Leader* implored readers to "have one big day of jollification."[62]

Potent cohesiveness also stemmed from shared song and verse. In an era before radio and television, agrarians made their own entertainment. Rural folk traditions long encouraged the rewriting of music and poetry to reflect

FIGURE 18. NPL picnic and rally at Louis Johannes's farm, near Steele, North Dakota, in 1920. Arthur Townley, who spoke at the event, arrived by airplane. (State Historical Society of North Dakota, C3266-004)

contemporary concerns. Enjoyed at picnics and precinct meetings as well as in fields, homes, or schoolhouses, rhymes and refrains reflected the needs and wants of League farmers. In refashioning familiar standards, farmers identified their problems, their perceptions, and their power. Created by participants, NPL-inspired poetry and song articulated shared sentiments and touched people's emotions.

One farmer-penned poem responded to the critique that the NPL simply took farmers' hard-earned cash. A song titled "Where Is the Money?" told skeptics, "It did not go to Minneapolis / Where most of our earnings go / . . . / We're tired of long hours of rest and toil / We've stopped to rest and THINK." Another agrarian used verse to proudly celebrate the ability of the NPL to bring farmers of many different sorts together:

> Three thousand Swedes came through the weeds
> And threw their votes for Frazier
> And Norse and Dane in a pouring rain
> They did the same with pleasure
> And French and Dutch, they did as much
> To take the Old Gang's Measure
> With Brit and Scot—just as they ought
> All voted for Lynn Frazier
> And Pat and Mike and Jewish Ike
> That gentleman of leisure
> With Hans and Fritz and Slavonitz
> They pinned their faith to Frazier.

A third poet suggested, "We'll stick until we're freed / And then we'll stick together through all eternity / While we are farming North Dakota."[63]

Early League songs pointed to the soon-realized occupation of North Dakota's state capital. "On to Bismarck," sung to the tune of the popular British ballad "It's a Long Way to Tipperary," suggested it was "time for farmers now to stand together for their cause / For farmer Legislators to make the laws." A month later, a member devised "Marching to Bismarck." The tune used the melody of "Marching to Georgia" and claimed the farmers would " 'radicate / The bosses that used to go to Bismarck."[64]

Song and verse rallied Leaguers to action. One agrarian from Saskatchewan wrote an anthem of anger: "But the bankers ride 'round in their grand motor cars / They are clad in swell clothes and they smoke fine cigars / Bought with wealth sucked from us and the thought of it jars." Like his neighbors, this farmer noted that "Now of this unfairness I'm heartily sick / And it's up to all of us to get busy and kick / And we'll get all our own if together we stick." In

Colorado, a Leaguer cleverly invited neighbors to an NPL picnic by writing a poem: "Everybody come, especially you / and the ladies / and the children / . . . / Let's all get together / and get acquainted / and have a good time talking politics."[65]

Outside Glencoe, Minnesota, the children of NPL members made refrains like this come alive. Forming a choir, "the young League folks . . . have good times practicing by themselves and especially good times singing at the League meetings." Wherever and whenever Leaguers gathered, song soon broke out. In Idaho, advertisements for NPL picnics in 1919 noted that "J. D. (Big) Brown, who lead [sic] the singing at the state convention of the Nonpartisan League in Boise in 1918 will enliven the meeting with some of his original songs." One of the songs he likely sang at these gatherings—penned by Ray McKaig's wife—claimed, "The League has come to rescue Idaho / . . . / It's goodbye to politician bosses / Henceforth united we will vote / Workers both from town and from the country / Thru the League now will make the Boss the 'goat.'"[66]

The reference to a goat reflects the most significant way League members powerfully refashioned their collective identity. In the earliest days of the NPL, North Dakota agrarians turned to the goat as an emblem of political power. During the Society of Equity convention in Minot in February 1916, whenever farmers (most of whom were also NPL members) angrily critiqued the state's governor, Louis B. Hanna, one man brought out "a real live goat, kept in the basement." It wore a sign: "The Rubes Have Hanna's Goat." This self-disparaging reference poked fun at the assumptions politicians made about farmers. But the phrase "got your goat"—prison slang that emerged in early twentieth-century America to describe making someone angry—meant much more. It used a farm animal to assert that clever and collectively organized citizens possessed more control over unresponsive politicians than they knew.[67]

As a symbol for agrarians exasperating indifferent elected officials, the goat took off. A goat made an appearance at a large NPL Fourth of July picnic later that year near Deering, North Dakota. Two farm boys, dressed as clowns, led the goat around the crowd. One report described how the animal took "a prominent place on the speaker stand and furnished all kinds of fun for the assembled farmers." This grassroots enthusiasm ensured that the goat persisted in NPL discourse long after Hanna's defeat in 1916. When League cartoonist John Miller Baer left North Dakota to serve in the US House of Representatives, the *Nonpartisan Leader* began featuring cartoons drawn by others. One cartoonist, who signed his images "Billican," often deployed a goat to symbolize the Nonpartisan League.[68]

FIGURE 19. Two boys at an NPL picnic outside Deering, North Dakota, on July 4, 1916. Dressed as clowns, they hold a goat with a sign that reads, "We have got Hanna's goat." Louis Hanna, then US senator from North Dakota, opposed the NPL's platform. (State Historical Society of North Dakota, 00032-WD-13-08)

Farmers loved the connection. In October 1919, a mass meeting of Nonpartisan Leaguers in Fargo recognized that "political parties have each adopted a certain animal as the emblem of their . . . organization." Because the NPL "built up a national force for order and justice against human rapacity," they needed one too. So "in order to enter a decided kick against the manifest wrong done them and the great cause that we do hereby adopt as our permanent emblem— THE GOAT that can't be got." One observer noted "the black and white enamel replica of Capricorn" hanging "from the chains and button holes of League members" throughout the crowd. Across the country, Leaguers roared with approval. Soon thereafter, NPL governor Lynn Frazier told a reporter from the *Nation* that agrarians "adopted this animal" as an emblem because "a goat is an animal that works with its head when attacked."[69]

One commentator further illustrated the differences between Republicans, Democrats, and Nonpartisan Leaguers. Paraded around "for the purposes of luring the unsuspecting to the circus ground where every sort of thieves abound," the Republican elephant was "ever under the control of one puny individual who leads him about by the trunk." The equally deficient Democratic donkey, "a stubborn, unruly beast whose usefulness is offset by his selfishness," offered only "senseless braying" that "annoys and destracts [*sic*] the people when they are trying to find a solution to the problems of daily life." In contrast, "the FARMERS and WORKERS of North Dakota adopted an animal, a breezy, alert fellow, interested in everything." The Nonpartisan League goat "looks at you quizzically, steadily, and without embarrassment . . . strong and hardy, capable of defending himself with his heels, butt [*sic*] preferring to USE HIS HEAD." Furthermore, the goat "is a producer but [one] who objects to being exploited as his relatives the sheep have been . . . a worker not a slave . . . an animal whose only objectionable feature is an odor, more or less offensive, which is really a protest against the conditions under which he lives."[70]

These analogies proved popular. Soon after adapting the goat as the organization's symbol, an NPL member produced "A One-Act Nonpartisan League Sketch" titled *The Donkey, the Elephant and the Goat*. Leaguers carefully designed the play to be "Presented on Any Platform or stage" by "High School Boys or Grown-ups." In it, a goat informs supporters of the two major parties that "the methods you employ are dead." Instead, "We make the law the advocate / Of Common people mainly / By placing it beneath the state / Where it can serve humanely." NPL newspapers claimed that the script made for "Fine Campaign Material and Good Reading" for "25 cents per copy."[71]

To meet the insatiable demand for goat-themed material from League members, the NPL's Progressive Feature Service Bureau even began publishing a monthly magazine titled the *Goat*. An editor promised that "THE GOAT will butt in periodically in the interest of the organized farmers and workers of the Northwest, using head, heels, and digestive powers to uphold and further the principle of Popular Government and Industrial Democracy." During its brief run, the journal sold goat buttons and pins and featured numerous cartoons depicting goats defeating politicians and other NPL enemies. As a symbol derived from agrarian vernacular, the goat showed more sticking power than any top-down representation of the collective public work done by League members.[72]

As farmers first took advantage of and then began to take ownership of the spaces fostered by the organization, they transformed the NPL from a membership-based group into a full-fledged movement. Taking advantage of the ongoing latency of the always-hard-to-define term "the people," agrarians refashioned their relationship with government and asserted their agency. The hard work of moving beyond mobilization ended up being done informally in rural spaces by farmers themselves. Everyday speech and action rendered a new vocabulary for power. Producing a meaningful culture that provoked and promoted collaborative civic work, farmers—not NPL employees—devised languages and emblems to describe their own transformation as well as their hopes for the future. League success depended largely on this member-driven shift from mobilization to organization.[73]

*

More than any other constituency, rural women took advantage of this self-empowerment to remake and expand their own roles as citizens. In the earliest days of the NPL, organizers focused all their attention on men. As the organization grew, League employees tended to overlook the many contributions made by women to the movement. But as full suffrage became the law of the land and critics openly attacked the League over gendered issues, the organization's nonpartisan orientation attracted women. For decades, politically active women saw party politics as distasteful and crooked. They seized on the opportunity to participate in electoral politics in a new way. As a result, farm women turned the organization into a locus for civic agency.[74]

From the beginning, farm wives and farmers' daughters made agrarian self-sustainability possible. Through their work both in the fields and the home, they brought in money, sustained households, supervised the labor of young children, and made it possible for men to focus on cultivation and markets. Beyond the many economic contributions they made to rural households, these women worked to foster and sustain the social ties that helped families survive the isolation of farm life. Women's labors even made the NPL's picnics and solicitation of song and verse possible. Annie Pike Greenwood, an Idaho farm wife, remembered a picnic featuring Ray McKaig as a speaker as well as tables "adorned with the pride of accomplishment of every farm woman in the county." These socially acceptable forms of political action for women in rural districts grew directly out of their existing roles in creating community bonds.[75]

The deep-seated interest in creating community sprang from the need to cope with the long-standing issue of loneliness on the farm. One North Dakota woman admitted to fellow *Nonpartisan Leader* readers in December 1916 that she "was very homesick and lonesome" and that the presence of her tiny children, who "are so much company," provided one of her only sources of comfort. NPL rallies and picnics provided opportunities to break the monotonous routines of daily life on the farm. These events offered a chance to meet with neighbors and share stories as well as solutions to daily troubles.[76]

Like men, rural women drew on existing affiliations as they considered involving themselves in the Nonpartisan League. In particular, the Grange had long pulled together women in common activity that included politics. Wherever the Grange supported the NPL, Grange women joined in the work of boosting the League. Alfa Salmon Ventzke, a Granger in Okanogon

County, Washington, defiantly distributed League literature even after locals forcibly removed her from the county board of education. After all, as Grange member Annie Pike Greenwood put it, "The cost of wheat, to be just, must take into consideration the world's greatest, as yet unfreed slave, the farm woman." Like Greenwood, many women even came to see politics as "the most worth-while interest of an American citizen."[77]

It took the NPL many years to envision the significance of farm women. After more than a year of focusing their attention on organizing farm men, the NPL finally provided a space for rural women to express themselves and learn from each other. In May 1916, the *Nonpartisan Leader* acknowledged that "the wives, the mothers, the daughters and the sisters of North Dakota's farmers have not been getting their full share of space in this news-magazine, in which they have taken as much pride, and for which they have 'boosted' just as hard as the men folks." The editors promised to rectify that situation by starting a semiregular women's page. The first editor of the page, M. M. Hollis, a home economist who worked with the North Dakota State Agricultural College's extension service, shared her expertise in "foods, in cooking, in sewing, and in general household planning." Envisioning women's roles as merely domestic, League leaders revealed their own gender bias. They also revealed their belief that full-blown suffrage seemed distant. Unable to anticipate a more powerful role for rural women, editors pushed Hollis to provide week after week of recipes and household tips.[78]

Yet readers saw the clear limits of the League's vision for farm wives, mothers, and daughters. In letters to Hollis, agrarian women repeatedly tried to point out that the organization might offer much more. Grace Johnson wrote the *Leader* in June 1916, noting that since "the fathers have banded together . . . to clean up the politics of the state . . . why can not the mothers of North Dakota band together as an organization to clean up the social life of our state?" Johnson called for the journal to "put this question before the wives of the Nonpartisan League members." Hollis happily agreed to do so.[79]

A lively discussion ensued, turning the *Leader* into a place for women to communicate and try to find common ground. In August 1916, "Mrs. E.D." wrote the *Leader* to suggest that organizing such a body meant little without full suffrage. She implored farm women to join the state's already organized voting rights movement—centered in North Dakota's larger towns and cities—and then asked whether "the Leader intends to come out for suffrage." Grace Johnson responded a few weeks later, agreeing that achieving suffrage was crucial, but wondering "what good would our vote do without an organization to demand that our votes be counted?" She believed the particular struggles of rural women demanded their own cohesion to better the schools, churches, and social mores that served farming communities. The very presence of debate suggested that these women thought long and hard about what the political empowerment of their husbands, fathers, and sons meant for them.[80]

News from north of the border further encouraged farm women to consider their options. In September 1916, Zoa Haight—daughter-in-law of Saskatchewan NPL founder Silas Haight and a League-endorsed candidate for the province's parliament in the June 1917 elections—weighed in on the debate. She told *Nonpartisan Leader* readers the story of how farm women belonging to the women's section of the Grain Growers' Association worked alongside colleagues in the Women's Christian Temperance Union to bring full suffrage to Saskatchewan. Haight made a powerful case for farm women's organizations that focused on achieving the vote. Equally important, the rural women of Saskatchewan attended to an "educational program, that we women may be able to vote intelligently, now that we have the right."[81]

Farm women's engagement with politics depended on both a space for action and the desire to deliberate. Rural women took advantage of the former and showed their acuity for the latter. Though women poured their energies into mulling over the options, the NPL itself largely ignored the dialogue. To be sure, the powerful emotions loosed among farmers by League organizers and the NPL's electoral victories encouraged farm women to pursue their own empowerment. But besides backing municipal and presidential suffrage for women and then pushing that bill through North Dakota's state legislature in early 1917, the League remained more focused on mobilizing new male members than developing more politically active farm women.

*

In its consistent appeals to agrarian manhood, the NPL overlooked the potency of organized and empowered women. They focused on motivating farmers by appeals to the patriarchal protection of families instead of imagining every member of the family as a productive agrarian and potential citizen. To be sure, the electoral system—which until 1920 generally depended on reaching male voters—encouraged this approach. But in identifying and expressing the latent economic ideology of the lower-middle-class farmer, the Nonpartisan League deeply vested itself in the gender ideology of the same.[82]

Rural women who wanted to formally contribute to the NPL often got nowhere. In May 1917, Minnie Boyer Davis, a farm journal columnist from Josie, Nebraska, offered her services to the *Nonpartisan Leader*. As "a good deal of a radical myself," she asserted that she could "write and speak" and might be of some use to the organization. The reply came quickly. The *Leader* needed no new writers. Noting her talents, the response suggested that there might be a job for her in the months to come. But a request for Davis's services never came.[83]

Some farmers challenged the organization to envision the political power of farm women. In April 1917, James McCullough told *Nonpartisan Leader* readers that "there is one phase or factor in our organization which . . . has not received the attention is should have, and that is the ladies." He urged the League to issue women "a cordial invitation" to attend NPL meetings. McCullough argued that "in the near future our wives will be voting with us." He even encouraged fellow members to involve their children in all League activities, because they would "be the voters of the near future." Henry Steinberger, an agrarian in north central North Dakota, remembered that when the NPL came, "it got a foot hold among men and women alike." The hardworking farm women he knew "learned of politics through politics and formed opinions" because couples discussed "politics at home."[84]

By dint of their actions, farm women kept making their significance to the NPL clear. In March 1918, the editor of an NPL-affiliated newspaper in Alexandria, Minnesota, found himself charged with sedition for supposedly discouraging enlistments. When local agrarians turned out to protest, "two old women, outwardly of the comfortable, motherly sort, wives of farmers," stood at the center of the crowd and "gave voice to songs of discontent." The anti-League reporter told readers that these matrons also discussed the possibility of boycotting stores in town. "The town needs us farmers," but, one said to the other, "we farmers don't need the town." Mrs. Edwin A. Schact, from Elgin, Minnesota, declared to readers of the *Minnesota Leader* that same year, "I am proud to think that we have an organization of our own to belong to and if the League members are entitled to a button then we want two."[85]

Such displays of loyalty garnered little attention from NPL leaders. They did recognize that wives, daughters, and mothers of League members wanted women to speak at NPL picnics and hired at least one woman, Hulda Bain, to speak to audiences in Minnesota. But League appeals to agrarian manhood continued unabated. In the late-1918 letter sent to farmers renewing their memberships, Arthur Townley insisted that "our own future—the future of our children—the future of the country—the fate of democracy depends upon us. Let us face our duty like men and fight like our fathers fought for American independence." He persisted in playing up masculine obligations to family to motivate and entreat voters. This rhetoric reinforced the patriarchal domesticity that structured lower-middle-class farm households.[86]

Not until April 1918 did the League begin to formally consider farm women as potential voters. Spurred by Woodrow Wilson's vocal support for suffrage as well as the US House of Representatives' passage of a women's suffrage bill that January, the NPL finally saw that women might soon be casting ballots. They began by highlighting the work done by agrarian women in the Populist movements of the 1880s and 1890s. Publishing a piece by an elderly Nebraskan named Mrs. Otto Mutz, Leaguers looked to reinforce the already existing notion among farm women that they could make a political difference. Indeed, Mutz's piece showed that the Populist movement "derived much of its power from the active part taken by women." A week later, the *Leader* published a story highlighting rural women's wide-ranging knowledge of farm economies and agrarian problems. It noted that "women throughout the West are rallying to the support of the Nonpartisan League."[87]

Lulu Miller, who lived on a reclamation project outside Kimberly, Idaho, believed that "the sun never shone so bright as on that day" an NPL organizer came to her farm. Though "the 'men folks' were not at home," she excitedly read through the *Nonpartisan Leader* left by the League employee—and then insisted that the men find the organizer to sign up for a movement that "gives us hope." Miller also showed concern not only for the family's crops but also for her own small-scale production. Even her garden-grown strawberries and barn-bound chickens fell prey to the vicious pricing of an unfair local middleman. The editor noted that the women clearly "have the hardest of all the economic burdens to bear."[88]

But farm women insisted on more than mere recognition of their hardships. Winning a *Nonpartisan Leader* contest in which women were asked to answer the question "What does the Nonpartisan League mean to me?," Mrs. H. L. Peterson argued that the League offered "an administration which will enforce the laws, clean up our towns, teach insurance companies what the meaning of the word 'insurance' is, give farm children a better school, and make the farm a place to live as well as to work." Most important, she felt that the NPL "means the renewal of hopes lost during the wageless years I have worked as a farmer's wife." Rooted in the significant contributions she made to the household economy, Peterson displayed thoughtfulness as well as ambitions for herself, her family, and the NPL.[89]

The 1918 elections in North Dakota further clarified the power of women voters. Settling on the proven incumbent, the NPL ran Neil Macdonald for the post of superintendent of public education. Macdonald had a national reputation for rural school reform. But his opponent in the general election, an anti-NPL county school superintendent named Minnie Nielson, won more votes. She successfully pointed out his weaknesses as well as the pro-League bias he inserted into the state's school curriculum. Despite losing, Macdonald refused to surrender his post. He rightly claimed that Nielson's overblown credentials rendered her unqualified for the office under state law. Only when Attorney General William Langer sided with the voters and insisted on Nielson's right to take over as state superintendent of education did the tumult end.[90]

Lost in the details of the dramatic controversy, many overlooked two crucial trends. First, voters in small towns went overwhelmingly for Nielson and the other League opponents. A clear town-country split now firmly defined the state's politics. Indeed, the election, in the words of one farmer, lined up "town and county in two hostile camps." Minnie Craig, an increasingly rare small-town NPL supporter, remembered that in the wake of the Nielson controversy "neighbors wouldn't speak if they were on opposite sides of the Political Fence."[91]

Second, because state superintendent of education was the only statewide position for which North Dakota's women were eligible to vote, female support for Nielson proved crucial to her victory. The former made the latter really matter. The mobilization of urban women made Nielson's win possible. Even more problematic for the NPL, many farm women "voted for their sex rather than for principle" and backed Nielson. Supported by the state's growing suffrage community and a network of women's clubs in cities and towns, Nielson represented a gendered repudiation of an otherwise entirely male slate of candidates. The mostly urban state federation of women's clubs and the local chapters of the Women's Christian Temperance Union not only sustained Nielson—the president of the North Dakota Federation of Women's Clubs—but also threw their general support to most NPL opponents. Within a year, they began actively working alongside the IVA to defeat the League.[92]

Nielson's victory also gave her a public position from which she attacked the NPL over gender issues. Responding to her success, the League put forward a statewide referendum on taking the "penal, charitable, and educational institutions of the state" out of her hands and placing them in those of an NPL-controlled board on which she sat as the only League opponent. The measure passed. In turn, Nielson accused the NPL of deliberately ensuring that "the women of the state would be unable to vote on the issue" because the measure included prisons and charities. These riders eliminated the application of women's suffrage to the referendum. "Yet these leaders claim that they are in favor of woman suffrage," Nielson trumpeted.[93]

This seemed to confirm urban women's suspicions that rural issues would always trump concerns for women in the NPL program. Furthermore, in an effort to undercut the League elsewhere, NPL enemies widely distributed pamphlets describing Nielson's claims. These reached readers well beyond North Dakota. As a result, even though the League pushed for and passed municipal and presidential suffrage for the state's women in 1917, club women in small towns turned against the League. In the wake of the Nielson fiasco, one pro-NPL journalist even suggested that the League was "not keen to see the arrival of equal suffrage, not because of disbelief in the principle but because of fear that a large feminine vote will be cast in the towns."[94]

League opponents also deployed Nielson's virtuous womanhood to cast the NPL as a hotbed of immoral thought. Accusing the NPL of introducing socialist and anarchist books into public circulation—especially Swedish feminist Ellen Key's *Love and Ethics*—League enemies cast Nielson as the lone morally upright member of the board. Key's book, which looked to unloose romantic love and reproduction from patriarchal visions of marriage and motherhood, seemed to critics to advocate "free love."[95]

Taking advantage of general fears over the Bolshevik Revolution in Russia, the accusations of NPL support for such immorality resonated deeply in rural communities. Though the largely trumped-up charges had little to do with any platform or political stance, Nielson's womanly and upright virtue stood in stark contrast to the League's supposed embrace of anarchic values. Without well-organized rural women actively laying claim to their own virtuousness to counter these claims, the NPL exposed itself to vicious and damning critique.

✳

By early 1919, Nielson's successful attacks highlighting the gendered weakness of the organization—as well as the imminent passage of universal women's suffrage—finally shifted the NPL's stance. In a speech at Forman, North Dakota, Arthur Townley boldly admonished farmers for overlooking the talents of agrarian women. "I have heard men make discourteous remarks about the women—that they talked too much," he began. In fact, the NPL leader informed the assembled farmers that "you will never be as good looking; you will never be as intelligent; you will never be as progressive, you will never be able to defend the League program and the people's measures as well as the women, until you learn how to talk as well as they do." The audience erupted with applause.[96]

The largely ignored farm women—fired up by the emergent participatory civic culture fostered by the NPL and fearing the power of city women aligned with opponents—took it upon themselves to organize each other. Recognizing that "the town women outvoted the country women" in the Minnie Nielson race, the wives of League legislators in North Dakota launched a formal

auxiliary to the Nonpartisan League in February 1919. Amy Edmunds, secretary of the NPL's new women's clubs, told *Leader* readers that they simply
could not assume that "the woman always votes as the husband votes." In order to face "so serious a menace to the League organization, not only in North
Dakota, but everywhere else the League is working," farm women needed to
come together. For a one-dollar membership fee, rural wives, daughters, and
mothers could work alongside menfolk and join together in an organization
dedicated to League goals.[97]

Edmunds excitedly noted that "THE WOMEN ARE VOLUNTARILY OR-
GANIZING THEMSELVES." Often they circulated among neighbors to find
members. Sometimes they even got an NPL employee organizing in their
district "to take them on his route." The long-standing exclusion of agrarian
women from small-town clubs explained some of the excitement. The strong
women's club movement that swept the United States in the early twentieth
century largely left farm districts untouched. "The wives and daughters of
the farmers are, as a rule, excluded" from such groups, she reminded readers, "owing principally to their isolated situation." A gap created by "a certain
antagonism between the farmers' wives and the women of the towns" might
now be filled. Almost immediately, the auxiliary proved its worth. Before the
successful June 1919 referendum vote in North Dakota, members of the nascent NPL women's clubs "distributed thousands of leaflets."[98]

Clearly impressed, Arthur Townley sent a letter to all the wives of NPL
members in which he extended "the hand of good fellowship in this great
fight for JUSTICE in the market places of the nation." He believed that "with
the help of the women, we feel that our strength is more than doubled." Farm
wives, in his view, "not only . . . influence the life of the present generation"
but also proved "instrumental" in "molding the thought and purpose of the
children in the home, thus laying the foundation of a better state of society."
Openly worried about societal discontent and the possibility that "revolution by violence" might overwhelm the work done by "fair-minded men and
women" like those found in the League, Townley encouraged rural women
to instead seek change "by the ballot," which would "do much to extend the
influence of this organization." To further encourage the growth of women's
clubs, the NPL's official policy also changed. The *Nonpartisan Leader* blared:
"The wife of every League member is also a member of the League, entitled
to equal voting privileges with her husband at all League meetings."[99]

Finally embraced by the League, farm women responded in turn. Maude
Pettyjohn, a farm woman from Dayton, Washington, appreciated the way
agrarians came "together against our common enemy that has its grip on all the
states of our union." Calling out "corporate greed" in most other newspapers,

Pettyjohn read League publications with "a feeling of gladness." "If the Non-partisan league, collectively and individually bring themselves to work and self sacrifice as they never have before to enlighten the people," she believed, "the tide will turn." A farm wife in Nebraska asked, if "the bankers, grocers, merchants, hardware men, doctors, laborers, all have their associations . . . then why in the name of all that's fair shouldn't the farmers too have their unions?"[100]

Within eight months, more than two thousand women had joined League women's clubs in North Dakota. Fearing the power of urban women as an anti-NPL voting bloc, women in other states rushed to copy the effort. Mid-1919 saw Minnesotans begin organizing similar auxiliaries. By early 1920, farm women in Colorado, Montana, and Nebraska came together in clubs.[101]

The auxiliary idea proved less popular in South Dakota and Idaho. In the former, League organizers put more energy into integrating women fully into the membership than creating separate clubs. In fact, in January 1919, Mrs. B. T. White already served as the state NPL secretary and treasurer. By then, the South Dakota NPL's bylaws declared that "women shall be eligible to this organization" and charged two dollars for members' wives and daughters "living on the same farm" to join. In Idaho, where the state League operated somewhat independently of the national leadership, women became NPL members as early as 1917. Because of that organization's focus on women who owned their own farms, they did not join in especially large numbers.[102]

Everywhere else, however, the idea of NPL clubs actively engaged rural women. Finally, farm women could work together to create a gender-specific space for uplift and politics. "How good it is," one relieved farm wife declared, "to find a friend with a common point of view in a common meeting place in the Nonpartisan League!" Tapping into the excitement, leaders of the auxiliaries immediately began orienting women toward political and social issues. The latter proved crucial because many rural women saw them as interconnected.[103]

Ruby Kraft, an active leader in the Turtle Lake, North Dakota, NPL women's club, believed that "a family is the 'Heart of Politics'" and asked, "who should be interested in the affairs of the world if not the mothers?" Firmly rooted in maternal expectations, Kraft implored others to support the League because it was "for an educational system that will give our country children an equal chance with the town children." Kraft firmly believed that "our duty as women of N.D. is to organize and make the womens vote as strong as the mens."[104]

The work of a farm mother, however, often made politics difficult to pursue. A few months later, Kraft apologized to a Leaguer friend. She admitted

FIGURE 20. The empowerment of farm women is reflected in this photograph of members and an invited speaker at an NPL Women's Club picnic in Conrad, Montana, on June 26, 1920. (Minnesota Historical Society, J1.4 r41)

that "this has been our busiest time, all three girls have been in the harvest field . . . it has not left me one minute to write." Her own difficulties helped her recognize the more pressing limitations of poor women in the township. For "families needing clothing," Kraft told her friend, she had "some here for distribution." Taking care of children and households and gardens and animals left little time for anything else. For "the woman who has only a very limited time for reading," Susie Stageberg, secretary of the Minnesota NPL women's clubs, suggested that she "divide her time between the Bible and political editorials."[105]

Nonetheless, farm women made time for meetings. Free from interference, they focused on their own empowerment. Many agreed with Margaret Hannah, secretary of the NPL women's clubs in Montana, who identified "the question of the contentment of the farm woman" as "one, among the many tremendous national problems." A colleague in Montana declared that "we are not going to talk about recipes for rhubarb conserve or how to remove rust from the stovepipe." Instead, she wrote, "we want to know about the great battles for human rights so that we can vote straight when the time comes." Women in Crook, Colorado, spent the summer of 1921 doing "civic work." In Dalbo, Minnesota, members of the NPL club began studying government, with "different members taking turns in leading the lesson." The female leaders of the clubs hoped "this great body of women workers will become a prominent factor in helping to improve conditions and make better citizens." The auxiliaries centered women's attention on politics, to be sure. But they also used the clubs to enable a wide range of community and civic activity.[106]

Although many women, as one wrote, were "tired of working for nothing, as the farmers' wives do all the time," the NPL auxiliaries promoted work as well as study. During monthly meetings, club members hosted suppers and dances to raise money for their activities. Women made quilts, knit mittens, and deliberated on pressing issues of the day. They read and studied together, focusing on civic tracts such as Charles and Mary Ritter Beard's *American Citizenship* (1914). One club started a lending library. NPL club women actively pressed local and state officials to support global disarmament and debated issues raised in local newspapers, such as taxation. The Franklin, Minnesota, NPL auxiliary even sold candy to raise money for union strikers in South Saint Paul.[107]

To coordinate all this work, the NPL hired Kate L. Gregg—an English professor who held a PhD from the University of Washington—to serve as manager of the Women's Nonpartisan League Clubs. The need to administer the growing movement led to annual county-wide and statewide conventions of the auxiliaries. This allowed women from the most remote farms to meet and learn from each other. A new statewide network of rural women—one that started to match the already connected women in small towns and cities— transcended the bounds of domesticity so central to rural middle-class life. In central Minnesota, women's club leaders even briefly published a monthly newspaper. One rural woman—"75 years old and lame"—responded joyously, saying, "Just think of it, we women have a paper, long may it live." She promised to "try and get subscribers as soon as I can get out."[108]

This "training of our mental powers raises us above the slave level," as one NPL club woman put it. Rural women began acting with new boldness. At a picnic featuring NPL candidates—including Charles A. Lindbergh—in Rock Creek, Minnesota, a farm mother "tied a banner, VOTE FOR LINDBERGH" on the county sheriff's car. The anti-League sheriff did not notice the sign until "it was too late." This proved especially dangerous. Just days later, NPL opponents "hung an effigy of Lindbergh on a telephone pole by the railroad crossing in town." But empowered by new community, this woman felt free to directly challenge even the most violent of NPL enemies.[109]

Aldyth Ward, a female farm owner from Raub, North Dakota—and the first president of North Dakota's NPL women's clubs—did more. She looked for allies, noting that the League might be a good vehicle for putting "the farm women and the city working girl in closer touch." After pushing a state legislator to join her in an investigation of women's labor conditions in North

Dakota, Ward found herself on a new investigatory commission. Its report pushed Governor Frazier to promote and then sign new eight-hour-day and minimum-wage laws in 1919.[110]

Some League men took great pride in newly empowered farm women. One local pastor rejected the notion that "these public spirited mothers neglect their homes." Instead, he believed that "the women, anchored to the home as they are, will by their devotional work and vote save this our beloved country from the dangers hanging over us." Even those men troubled by active NPL women began to recognize that their vote would be crucial to future electoral success.[111]

By breaking open the League's commitment to protecting and promoting a particular vision of agrarian manhood, the rise of women in the League expanded the civic sphere in the rural West. No longer shackled by a discursive and structural emphasis on agrarian manhood, the broader potential of the NPL's nonpartisan tactics showed. Most notably, organized laborers saw the gains made by farmers, and began imagining their own politics in new ways. As agrarians made the League their own, union workers also began to see the possibilities in nonpartisan electoral action.

*

For decades, rural organizers and labor activists alike had hoped to bring agrarians and workers into a farmer-labor-rooted reform alliance. Deep mistrust on both sides too often trumped attempts to do so. Throughout the early twentieth century, industrialization and the rise of corporate capitalism tended to pit farmers and laborers against each other.

From the start, League leaders hoped to attract labor unions as allies. Because so many had spent years supporting the Socialist Party, this coalition seemed natural. That party's platforms, however, proved too labor centered to gain any traction among agrarians. But farmer empowerment through the NPL gradually encouraged agrarians to think more broadly about political economy. Growing secure in their own electoral acumen—and realizing the limits of class-oriented organizing in states where industry coexisted alongside agriculture—League members grew to see laborers as potential allies.

In states where organized labor represented a tiny fraction of the adult population—such as North Dakota—union chiefs quickly allied themselves with the NPL in 1916 and 1917. But in locales with already influential labor movements—such as Minnesota, Montana, and Washington State—wartime repression that targeted both Leaguers and organized labor helped orient union workers and farmers alike not just to think about alliances but to actually form them. This helped the NPL extend its reach into cities.

The first stirring of real coordination between Leaguers and laborers out-side North Dakota emerged in late 1917. Wartime repression orchestrated by Minnesota's Commission of Public Safety—which targeted the NPL and labor unions alike—ultimately created the conditions under which farmers and workers joined together. In Saint Paul, the commission's thinly veiled attempt to destroy the streetcar union not only led to violence and martial law in December 1917 but also drove the labor movement there away from the Democratic Party and into the arms of the NPL. In 1918, the city's Trades and Labor Assembly backed the NPL in the summer gubernatorial primary and put up a slate of independent union candidates for city offices. Labor leader William Mahoney seized the moment and began strategizing about how to formally connect workers in an alliance with the Nonpartisan League, con-veniently headquartered in the city. Union votes alone would not be enough. But when joined with those of farmers, they might influence state politics in powerful ways.[112]

Minneapolis's socialist mayor, Thomas Van Lear, came to similar conclu-sions. He broke with his national party's stance on both the war and the NPL. Noting his shared concern about civil liberties and concentrated corporate capital, Van Lear told League farmers in 1917 that "we ought to have just a little bit more democracy in America." The city's less radical local federation of labor followed Van Lear's lead once he declared himself for US involve-ment in the war. Courted by League leaders, all began encouraging workers to organize themselves to join with the NPL in electoral politics.[113]

In February 1918, Minneapolis union leaders "decided to organize for "'Non-Partisan' political action." Arthur Townley spoke at Minneapolis labor meetings that summer, backing the effort. NPL enemies decried the emerging farmer-labor coalition by labeling its advocates "Municipal Townleyites" and trying to convince agrarians that the NPL supported urban Socialists. But the opposition failed. Most Minnesota farmers appreciated that workers voted for NPL state gubernatorial candidate Charles Lindbergh—whose affinity for organized labor as well as farmers led the New York Times to dub him "a Go-pher Bolshevik"—in June 1918.[114]

With Lindbergh's defeat, both Van Lear and Mahoney worked with the NPL to craft a farmer-labor slate of candidates for the general elections in November. Though most of those candidates went down to defeat, the effort showed the potential power in a formal farmer-labor alliance. In order to better coordinate labor unions as a cohesive presence in electoral politics, in early 1919 Minneapolis's Van Lear and Saint Paul's Mahoney pooled their re-spective influence in the state's labor movements. Meanwhile, an undercover detective working for Twin Cities businessmen reported that NPL leaders privately recognized the "difficult and delicate proposition" of bringing together workers and agrarians but nonetheless did whatever they could to encourage "a closer relationship between the laborer and the farmer."[115]

Encouraged by the League's victory in North Dakota in 1918 as well as the establishment of the NPL platform in that state, Minnesota labor leaders saw a rising tide of farmer electoral power engendered by the League. They hoped to attach themselves to it. Van Lear and Mahoney's combined work convinced the Minnesota State Federation of Labor to commit itself to direct political action in July 1919. That month, the federation created a Working People's Nonpartisan League (WPNPL) in Minnesota. Drawing directly from the example of the NPL, Van Lear and Mahoney crafted the WPNPL to serve as a nonpartisan vehicle through which farmers and labors might work in coalition against the two major parties. The WPNPL and NPL began coordinating plans for the 1920 elections.[116]

Inadvertently encouraging cooperation, the Commission of Public Safety's reckless disregard of constitutional rights spawned an emerging farmer-labor alliance in Minnesota that meant much more to League leaders than the perfunctory and easily arranged alliance between the NPL and organized labor in North Dakota. In Montana, wartime repression also pushed laborers and farmers to work together. The nascent affinity of the state's Society of Equity and labor unions blossomed in the context of state-sanctioned suppression. During the 1917 copper miners strike in Butte, League farmers invited union ally and US House member Jeanette Rankin to speak about the unrest. Telling them that "you have gone a step farther than any farmers in the past," Rankin thanked them for realizing "that the handicaps that confront the producer, whether on the farm or in the mine are very closely related." She further assured the worried agrarians that "out of the 15,000 strikers . . . not more than 300" carried "I.W.W. cards." The miners could be trusted.[117]

Some farmers still worried. When the 1917 strike failed, private agents masquerading as IWW members organized another brief but violent strike in September 1918. The second walkout brought the heavy hand of law enforcement down on labor organizers in Butte. It shattered the union movement there. Desperate for new directions in organizing, Butte's labor leaders became intrigued with the NPL's nonpartisan and workplace-oriented model for electoral politics. They established connections with excited League leaders in their new headquarters in Great Falls. Copying the NPL's nonpartisan approach, labor leaders soon formed the Nonpartisan Club of Silver Bow County, headquartered in Butte.[118]

Preparing for the 1918 contest, the Nonpartisan Club ran candidates in the Democratic primary. One of them, the controversial editor of the *Butte Bulletin*, William Dunn, won his contest to run for the state legislature from that city. He also won in the general election. This victory allowed Dunn to support farmers' initiatives during his term in Helena. The farmers and laborers together also endorsed Jeannette Rankin's run for the US Senate seat. Though she lost the election, her campaign further fostered a firm farmer-labor alliance in Montana.[119]

From the start, Washington State's farmers insisted on holding on to their hard-earned relationship with organized labor. Since 1911, a joint legislative committee had worked as a clearinghouse for the two groups to work together on political issues. But as in Minnesota and Montana, World War I provided an opening for business interests—under the guise of the state—to suppress organized agrarians and workers. This tied together farmers and workers more closely than ever. A concerted campaign against IWW members that often caught up less radical trade unionists turned the latter sour on working through established parties. In turn, the vitriolic denunciation of the Washington State Grange and the persecution of NPL organizers convinced farmers of the need to finally and formally cement their alliances with the state's workers for direct electoral action.[120]

Finally, back in North Dakota, the NPL further cemented its good relations with the tiny union presence in the state. In October 1919, a looming nationwide coal strike roused NPL governor Lynn Frazier into prolabor action. Knowing that North Dakotans depended on coal to survive the frigid winters, Frazier tried to broker a deal between the state's coal miners and mine owners to ensure a steady supply during the months of greatest consumption. But the owners remained recalcitrant. Then a brutal blizzard struck the state. Motivated by public safety as much as a desire to support organized labor, Frazier declared martial law and ordered the National Guard to keep the mines open. The governor slyly noted that "the Washburn Flour Mill people at Minneapolis" owned one of the largest operations affected. The action weakened the mine owners' hand in the subsequent contract negotiations. When the crisis finally abated, the governor's actions confirmed that the NPL hoped to help unions whenever possible.[121]

By 1919, the League followed through on its long-stated desire to work in coalitions with labor unions. On the cusp of realizing what many union and farm leaders had aspired to for decades, the NPL showed itself ready, willing, and able to reach out to laborers in electoral coalitions across the West. Indeed, such coalitions emerged in time for the 1920 election season.

As entrenched politicians invented new ways to undercut NPL power, these alliances proved crucial.

✶

The League's expression of power in North Dakota and elsewhere gave it new weight with workers across the West. It also meant that journalists, intellectuals, writers, and artists around the country paid close attention to the NPL. Farm publications, of course, chronicled the rise of the insurgency. Their editorials offered a mixed response. The mainstream press—including the *New York Times*, *Chicago Tribune*, *Saturday Evening Post*, and *Literary Digest*—likewise kept tabs on the rise of the League. Usually, their reporters belittled it. Smaller progressive outlets such as the *Nation*, the *New Republic*, the *Liberator*, and the *Masses* proved especially interested in what many characterized as the "League experiment." Many of their articles on the NPL sang its praises or at least offered evenhanded commentary. Even academic journals, such as the *American Journal of Sociology* and the *American Economic Review*, raced to analyze the political movement growing in the West.

One of America's leading intellectuals, Thorstein Veblen, offered up his own take on the League in 1918. Famous for describing the rise of "conspicuous consumption" in American life in 1899, the irascible Veblen courted controversy throughout his career. Veblen's stance on the NPL proved no less contentious. Raised on a farm in southern Minnesota, the son of Norwegian immigrants espoused agrarian-based political action. In a memo prepared for federal Food Administration head Herbert Hoover, Veblen pointed out that the persecution of Leaguers disrupted agricultural production at a critical moment in the war. Veblen rightly noted that "the bond of union among" farmers "is a felt antagonism between their own material interest, on the one hand, and the interests of the commercial and other business elements of the community, on the other hand." But because "the party of the commercial clubs . . . is in control of the legally constituted administrative apparatus, police and judiciary . . . the federal administration is coming in for a share of the distrust" from disgruntled agrarians. Veblen believed that only by offering the League "generous treatment and fair dealing" could the government ensure a steady supply of grain and meat to feed the troops and the nation. Suspicious of his sympathies, the Bureau of Investigation opened an investigation into his political affiliations.[122]

A year later, Veblen further outlined his appreciation of the NPL's position. In *The Vested Interests and the Common Man* (1919), the economist described the League as "large, loose, animated, and untidy, but sure of itself in its settled disallowance of the Vested Interests, and fast passing the limit of tolerance in its inattention to the timeworn principles of equity." As a biting attack on corporate capitalism, *The Vested Interests* suggested that the NPL and other outlets of discontent resulted from "the time-worn rules no longer" fitting "the new material circumstances." Defending the NPL's call for "common honesty" in economic relations, Veblen posed the question "And why should it not?"[123]

Political thinkers also paid close attention to the League. In Harlem, the future labor leader A. Philip Randolph suggested in 1919 that African American radicals might forge an alliance "with white radicals such as the I.W.W., the Socialists and the Non-Partisan League, to build a new society—a society of equals, without class, race, caste or religious dimensions."[124] Upton Sinclair, noted socialist, muckraker, and author of the best-seller *The Jungle* (1906), openly supported the farmers' organization. In an editorial, Sinclair told readers that "if you want to see the joyful spectacle of farmers and their wives thoroughly awakened, class-conscious and alert, discussing their own interests and acting upon them, repelling the slanders and evading the snares of their cunning exploiters, send to St. Paul, Minn., for a copy of the Nonpartisan Leader, and read!" He also defended the League in subsequent books, such as *The Brass Check: A Study of American Journalism* (1920) and *The Goslings: A Study of America's Schools* (1924). In turn, the NPL promised Sinclair it would sell a thousand copies of *The Brass Check*.[125]

One writer did more than analyze and comment on the League. He joined in its work. After publishing the influential and popular *Our America* (1919), novelist and journalist Waldo Frank looked for a new challenge. Excited by news of the NPL, the New Jersey–born Jewish intellectual (who soon thereafter became a regular contributor to the *New Yorker* and editor of the *New Republic*) relocated to Ellsworth, Kansas, in 1919. There, he began reporting for the *Ellsworth County Leader*, the leading League newspaper in the state. He hoped to become an authentic and engaged voice of the people. Living with Milton Amos—Kansas's leading Leaguer—he encountered agrarians fearful of losing their land. In letters to his friend Sherwood Anderson, then garnering praise for his new novel *Winesburg, Ohio* (1919), Frank poured out his appreciation of the "poor dull sweet brothers" who turned to the NPL to defend their livelihoods and families. Indeed, Anderson himself had just turned down a tempting $5,000 offer to write political commentaries condemning the NPL.[126]

Yet Frank's idealized vision of agrarians could not disguise his growing disgust for the League itself. After attending the NPL national convention in Saint Paul in December 1919, Frank decided he could no longer engage in the dirty work of politics. He wrote Anderson that "as I went up in the scale of authority, there gradually faded out all that was sweet and gentle and lovable in the farmers' world—: the farmers' world—its life—faded away." "In its place," Frank found nothing but "doctrine, dogma, figures, political manipulation, words." His anger rose to a fever pitch as he pronounced that "political representation will *always, must* always be a game for the tricky, the brutal . . . the shallow." Frustrated by League infighting, the ever-romantic Frank gave up on the NPL.[127]

The internationally renowned sculptor Gutzon Borglum also threw his support to the League. Long before he began working on South Dakota's magisterial Mount Rushmore, Borglum raised money and spoke for the NPL on the East Coast during the 1918 campaign season. He envisioned the farmers' movement as a vehicle for reforming his beloved Republican Party. In 1920, Borglum attempted to broker national GOP support for the League in North Dakota with presidential candidate Warren Harding, but to little effect.[128]

The artist also solicited funds for the League. Portraying the NPL as a wing of the party that might eliminate Republican reactionaries once and for all, Borglum told wealthy patrons that "I am heart and soul with these people of the Northwest and believe fully in their grievances. . . . But I want to keep them in the Republican Party." Anxious for any financial help they could get, League leaders carefully cultivated Borglum. They assured him of their desire to work with the party even as they publicly rejected such a stance. He repaid their interest with a deep commitment to the NPL's success. He even toured North Dakota boosting for the League.[129]

Sinclair Lewis, the Minnesota-born novelist, emerged as the NPL's most important literary supporter. Ten years before becoming the first American to win the Nobel Prize for Literature, he published *Main Street* (1920)—the best seller that made him famous. Research for the book brought Lewis back to Minnesota in 1917. In his rented Summit Avenue mansion in Saint Paul, Lewis and his wife welcomed Nonpartisan League leaders. This fed local gossip about the couple's political associations. Ostracized, the novelist and his family fled to a summer home on Cape Cod. That fall, they returned to Minnesota, renting a house in Minneapolis. There, his wife remembered, "we were seen with disapproval at a meeting of the Nonpartisan League." Again, neighbors whispered. Moving to Mankato in 1919, Lewis attended nearby Nonpartisan League picnics and rallies without further incident. Soon thereafter, his editor at the *Saturday Evening Post* asked him to pen an anti-League story. Lewis delicately declined the offer, noting that the League remained "too controversial."[130]

Main Street, released the following year, depicted life in the town of Gopher Prairie, a thinly veiled version of Lewis's hometown of Sauk Centre,

Minnesota. The book offered a powerful critique of the provincialism that defined early twentieth-century small-town life in America. It also subtly defended NPL farmers. Its main characters—Carol and Will Kennicott—fall in love, marry, and then drift apart. The rift in their relationship grows from Will's inability to see past the narrow interests of businessmen in a tiny burg as well as his denigration of farm families. In fact, the narrator suggests that Carol comes to understand that "the prairie towns no more exist to serve the farmers who are their reason of existence than do the great capitals; they exist to fatten on the farmers." Indeed, she soon realizes that "the wheat money did not remain in the pockets of the farmers; the towns existed to take care of all that."[131]

In the novel, Nonpartisan League agitation among local agrarians—and Will Kennicott's disgust for it as a local land speculator—proves the narrow-mindedness of Kennicott and his neighbors. It also sets the scene for the ultimate failure of the couple's marriage. Learning of an NPL speaker willing to defy an order not to speak in the county, a local mob "of a hundred businessmen" had "taken the organizer from his hotel, ridden him on a fence-rail, put him on a freight train, and warned him not to return." Will applauds the act. For Carol, it is the last straw. She decides on separation. Basing his story on actual events allowed Lewis to offer a convincing plot device to display Carol's building anger budding into action.[132]

Lewis's stance on the League became clear to anyone who perused *Main Street*. In his mind, farmers possessed real grievances, and small-town dwellers problematically exploited them. Indeed, the latter unfairly denigrated agrarians as simpletons and rustics. One impressed reader—League leader Joseph Gilbert—believed Lewis got it right. After losing his appeal in the US Supreme Court, he began serving his sentence in the Goodhue County, Minnesota, jail. With plenty of time on his hands, Gilbert devoured *Main Street*. He later told an interviewer that it portrayed "with camera-like fidelity the superficial, mediocre, but dominating characteristics, not only of Gopher Prairie, but of American urban life in general." For Gilbert, those characteristics were "love of power as measured in terms of money, ostentatious display, and a corresponding lack of knowledge, together with intolerance towards all progressive ideas."[133]

Sinclair Lewis did more than confirm Leaguers' animosity toward small-town businessmen. He also proved instrumental in helping far-removed readers better understand the farmers' insurgency. In fact, Lewis convinced his dear friend and New York–based editor Alfred Harcourt to solicit a book on the Nonpartisan League as he launched a new publishing company. Using his contacts among NPL leaders, the novelist even arranged for Herbert Gaston, the director of NPL publications, to write the book.

Upon the publication of Gaston's sympathetic but evenhanded account of the League's rise in 1920, Lewis demanded a free copy from Harcourt. "Don't forget I'm the father and the mother of that book," the Minnesotan reminded his colleague. "You better send me [a] copy of [Gaston's] *Nonpartisan League*." All told, Lewis embraced the farmers of the Nonpartisan League because in his mind they embodied the best of lower-middle-class ideology—accumulation without concentration—in the face of what he would stereotype as an emergent consumer-oriented conformity in his next novel, *Babbitt* (1922).[134]

⋆

In December 1919, the NPL numbered 208,800 paid members, including 50,000 in Minnesota, 40,000 in North Dakota, and 25,000 in South Dakota. Farther west, 20,000 NPL members lived in Montana, 12,000 in Colorado, 12,000 in Idaho, and 10,000 in Washington. Though not yet a presence in political life there, the League also had 15,000 members in Wisconsin and 13,000 in Nebraska. The remaining 11,300 adherents lived in Iowa, Kansas, Texas, and Oklahoma. Based on those who still received the *Nonpartisan Leader*, the League estimated that another 30,000 to 40,000 farmers remained staunchly behind the organization. That made for 250,000 dyed-in-the-wool Leaguers. Thousands more voted for NPL candidates in elections. As a voting bloc, the NPL had no peer. A simple idea to bring North Dakota voters' dream of a state-run terminal elevator to fruition blossomed into something much bigger than anyone had expected.[135]

While electoral success outside of North Dakota remained limited, the seizure of one state government showed citizens of every stripe that the NPL could effectively govern when given the chance. Instituting their complete program of state-owned industries, the League fulfilled the promise it made to voters. Despite some failures and many challenges, the NPL ably reflected its constituents' desires.

Equally important, farmers' allegiance to the Nonpartisan League peaked in the months following World War I. A culture of empowerment, created by agrarians themselves, flowed from the machinations of League-organized publications and gatherings. Instead of merely mobilizing voters, the NPL started to become an organic organization that drew on the power of transformed members. The newfound power of the rank and file tapped into sentiments and community building that transformed the League into a full-blown movement.

More than any other constituency in the emerging movement, farm women created a space for themselves. Turning the culture for change created by members into a potent transformational force, they empowered themselves and their families. Transforming the NPL, farm wives, daughters, and mothers reoriented the organization away from appeals to agrarian manhood and toward a broader vision of economic equity for all. This made for new possibilities in politics. Breaking free of self-imposed gendered assumptions, the League then proved able to mobilize urban workers as potential allies. In multiple states, organized laborers joined the League in coalitions.

Finally, the widespread sympathy for the NPL among intellectuals, writers, and artists across the United States symbolized the cultural reach of the League. The League's commitment to government competition in the marketplace as a counterweight to corporate capitalism broadly appealed to many of the nation's leading thinkers. The NPL idea seemed to offer something new in the tumultuous postwar climate.

But poised on the cusp of broader success, the NPL faltered. Facing the constant stress of powerful and wily opponents, the limits of the NPL's model and tactics began to show. As the 1920 elections approached, the very forces that engendered the League—as well as those that the organization engendered—threatened to tear it apart. The moment of triumph also proved to be the moment when everything began falling apart.

Reverses

I would state rather that the league was helped by this opposition and killed by its friends.

      A. B. GILBERT (former editor of the *Nonpartisan Leader*), July 1926

Anxious to understand the scope of the ascendant farmers movement, an ambitious young *New Republic* reporter named Charles Merz traveled to North Dakota to investigate. In May 1920, he observed that the NPL, a "political force of substantial power," courted constant controversy. With this in mind, Merz focused on being as objective as possible. His essay carefully addressed the League's weaknesses as well as its achievements. In "an institution in which the few decide and the many follow," the question of democracy within the NPL proved chief among the former. Merz found that authority in the League—after five years—still ultimately rested "in the hands of a very small group of men."[1]

This concern permeated his analysis of the NPL. "Every political organization operating on the modern scale—whether it be the Nonpartisan League, the Republican party, or the United States of America" shared the "danger that too much power is centering upon a few leaders." Merz argued that "no ordinarily human leaders are both able enough and honest enough to manage it rightly." Noting the pervasiveness of the dilemma, he reported that "the problem of democratic control within the Nonpartisan League is only a miniature of the problem of democratic control within the modern state."[2]

Merz, a close confident of *New Republic* cofounder Walter Lippmann, cared deeply about the state of democracy in America. In the wake of World War I, many intellectuals and reformers—such as Merz and Lippmann—gave up on participatory democracy. They had watched George Creel and other war propagandists work to create an irrational and destructive public that ran roughshod over civil liberties. They saw average citizens, often working in voluntary associations founded by reformers, easily swayed from core American beliefs. This suggested that the people could no longer be trusted.

Commentators lost faith in the public as the foundation of political life in the United States. Just a few years after Merz published his analysis of the Nonpartisan League, Lippmann decided that "the public must be put in its place." Only then would the nation "live free of the trampling and the roar of a bewildered herd."[3]

NPL leaders admitted to a tight hold on the organization. Walter Thomas Mills, a prominent former Socialist and Arthur Townley ally, published a pamphlet addressing the issue in 1918. He acknowledged that the League was not a "democratic organization." "It is impossible to fight the political machines built, financed, and managed by the great private monopolists," Mills asserted, "except by the building of a machine with which to fight them." He went on to argue that "battles can be fought only with someone in command. Townley is in command." Finally, Mills suggested that the object of the battle—"to obtain democracy"—ultimately made the League democratic. Carefully distinguishing between a democratic movement and a movement for democracy, Mills argued for the latter.[4]

Agrarians, then, knew "the organization to which they belong is neither entirely efficient nor ideally democratic." Nonetheless, Merz told the *New Republic*'s readers, the NPL remained "the instrument of a living faith for many of the hundreds of thousands of farmers." When the young reporter concluded that if the League's "leaders fail because they listen to the call of power" they "will destroy a useful thing . . . a faith which the next new force will find it difficult to recover," he got it only half right. To be sure, the "call of power" convinced Arthur Townley and a small core of dedicated lieutenants that victory depended on resisting the siren song of democracy. The strength of their opponents cemented this belief even as many farmers developed a new confidence in their own clout. The tensions between the two visions— one insistent on tight control from above, the other equally insistent on expanding participation in the organization's decision-making—broke the NPL wide open. Farmers' emerging commitment to citizen-centered democracy fueled challenges to Townley's tight control of the League. These agrarians belied the emerging perception that popular politics threatened the nation's stability. At the moment when the public seemed so easily led astray, many Leaguers emerged with a stronger commitment to participatory democracy than ever.[5]

Pressure applied by powerful enemies convinced Arthur Townley to retain his iron grip on the League. Desperate to keep the coffers full, NPL leaders devised and promoted commercial enterprises without consulting the membership. Unilaterally extending the League beyond politics, NPL leaders engaged in misguided and even illicit entrepreneurship. Angering allied farm organizations, these initiatives failed. Allegations of corruption ensued. The dissonance between the NPL's calls for democracy and its organizational practice spawned internal opposition. Organizers began to challenge Townley's insistence on top-down structures. Rebellions by staff, coupled with concerns raised by agrarians, forced Townley to make a show of democratizing the organization. His power began to slip away.

Struggles over democracy inside the organization took on a new intensity when combined with the struggle for democracy outside it. Established political parties responded to the threat posed by the NPL by trying to reverse years of popular electoral reform. Attacking the strategic innovation that made League success possible—the open and direct primary—politicians in seven states introduced legislation to alter or abolish the practice. Of the four states with the most Leaguers, three adopted less inclusive election laws. These put new boundaries on hard-won democratic practices.

The emergence of dissatisfied Leaguers also encouraged adversaries to engage in vicious smear campaigns and to cultivate controversy. Efforts to expand the League's power in new places fostered further backlash. The recently formed American Legion touted 100 percent Americanism and envisioned the NPL as a rural manifestation of Bolshevism. NPL members and employees alike faced mob violence. This breakdown in civil society—in the name of protecting democracy—denied some Leaguers their rights to free speech and assembly.

The NPL could not bear these internal and external pressures. The League held on to power in North Dakota and instituted its program there even as new primary laws made it more difficult than ever to find electoral victories elsewhere. The faith that fueled the fight against the NPL's enemies also led farmers to speak out for more power in their own organization. Fissures in the unified farmers front began to show. The struggles that ensued led to the League's slow fragmentation. Failure in the 1920 election season made this fragmentation apparent. Even though the NPL did not disappear in the wake of these electoral losses, it never fully recovered.

✶

In 1918 and 1919, ill-advised business ventures provided League opponents with fodder for critique. The League's interest in forming new farmer-oriented companies proved consistent with their deployment of innovative approaches to sales and credit in organizing. The organization criticized corporate capitalism even as it drew on many of its practices. It firmly planted one foot in popular politics and the other in contemporary efficiencies derived from corporate values.

The resulting commercial initiatives—which included a bank, a fishing company, a binder twine business, and a chain of retail stores—all ended up ensnaring the NPL in complicated legal proceedings and bad press. The origins of these enterprises remain murky. In many cases, even contemporaries could not be sure about who, exactly, was involved. Furthermore, poorly documented and poorly explained initiatives revealed the limits of the League leadership's strategic thinking. All told, the NPL's entrepreneurial streak undercut its own stated aim of bringing democracy to the rural West.

From the start, monetary difficulties dogged the League. Building a movement on credit fueled the NPL's rapid expansion. Yet it also created keen organizational challenges. Small-town bankers often refused to honor post-dated checks given to the League by farmers. In early 1917, the NPL purchased the Scandinavian American Bank in Fargo. This ensured a sympathetic clearinghouse for its own transactions. Run by John J. Hastings—the NPL's bookkeeper—the Scandinavian American Bank handled all League accounts as well as those of individual farmers.[6]

Shoddy record keeping and dependence on credit led to scandal. Hastings took unusual liberties with large amounts of money. He purchased small-town banks in an effort to further capitalize what the League described as "a farmers' bank." Hastings also freely extended loans to prominent Leaguers and even to the NPL itself. For all these activities, he charged a commission. Meanwhile, much of the bank's own collateral rested in postdated checks. One leading financial journal sarcastically suggested that "it remains to be shown that this practice of extending credit is better than the old banking practices."[7]

Rumors swirled about Hastings's financial mismanagement. League enemies suggested that Townley personally profited from funds derived from farmers. Because the state institution proved "partial to" the Scandinavian American Bank "above all other banks in the state in the matter of redeposit and loans and rediscounts," some even suggested that money might be nefariously redirected from the new State Bank of North Dakota to the NPL's own institution.[8]

An investigation of the Scandinavian American Bank's finances by a North Dakota state assistant bank examiner in September 1919 revealed a "vast, unwieldy financial monstrosity unable to take care of its obligations." In October, the state shut it down. The League responded with a mass meeting in Fargo. Speakers insisted that the report represented a political attack. Newly inspired, hundreds of NPL members deposited their own hard-earned money in the bank. Three weeks later, after a second report by the examiner's League-friendly supervisor, the NPL-dominated state supreme court found no wrongdoing and ordered the bank reopened. The troubles, however, continued. In 1921, the state again shuttered the bank, this time for good.[9]

Further controversy over the bank's practices stemmed from the League-related businesses that used the institution. Formed by the Scandinavian American Bank's president, H. J. Hagen, the Bering Sea Fisheries Company derived from a loan granted to Hagen by the bank. His collateral consisted of little more than "uncaught salmon in Alaskan waters." Hagen then tried to attract shareholders. In the wake of the investigation of the League's bank, the only thing Hagen landed was a jail sentence.[10]

Other enterprises emerged. During the 1910s, supplies of the sisal-derived twine that wheat farmers across the United States and Canada depended on grew tight. Cultivated by Mexican farmers, sisal offered the best fiber for the cord used by the thousands of mechanical binders that harvested and bundled grain. The corporations that built binders—such as International Harvester—determined twine production and prices. In combination with instability in Mexico and war-induced demand for wheat, this left farmers longing for a cheaper, agrarian-owned supply.[11]

Promoters from Florida approached Arthur Townley. Sisal, they argued, could be grown in South Florida. As the biggest organization of farmers in the United States, the Nonpartisan League seemed like a logical investment partner. An agrarian-controlled twine supply held great appeal. Though the NPL did not officially endorse the venture, numerous League leaders, NPL-affiliated officials, and the Publishers' National Service Bureau (the company created by the NPL to foster and support farmer-owned newspapers) did invest.[12]

The company that emerged—the United States Sisal Trust—involved a number of prominent NPL members. James Waters, the League-appointed head of the newly organized Bank of North Dakota, served as president. Prominent shareholders included Job Wells Brinton, a Townley lieutenant; Alex Stern, the mayor of Fargo; and Obert Olson, the state treasurer of North Dakota. Farmer-owned banks across North Dakota also purchased sizable shares. Together, they raised $1 million and purchased 22,400 acres of sisal-growing land in Dade County.[13]

In the end, the company did little more than bilk hundreds of investors out of thousands of dollars. In February 1920, Brinton criticized the company's leader, James F. Jaudon, a noted Miami landowner and entrepreneur. "He simply lacks business ability and business judgment," the Leaguer noted. Brinton later accused John J. Hastings of investing thousands of Townley's own dollars (themselves stolen from League coffers) in the scheme. James Waters corroborated this claim, noting that Townley owned "a controlling interest in the concern." But the one who lost the most in the venture proved to be Brinton. He forfeited all of the $25,000 he invested in the trust.[14]

Even so, the Consumers United Stores Company proved to be the League's most spectacular business failure. Soon after joining the League, Brinton— the former mayor of Beach, North Dakota, and longtime friend of Townley— hatched a plan to harness the political power of farmers to their purchasing power. He envisioned "a co operative chain store system established for service" but "not profit or dividends." Rejecting long-accepted cooperative principles practiced by farmers for decades, Brinton blended corporate centralization with a membership model. Townley and other League leaders assented to the plan.[15]

Brinton called the Consumers United Stores a "North Dakota Company for North Dakota People." Organizers for the stores sold memberships—not shares of ownership and control—that allowed farmers to buy hardware, clothing, shoes, furniture, food, and other offerings at 10 percent over wholesale prices. This discount appealed to many farmers, as it was larger than the one proffered by most small-town merchants. By the beginning of 1917, organizers had established three stores. Farmers paid one hundred dollars for a ten-year membership. For those so inclined, Consumers United also paid "cash based on the market prices, less the cost of handling" for "produce, cream, butter and eggs from the members." Once one hundred agrarians in the same area joined, the company opened a store.[16]

Playing on farmers' loyalty to the NPL, the scheme proved popular. In Ward County, despite "three successive crop failures" and "cases of poverty . . . that are pitiful," one organizer convinced "about 25%" of local farmers to buy a store membership. The local store did a "better than $200 per day" business. By June 1918, Brinton raised more than $1 million. Eight North Dakota towns had Consumers United stores, and the company already counted enough members across the state to start twenty-one more.[17]

Representatives of the NPL-organized Consumers United Stores Company insisted that it was a separate entity from the League. They rejected claims that the League operated or ran the company. Indeed, brochures noted that the company "is entirely independent of the Nonpartisan League, which is a voluntary political organization." Yet the same leaflets sported the endorsement of Arthur Townley. Furthermore, most Consumers United Stores Company organizers carried letters on League letterhead authorizing them to solicit NPL as well as company memberships. In fact, the League turned to its best organizers to solicit new members, who envisioned the job as a plum assignment. And, like NPL organizers, they earned a commission on each membership.[18]

Deploying the familiar language of agrarian cooperation, organizers promoted the company as a way to make "democracy a real factor in our political and economic life." But the stores worked "on the same principle as those operated by the big chain stores organizations in the United States; with centralized management and control." As a self-described "corporation," the company focused on bringing new efficiencies and a much larger scale of purchasing consumer goods to rural communities. It envisioned both existing farm cooperative stores and privately owned businesses as too small and inefficient.[19]

Further complicating matters, salesmen used at least two different contracts to enlist members. One insisted that all membership dollars would be used as capital to establish and operate the stores. The other declared that "90% of all money subscribed herein shall be used by said company to establish" a local store. The remainder would "be used by the company either to establish and maintain a central buying agency or wholesale establishment, or to carry on educational work or propaganda . . . in the interest of or beneficial to farmers."[20]

Invariably, the poorly defined scheme led to confusion and controversy. Despite claims that Consumers United Stores did "not interfere with the local merchant or local market by selling goods to the public for less than the market prices," the outlets did exactly that. Local store owners, already attacked by NPL newspapers, now saw their very livelihood directly threatened. One responded viciously, referring to the venture as "*The United Easy Marks Association.*" Meanwhile, League leaders used the company's coffers as a fund, transferring monies directly to the NPL at their own discretion. Finally, in its critique of existing stores of every sort, the company inexplicably defied established farmer consumer cooperatives connected to Leaguers and their allies.[21]

Deeply concerned about the Consumers United Stores Company, the leadership of North Dakota's Farmers Union as well as the state Grange sought redress. In January 1918, an all-night meeting with Arthur Townley revealed the NPL leader's hubris. Townley insisted on the soundness of the Consumers United Stores Company plan, refused to discuss the use of member-derived funds for political work, and rebuffed all criticism. In response, the other farm leaders publicly declared that democratic cooperation—not a member-based corporation—should be the guiding principle of any farmer-owned enterprises. They and their members "would have nothing to do" with the company. Furthermore, some grumbled that "there would be war to the knife if the league sought to disrupt" cooperatives supported by the Farmers Union and state Grange. The *Bismarck Tribune*, firmly against the NPL, reported that Townley met his match among these "sensible, slow-going, thinking farmers." Three days later, the *Tribune*'s editor predicted: "That meeting was the beginning of the end in North Dakota."[22]

Increasing criticism—and internal squabbling over the appropriate use of Consumers United Stores funds—gradually weakened the company. As early as 1917, when Brinton expressed concern about the use of company money in political campaigns, Townley assured him that "we can re-write all the books if necessary." The cavalier attitude caught up with the leader. Ensnared in controversy, Townley eventually devised a different strategy. In late 1919, the NPL reorganized the Consumers United Stores around more typical cooperative principles. The company transformed memberships into shares and laid out a dividend plan.[23]

But the damage—to the Consumers United Stores Company and the League itself—had been done. The stores went out of business. Like the other League-associated enterprises, the effort ended in litigation and failure. The poor record of these businesses hurt the NPL's expansion effort everywhere. League organizer Alfred Knutson later admitted that "would-be friends and allies were alienated . . . to a great extent by reckless business and shortsighted policies." Much as the political victories in North Dakota carried great weight outside the state, these missteps undercut the NPL's plans to organize more broadly.[24]

The misguided ventures in North Dakota, for instance, forced Leaguers in far-off Idaho to respond to charges that they, too, were selling "stock in various league enterprises." Opposition newspapers told agrarians to beware of such efforts. In response, the *Idaho Leader* assured readers that the NPL in that state would not ask "the farmers to lend their credit to the League for the purpose of establishing various enterprises." In fact, it "is no way sponsoring ANY promotion schemes." The editor even asked farmers to report any solicitors claiming an association with the League to the state headquarters.[25]

Ultimately, the NPL's entrepreneurial ventures convinced a growing number of farmers that the League did not represent their political or economic interests. J. W. Ingle, a member of the North Dakota Farmers Union who attended the January 1918 showdown with Townley over the Consumers United Stores Company, soon thereafter broke with the NPL. In a pamphlet published by League opponents, Ingle said he wanted "it distinctly understood that I am not fighting the Nonpartisan League. I am fighting Townley and his methods." Ingle, a League booster who initially helped an organizer secure Consumers United Stores memberships in his neighborhood, came to realize that he "didn't go into the league with the idea of building up any wild cat proposition such as this store scheme." He and the others at the January 1918 meeting asked Townley "to take the store proposition away from the league because we were afraid it would hurt the league." Ingle accused Townley of responding with obscenities. He claimed that the NPL leader called the Grange and union farmers "a set of G——damn hogs. We have got to drive you to the polls with a ———— elm club to make you vote." Put off by his purported profanity, insults, and derisive attitude, Ingle flatly declared that Townley was "not of our kind of people." As controversies over League-related business controversies grew, other agrarians began to agree with Ingle.[26]

∗

Arthur Townley's belief that farmers needed to be driven to the polls—publicly expressed or not—illustrated a central feature of his thinking. Even as the NPL fostered newfound democratic sentiment among the rank and file, he kept a firm hand on the organization. Along with entrepreneurial missteps, a growing sense of civic agency among members led to calls to democratize the organization itself. But Townley stubbornly insisted that only strong leadership could trump the growing challenges to the League. This stance gradually alienated advisors and members alike. Growing discontent left many disgruntled, fed challenges to Townley's power, and divided the NPL.

Townley instituted and fostered a clear hierarchy among advisors and organizers. He resolutely held the League "strictly in his own hands." As the inventor of the automobile-driven and commission-based League sales pitch, his reputation loomed large. Townley used that proven authority—and the NPL's success—to demand absolute loyalty. Recognizing money as the lifeblood of politics, he remained convinced that the organization's top-down structure kept agrarian funds flowing into NPL coffers. Townley also believed agrarians remained politically naïve. In 1919, he told one group of farmers that their propensity to be fickle made them "an awful hard bunch to handle." The NPL leader also presumed that ambitious politicians would take advantage of obscure parliamentary procedures to wreck the League.[27]

Thus Townley, as one colleague put it, "intended that affairs move, and move swiftly. He did not want to waste time with formalities, with parliamentary meetings and the machinery of conventional organization, little understood by farmers." He maintained this outlook even though detractors decried Townley's staunch commitment to his own unquestioned authority. Some believed it stemmed from his desire to make money. Others called his autocratic tendencies a form of "irresponsible power . . . contrary to the spirit of American institutions." Many—including a small but growing number of Leaguers—believed that his "rank Autocracy" rendered him little more than a corrupt "political boss."[28]

Townley acknowledged that he had never been elected president of the NPL. Speaking to agrarians in Grand Forks in 1917, the NPL leader observed that "40,000 farmers" subscribed to the League program, which "was a pretty fair election . . . about as good as we could accomplish at that time." Furthermore, Townley believed that farmers needed to be shown their own power. It might have been better to call for a convention to form a new organization, but he claimed that "we had to show you first that something should be done, before you would come." Townley remained convinced that only with a strong leader would agrarians realize their own political power.[29]

The immense and concentrated power held by League opponents served as another rationale for Townley's paternalistic rejection of democratic structures. Effective politics seemed to demand the concentration of power. The organization's top-down hierarchy no doubt made for swift decision making in the heat of electoral contests. But even his advisors became uneasy. Arguments about Townley's self-assured leadership distracted already busy men. Intense internal debates began in 1917, continued in 1918, and broke into the open in 1919.

Samuel Maxwell, an organizer for the League in Colorado, was the first to directly challenge Townley's tight control of the NPL. In October 1917, he penned a letter to Townley demanding that the money collected by organizers stay in the state. Maxwell also wanted members—instead of R. W. Morser, the state organizer—to control the Colorado chapter of the NPL. In Maxwell's mind, "the democratic control of the money is the main thing." If ignored, he threatened to use his influence with the leaders of the Colorado Farmers Union and state Grange to have them withdraw their support of the League.[30]

Townley immediately sent for Maxwell. When the Coloradoan arrived in Saint Paul, the NPL leader could not be found. Instead, Maxwell met with Joseph Gilbert and Arthur LeSueur. Confronted with the details, Gilbert admitted to the truth of Maxwell's charges. Yet neither Gilbert nor LeSueur would support Maxwell. In fact, Gilbert insisted that "such an arrangement is absolutely essential in the present stage of the growth of the League." LeSueur then traveled back to Denver with Maxwell. There, LeSueur agreed to remove Morser as soon as possible and to consult with Townley about keeping the membership dues in the state. Months went by. Nothing happened. One prominent Colorado Grange leader—state senator Agnes Riddle—broke with the League over its inaction. Townley tried to entice her to stay on with the offer of a job in the national organization. She refused. Meanwhile, League leaders invited Maxwell to serve as a national lecturer. He accepted, took a pay raise, and began stumping throughout Minnesota, North Dakota, and Nebraska.[31]

More resistance emerged a few months later. Organizer Walter Quigley, disturbed by Townley's refusal to share power with NPL members, allegedly started what he called "a one-man revolt." Securing an affidavit from an Iowa soldier who suggested that Townley had offered him a position as a North Dakota game warden to avoid the draft, Quigley confronted the League's leader. To back his bluster, the organizer showed Townley a series of articles to be submitted to news outlets alongside the accusation of corruption if Townley did not "reorganize the league."[32]

Threatened with blackmail, Townley met with Quigley and promised to call a national convention after the November 1918 elections. There, he would institute changes "and leave the running of" the League "to the members themselves." Recognizing that "a shakeup" before the elections might prove disastrous, Quigley surrendered the affidavit and agreed to wait. Again, nothing happened. Under pressure to quit, Quigley refused to do so until after the November 1918 elections. Finishing up as an organizer in Nebraska, he finally published his tell-all articles with the *Lincoln Daily Star* in April 1919.[33]

By May 1918, members as well as advisors expressed concern about Townley's iron grip on the organization. Will Wasson, a pro-NPL newspaperman in Kensal, North Dakota, admitted to S. A. Olsness, North Dakota's insurance commissioner, that he and others had decided to oppose "Townley's self-imposed rule." Wasson believed "that its [sic] time to break with Townley. . . . The League faces defeat with such leadership." Olsness's pleas to stay focused on the bigger issues facing farmers fell on deaf ears.[34]

Over 250 members of the League in Kidder County, North Dakota, agreed with Wasson. That same year, they began organizing "for the purpose of electing a president of the League." League leaders rushed there to dissuade them from their effort. Around the same time, NPL members from across North Dakota petitioned to elect the League's leaders and embed the initiative, recall, and referendum within the NPL itself. Ignored, at least one signee became a prominent member of the Independent Voters Association. All this grousing among members resulted in the publication of Walter Thomas Mills's ardent defense of the League's hierarchical structure as essential for establishing democracy.[35]

More serious divisions soon emerged. Joseph Gilbert, indicted in Goodhue County, Minnesota, in 1918 on trumped-up charges of sedition, came to be seen as a liability to the League. When Townley—over the wishes of most of his advisors—demoted Gilbert to a lesser post in North Dakota, the latter responded by drafting a lengthy petition demanding change in the organization. Signed by seventeen other League employees, the missive declared the signers' loyalty to the organization even as it called for internal democracy.

Gilbert's letter acknowledged the pivotal role played by Townley in leading the League to national prominence. It also noted that in the early days of the NPL "an autocratic policy was productive of the greatest efficiency in promoting rapid growth." But now that policy held the potential to undermine the League's "permanent success." The petition acknowledged that "already many of the farmers perceive that they have the form but not the substance of democracy." If the League leadership did not—on its own—democratize the organization, the results might be "disastrous."[36]

The appeal went one step further. Besides calling for state and national conventions of NPL members immediately after the 1918 elections, the dissenters demanded the democratization of the movement's management and finances. Departments—educational, legal, publicity, finance, legislative, campaign, public affairs—headed by trusted Townley lieutenants would be required to report to each other as well as a national executive board. They also proposed preapproved monthly budgets. The signers believed that giving League members more control as well as more carefully accounting for expenses would ensure a bright future for the NPL.[37]

The petition provoked Townley's wrath. He demoted those who supported it. Many stayed on with the League in a lesser capacity. Still dependent on the NPL to pay his ever-mounting legal fees, even Gilbert acquiesced. Nonetheless, those close to the inner circle came to expect intrigue at League headquarters. O. M. Thomason, another signer, wrote a close friend at League headquarters a few months later to find out about "the things on the 'inside.'" Wondering whether or not there were any "revolutionary tendencies," he clearly still yearned for change. The friend immediately assured a disappointed Thomason that "splendid harmony" reigned in Saint Paul.[38]

Years later in the *Nation, Nonpartisan Leader* editor Arthur B. Gilbert not only suggested that the signers of the petition "were woefully lacking in the ability to organize and agitate" but also believed "there was too much democracy in the league for a fighting farmer organization rather than too little." Like Townley, he came to believe that "farmers can't fight successfully unless they trust leaders who have proved themselves." This hindsight conveniently obscured Gilbert's own support of the petition. Apparently, he regained Townley's trust, because by 1920 he served as the editor of the League's primary organ. That—as much as subsequent events—changed his mind about Townley's grip on power.[39]

The petition did, however, impress upon Townley the need for a public affirmation of his leadership. He responded to the signers' pleas that something be done in the wake of the general election by calling for a national convention in December 1918. Delegates came to Saint Paul. Most served as state managers or League employees. The others were noted Townley loyalists. Together, they formalized centralized power in the existing executive board of three men—Townley, William Lemke, and Francis Wood. The assembly also audited the League's finances and found them sound. Finally, Townley told the assembled that he would continue to serve only if the rank and file supported him. The *Nonpartisan Leader* then solicited the opinion of members. A ballot published in the newspaper offered Leaguers the chance to vote for or against Townley. About half the members participated in the referendum. Counted by League employees in the Saint Paul office, the tally overwhelmingly fell in Townley's favor. The *Nonpartisan Leader* declared the matter settled.[40]

Combined with the seizure of power in North Dakota, this carefully manipulated show of support confirmed existing organizational arrangements. Yet soon thereafter, Arthur LeSueur—lawyer, longtime Townley confidant, North Dakota's best-known Socialist, and head of the NPL's Education Department—broke with the League. In January 1919, LeSueur expressed concern about the structure of North Dakota's Industrial Commission. Over his objections, the NPL-controlled state legislature (guided by Townley) decided to vest all oversight of the new state mill, elevator, and bank in a committee made up of the governor, attorney general, and commissioner of agriculture and labor. The longtime lawyer railed at this arrangement, believing that the legislation should "TAKE THESE INSTITUTIONS AS FAR OUT OF POLITICS AS POSSIBLE." Townley ignored his pleas and plunged ahead with this centralization of power. The rejection confirmed LeSueur's long-standing concern about "the known autocratic and irresponsible" NPL leader. Just three months later, LeSueur left the League. He publicly criticized it in the tumultuous years to come.[41]

Other longtime Townley associates departed around the same time. Leon Durocher—a former Socialist who worked for Townley and Bowen in western North Dakota in 1914 and followed them into the NPL as one of the first organizers—drifted away in late 1918 over concerns about democracy in the organization as well as the misuse of funds. He landed at a profarmer bank in Minot started by an old Socialist friend of Townley and LeSueur. When, in May 1919, the *North Dakota Leader* accused the friend's bank of selling out agrarians, Durocher sent Townley an angry letter disassociating himself from

the League. Anthony Walton—one of the first farmers to join the NPL in
North Dakota—joined him. A year later, Durocher claimed that Townley ran
the organization as a "dictatorship." By 1921, Durocher had joined with Job
Wells Brinton in a public lecture campaign against Townley.[42]

All of these defections formed the context for the most significant—and
public—break with the NPL. It came in North Dakota in March and April
1919. Even as the League's control of state government led to the implementation of its entire program, the organization lost three of its own endorsed
and elected politicians. William Langer, attorney general, broke with the
NPL over concerns about the United Consumers Stores Company, the Scandinavian American Bank, and Townley's refusal to relinquish power. Carl
Kositzky, state auditor, rejected much of the League's 1919 program during his
testimony on tax increases during the 1919 legislative session. Thomas Hall,
secretary of state, the only statewide incumbent endorsed by the NPL in 1916,
took issue with the tactics used by Townley and his lieutenants to guide the
legislature. A small number of NPL-affiliated legislators joined them.[43]

Kositzky departed the NPL in early March after Townley damned him for
his intransigence on the floor of the house of representatives. Within days,
the state auditor dedicated himself to defeating the League. He began debating NPL lecturers in small towns across North Dakota. Langer, frustrated by
what he saw as Townley's interference, grew more convinced of the leader's
corruption as well as the wrongheadedness of political control of the Industrial Commission. Early in April 1919, Langer privately assured a farmer that
despite concerns, "the original industrial program of the League is all right."
By the end of that month, he issued a public letter claiming that Townley remained "greedy for power, hungry for money, self-indulgent in [his] whims,
and with a mighty hate for all honest men who dare to counsel moderation." After gauging public reaction to Kositzky and Langer, Hall joined the
revolt.[44]

These high-profile partings fueled an intense debate. Many small businessmen applauded their departure. One applauded Langer's "Noble Challenge
to that Hypnotizer of our Blind population—the farmers—who followed
him like their own sheep." Langer's defection, in particular, convinced some
agrarians that they could no longer support the NPL. T. T. Jorstad, secretary
of a farmer's club outside of Harvey, North Dakota, told Langer that he applauded the League for its "economic reforms" but shared similar worries
about the "autocratic abuse of power, even if it appears in the garb of would
be deliverers from BIG BIZ."[45]

Even loyal Leaguers grew worried about what one called the "autocratic
elm club wielded over the heads of the legislators" by Townley. Many years

later, farmer Jess Joiner recalled that "in them days they kind of insisted you
vote the way they tell you," though "a lot of people didn't like that too well . . .
especially after they got on their feet a little bit." By 1920, a reporter for the
Nation interviewed farmers in North Dakota and Minnesota. Many told him
that "the League has grown to a point where control can no longer be so
centralized." Everywhere the journalist found "the stirrings of rank-and-file
democracy . . . in the Nonpartisan League."[46]

Even so, many agrarians remained convinced that Arthur Townley was
"one of the world's greatest, if not the greatest present day statesmen." North
Dakota State Insurance Commissioner S. A. Olsness—a committed Leaguer—
told one doubting friend that "no organization is perfect because no man
is perfect." In fact, he suggested that Langer's dissent (and that of others)
stemmed from "jealousy of power and ambition to be first and 'biggest cock
on the walk.'" The charges that flew among League leaders and politicians
created a morass that confused those less familiar with the details. Many members were not sure whom to believe. Mounting dissent—spurred by Townley's
unrelenting insistence on control of the organization—sprang from the confusion. Alongside efforts to force the League out of the primaries, this foment
fostered the gradual fragmentation of the NPL.[47]

*

Party politicians did whatever they could to further that fragmentation—
mostly by changing primary laws. A broad impulse for popular government
and direct democracy transformed state elections in the early 1900s. Growing
out of antimonopolistic movements, these Progressive Era reforms set the
stage for the emergence of the NPL. The open primary, in particular, made
it possible for voters—not politicians—to select a party's candidates. The
League depended on the primary to empower voters and sidestep political
parties. It entered them without any intent of engaging a party's machinery
or processes. It merely sought to use parties and primaries alike as an electoral vehicle.[48]

But election rules could be changed. The major parties used the rise of
the NPL to point out perceived problems with popular participation in the
political process. In an effort to defeat the movement, established politicians
tried to modify or even eliminate the open primary. In the states where the
League did not seize political control in 1918, officeholders worked together
across party lines to force the NPL to operate more like an easily defeated
third party. Leaguers fought them hard. They hoped not only to preserve
their own electoral advantage but also to ensure that democratic reforms did
not disappear.[49]

The adoption of open and direct primaries in most states in the early 1900s made the Nonpartisan League imaginable. This electoral innovation offered an avenue for nonpartisanship. A nonparty organization could assemble and organize voters around particular issues and then concentrate them at the polls to nominate the candidates that suited them. As NPL leader William Lemke later put it, the primary meant that, with careful strategizing, "the party machinery of both parties belong[ed] to the people who do the voting" instead of party leaders.[50]

Those who founded and led the League acknowledged the significance of a nonpartisan approach to the open primary as a central vehicle for their success. In 1918, Ray McKaig told a newspaper editor in Iowa that the organization used the primary "machinery for the good of the people." Indeed, he noted, "The primary system of those various northwestern states makes it possible for us." Identifying and backing "a nonpartisan ticket in the present parties" offered an avenue for electoral success that avoided the well-known limitations of third parties. McKaig reminded the Iowan that "the political thugs are waiting with stuffed legal clubs to kill any independent third party movement."[51]

But the law could also be used to thwart the primary-centered approach of the NPL. Frightened by the establishment of state-run industries in North Dakota and the League's control of government there, party politicians across the West sprang into action. They immediately began proposing modifications to primary law in states where the NPL threatened their hold on power. It proved the most effective way to reaffirm partisanship and defeat the League.[52]

In South Dakota, Richard O. Richards, a Republican Party insider, recognized the NPL's dependence on the primary earlier than most. He spent his entire career leading the forces agitating for a direct primary in that state. His elaborate and complicated solution—unlike any other in the country—required party affiliation, endorsements for party officials, formal recognition of intraparty factions, and party platform proposals. The state adopted it in 1912. Unhappy voters then rescinded Richards's closed direct primary in 1915. Soon-to-be governor Peter Norbeck, a rising star in the Republican Party, successfully proposed a more straightforward open primary law. Richards, however, continued to tweak his version and bring new proposals to the legislature.[53]

In 1916, with further reform in mind, Richards directly sought the primary-dependent NPL's support. He used his connections to help the League purchase automobiles for its organizers in South Dakota. Richards hoped to gain the NPL's backing for his version of the primary law. But Leaguers recognized that Richards's party-centric primary proposal would make it more difficult for the NPL to win in South Dakota. Even as the League took advantage of his connections for cars, it rejected Richards's offer.[54]

Rebuffed by the NPL, Richards turned to the organization's enemies. Minnesota governor J. A. A. Burnquist met with Richards several times in late 1917 and early 1918. Convinced that the League would sweep Minnesota's primaries in 1918, Burnquist considered calling a special session of the legislature to adopt Richards's unique primary law "in the hopes this would be the means of defeating them there." Back in South Dakota, now governor Peter Norbeck—another League opponent—saw similar potential in Richards's primary law. He threw his support behind Richards, and in November 1918, the open primary became closed. This made it impossible for the League to take over the Republican Party through the primaries. NPL organizers in South Dakota could do little more than "act as a separate political party."[55]

Politicians in Idaho took a different tack. There, the NPL took over the Democratic Party in the 1918 primaries but fell short in the general election. Democratic stalwarts, angry over the seizure of their party by farmers, turned to their Republican opponents for help. Shared interest in holding power trumped partisanship. Creating an unusual coalition in the statehouse, anti-League representatives from both parties even met in nightly secret caucuses. They defeated every NPL legislative proposal and decided to eliminate Idaho's open primary. Over the protests of both of Idaho's US senators, the legislature and governor did exactly that. The state became the first in the country to return to a convention system for all party nominations. The new legislation even insisted that candidates for office show sworn affidavits attesting to their standing in a party. This forced the Idaho NPL into more conventional third-party politics.[56]

Other states followed a similar course. Even progressive politicians who had championed the open primary in years past faced intense pressure to change their stance. Attempts to alter or repeal the primary in Colorado, Kansas, and Minnesota ultimately failed to reach the governor's desk, largely because of the opposition of longtime primary supporters. But during the 1919 legislative session in Nebraska, politicians eliminated the open primary for statewide offices below the rank of governor. Meanwhile, Democrats and Republicans in Montana came together to reinstate the convention system.[57]

The League responded by marshaling a massive petition effort to refer the new law to the people of both states. In Montana, NPL organizers used posters, picnics, and red, white, and blue automobile stickers with the slogan

"Save the Primary" to reach beyond the League's membership for support. Emphasizing the need to preserve a more democratic election process, Leaguers looked to reinvigorate their own rank and file as well as attract support from nonmembers. Besides drawing in labor unions, the NPL peeled off reform-minded progressives from both parties that otherwise distrusted the League. They recognized that the proposed change in election law represented "the death of popular government in Montana." Leaguers in Nebraska followed the Montanans' lead. The NPL's legal advisor in Nebraska, Christian A. Sorensen, turned to the state's leading progressives as well as labor unions for ideas and aid.[58]

In both states, the effort proved successful. Turning out large numbers of petitioners, Leaguers succeeded in putting the new and restrictive primary laws on the 1920 election ballot. Politicians in Montana countered the NPL effort by calling a special session in September 1919. Under the public guise of offering relief to drought-stricken farmers, they again approved a closed primary that forced voters to use a straight party ticket. To ensure success, they declared the law petition-proof through the unprecedented invocation of a statewide emergency allowed by the state constitution. Eventually, courts overturned the latter injunction. Even as the *New York Times* decried "the easy hospitality of the direct primary system" and the "alien and hostile" nature of the NPL, in November 1920, voters affirmed the earlier, open primary laws in Nebraska and Montana alike.[59]

Further efforts to modify primary law in order to keep the NPL from taking control failed—except in Minnesota. In January 1921, Governor J. A. O. Preus called on the legislature to change the primary law. He insisted that the NPL made a "mockery of the oaths of allegiance voluntarily taken by all candidates at the primary." Careful to not "make excellent League propaganda," state legislators made two crucial changes to election law. First, conventions would identify endorsed primary candidates for each party. Second, anyone defeated at the primaries could not run again in the general election. These changes cemented the NPL's future in Minnesota. From then on, the NPL needed to operate as a third party in the North Star State.[60]

The push to protect partisanship even emerged at the federal level. In late 1921, the newly elected president Warren G. Harding appreciated the need for "popular Government" but only "through political parties." In his first annual message to Congress, the new president kept the NPL in mind when he argued that "divisions along party lines" provided "vastly greater security" and resulted in "much larger and prompter accomplishment" than "to divide geographically, or according to pursuits, or personal following." By positing

FIGURE 21. This advertisement from the *Montana Nonpartisan*, published May 17, 1919, shows the great lengths the NPL went to in the effort to protect direct democracy. (Library of Congress, Chronicling America: Historic American Newspapers)

partisanship as the heart of a democracy, Republicans and Democrats alike looked to secure their own position and kill the League's chances for electoral victory.[61]

All told, changes to primary law worked both for and against the NPL. In every state, the attacks on direct democracy infused new life into the League. Working alongside laborers and progressives, the NPL skillfully framed attempts to eliminate or alter open primaries as an outright attack on popular government. Even as it drained away precious time and resources, this confirmed the organization's own commitment to democratic processes and outcomes in elections.

Yet in three of the four states with the largest League presence—Idaho, South Dakota, and Minnesota—the primary no longer served as an avenue to power. Indeed, parties redoubled their efforts to exclude any other form of political organizing for elections. For a movement described by one commentator as "outside of and wholly dissociated from any party or parties," this forced a retreat from the League's primary tactical innovation. While the NPL continued to put forward candidates in the primaries in states such as Montana, Nebraska, Kansas, and Wisconsin, Leaguers were left with limited choices in the states they most prized. Only as a third party—long acknowledged as a problematic path to power—could the NPL succeed in Idaho, Minnesota, and South Dakota. These two different electoral tacks required profoundly different tactics. After 1919, differences in state primary laws confounded any unified national effort.[62]

*

Ideological fracases fueled more problems for the NPL. League opponents took advantage of growing concerns about the Communist triumph in Russia in 1917. Across the United States, many worried about a similar revolution at home. Throughout 1919, economic dislocation, racial violence, and a newly ascendant labor movement fed those worries. Looming unrest seemed to be around every corner. The ensuing Red Scare channeled these concerns into a confrontation with perceived antipatriotic subversion. For many, loudly proclaimed anti-Bolshevism became a useful political credential. Enemies of the NPL quickly dubbed affiliated agrarians as Bolshevists. This charge sustained the swirl of controversy that enveloped the League.[63]

In Kansas, the firmly anti-NPL Farmers Union president Maurice McAuliffe questioned the League's loyalty throughout World War I. Even before the conflict ended, he characterized the NPL as revolutionary. In his mind, "the Bolshevists of North Dakota" represented a real threat. To promote the parallel, McAuliffe insisted that the League's program in its home state would lead to "state ownership of all property." Throughout late 1918 and 1919, the League faced analogous charges from businessmen, politicians, and other agrarians alike.[64]

In Minnesota, the Commission of Public Safety—which viciously attacked the League during wartime—lost its authority in early 1919 but continued to meet. Bankers and businessmen turned to a detective agency in Minneapolis. Infiltrating the League, the agency's employees sought evidence of direct connections between the NPL and Bolshevists or members of the radical Industrial Workers of the World (IWW). Instead, they found Leaguers carefully avoiding any such links. In one case, the League privately dismissed an organizer previously affiliated with the IWW "because his connection was detrimental" to the NPL.[65]

Despite efforts by the League to protect itself, the most forceful opposition to the NPL's supposed Bolshevism came from the American Legion. The new veterans organization called on members "to foster one hundred percent Americanism." Drawing on the authority of their own World War I service, these veterans became the unofficial arbiters of national loyalty in hundreds of small towns across rural America. Despite the Legion's firmly nonpartisan stance toward politics, the *New York Times* noted its "pronouncedly anti-Bolshevist" stance. At its founding convention in November 1919 in Minneapolis, the Legion hotly debated an anti-NPL resolution, but eventually tabled the matter over resistance from North Dakota veterans. Local posts elsewhere took the matter into their own hands. They generally envisioned the League as un-American.[66]

In response, the NPL accused the veterans organization of becoming the tool of party politicians. In September 1919, a Legion post forbade the band contracted to play to an audience of two thousand farmers at a League picnic in Lake Madison, South Dakota, to attend. The NPL insisted that Governor Peter Norbeck "used" the Legion to encourage the "terrorism, intimidation and persecution" of Leaguers. To be sure, established politicians embraced the Legion as an influential interest group. The veterans' political use of their sacrifice and patriotism—as well as their sway in Washington, DC—made them a crucial ally.[67]

Whenever it could, the League solicited the support of returned soldiers. In the *Nonpartisan Leader*, reporters pointed to the large number of veterans employed by North Dakota's NPL-led government. In July 1919, veteran W. H. Metcalf of Malad City, Idaho, trumpeted his support for the League. He told the *Idaho Leader*'s readers that he organized for the NPL in order to create "not a sharing of wealth, but a sharing of opportunities to make a home and a decent living." A month later, in Nebraska, veteran George Hornby publicly decried "the pseudo-democracy" of "the autocratic system" that left laborers and farmers on an uneven playing field. His dissatisfaction with "the type of democracy" veterans fought for drove him to organize for the NPL.[68]

Recognizing the power of such appeals, the League promoted an alternative veterans organization. Based in Minneapolis—and sustained by the NPL—the World War Veterans hoped to be a more sympathetic version of the American Legion. Minnesotan George H. Mallon, a trade unionist and Medal of Honor winner, became the most prominent member. League newspapers everywhere touted the need for NPL-affiliated veterans to join the little-known group. The American Legion largely ignored the group's anemic organizing efforts. Threatened only by the World War Veterans ideology— seen as less than "one hundred percent Americanism"—the Legion argued that it confirmed the NPL's lack of patriotism.[69]

In states with a strong League presence, the Legion and other opponents did little to openly intimidate the NPL. The League's passage of a favorable veterans bill in North Dakota, for instance, stemmed discontent in that state. In many rural Minnesota posts, returning soldiers joined both the Legion and the League. In such places, political and legal options better served those fearful of a "Bolshevistic" NPL.[70]

But the League found itself a victim of Legion Red-baiting elsewhere. In August 1919, rumors of a red flag on an automobile convinced returned soldiers to head an anti-NPL mob in Beatrice, Nebraska. The three-hundred-person crowd broke up a farmers' meeting and physically assaulted the League's attorney and two of "the d—n farmers." Despite pleas to local police, authorities provided no protection. Two months later, in Hartford, South Dakota, "hoodlums and rough-necks" fired guns and sent "a volley of missiles . . . through the windows of a hall" hosting an NPL assembly. Just a few days later, members of the American Legion interrupted a League gathering in Monroe, South Dakota, threatening to lynch any organizers and demanding an apology for the group's disloyalty. Local law enforcement did nothing. The next week, the recalcitrant Leaguers held a second meeting, signing up many local farmers—including two "REAL soldiers."[71]

The violence soon spread south and west. In November 1919, a mob convinced of NPL disloyalty seized a speaker at gunpoint during a League meeting in Stafford, Kansas. Others stuck around to throw "pieces of 2 x 4s, iron bars, and other miss[i]les at the men on the stage." The intensity of the Legion's animosity grew during the summer of 1920. Members forcibly broke up NPL gatherings in Macksville, Ensign, Dodge City, Kingsley, and Hoisington, Kansas. Further afield, a "delegation of ex-servicemen" in Walla Walla,

Washington, accused an NPL organizer of insulting the American Legion, forced him out of town, and insisted that he not return.[72]

Ellinswood, Kansas, saw the most infamous incident. In June 1920, NPL speaker Walter Thomas Mills and Milton Amos, chair of the League's efforts in Kansas, arrived there for an NPL picnic. Quickly seized by "300 men in the United States army uniform," assaulted with fists and clubs, driven to nearby Great Bend, marched to a local stockyard, locked in a pen, and egged by an angry crowd, the men barely survived. One week later, Leaguers published a missive—dated before the NPL picnic—on Great Bend's American Legion chapter letterhead declaring that "the Non Partisan League demonstration would not be tolerated at Ellinswood." Marching on the state capital for redress, Mills and Amos displayed their egg-stained clothing and demanded protection from the governor. The two Leaguers insisted that "every American Legion post in Kansas will pass resolutions against the action taken at Great Bend" because "it cannot remain a national organization and continue to recognize such unlawful action." In response, one member of the Legion bragged of his involvement.[73]

Much of the intensity of anti-League sentiment in Kansas's American Legion stemmed from a renewed effort by NPL opponent Phillip Zimmerman. When the Kansas State Council of Defense disbanded in December 1918, Zimmerman found himself fighting the NPL alone. His self-funded "Anti-Bolshevik Campaign" took up the fight against the League. By September 1919, Zimmerman ran out of money. He even stood to lose his home in Lindsborg. Desperate to raise funds, he sent out one last appeal for money in early October. This time he garnered the full attention of Governor Henry Allen.[74]

Allen, elected to the post in 1918, already took a dim view of the NPL. As the editor of the *Wichita Beacon* during the war, he generally opposed the organization. Just a few months into his tenure as governor, Allen queried his friend and fellow Republican South Dakota governor Peter Norbeck about the NPL. Norbeck assured him that the League remained dangerous. Then the summer of 1919 saw League organizers active all over Kansas. Allen took notice. Even before the NPL publicly revealed its intent to intensify the enrollment efforts in Kansas, the governor began to quietly support Zimmerman. His advisors drew up a plan to organize the state's businessmen in support of the anti-NPL crusade.[75]

Trying to cultivate a new infusion of backers and cash—and sensing the governor's interest—Zimmerman told Allen of his effort to "line up the American Legion against our favorite enemies in Kansas." The longtime NPL opponent believed that with "the Legion . . . concentrating their fire on the

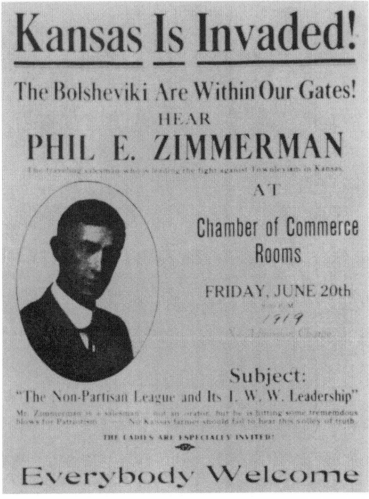

FIGURE 22. Phillip Zimmerman's anti-NPL campaign in Kansas in 1919 tried to connect the league to Bolshevism. (Kansas State Historical Society, JK.C 1919 *1)

red-flaggers," the League would ultimately fail. At the very moment he finally began to attract allies, however, Zimmerman could no longer support his family. Unable to raise money, he decided to accept a long-standing offer from the Greater Iowa Association to coordinate its work against the NPL.[76]

Despite this defeat, Zimmerman ultimately succeeded. At least one advisor to Governor Allen believed that Zimmerman had "done more to head off radicalism and Townleyism in Kansas than any ten men have done in any of the other league-infested . . . states." In the Sunflower State, the American Legion responded to Zimmerman's repeated accusations by targeting the

League. Meanwhile, Allen's fear of the NPL's potential challenge in the 1920 election season led him to tacitly support this repression, which continued all summer.[77]

Unlike Zimmerman, Theodore Nelson—leader of North Dakota's anti-NPL organization, the Independent Voters Association (IVA)—found plenty of funding. In the wake of the pathbreaking 1919 legislative session, the IVA attracted more attention than ever. Emotions ran high. Bitter businessmen and establishment politicians looked for any way to rid the state of the NPL. One, Oscar Sorlie, publicly joked that he left issues of the *Nonpartisan Leader* in the outhouse in lieu of toilet paper. His mother-in-law, though, "said she had never been troubled with the piles" until he did so. She "asked [him] kindly not to leave it there anymore." Vitriol became a common feature of political discourse across the state.[78]

IVA leaders knew, however, that money and allies—not anger—fueled successful political campaigns. As a membership-based political organization, it charged twenty-five dollars for a two-year affiliation. An excellent organizer with extensive experience in the state's equity movement, Nelson effectively targeted North Dakotans disheartened by League advances. He also reached out to fellow Norwegian American Nicolay Grevstad. Grevstad advised Minnesota senator Knute Nelson, served on Minnesota's Commission of Public Safety, and worked as the national Republican Party's liaison with the foreign-language immigrant press. Grevstad quickly lobbied the National Republican Committee for $10,000 for the IVA's fight.[79]

Finally, the IVA garnered the support of the two major political parties in North Dakota. In February 1920, Theodore Nelson adroitly convinced anti-NPL politicians in both the Democratic Party and the Republican Party to join together to back the IVA. Insisting that there "must be cooperation between the Republicans and Democrats opposed to Socialists," one prominent Republican politician agreed with many who believed fusion to be the only option. In April 1920, the three groups formalized their alliance. Nelson and the IVA would coordinate the anti-NPL campaign. Equally important, both parties agreed to raise money for the effort. With funding secure, the IVA wisely began surveying the state's population to determine the number of potentially sympathetic voters. It also began publishing a weekly newspaper—the *Independent*—as well as a series of pamphlets that linked the League to Bolshevism.[80]

As for more vicious responses to NPL rule, the League's hold on North Dakota's government and the large presence of its members in the state's American Legion ensured that veterans did not engage in extralegal repression. Nonetheless, Red-baiting remained a favored tactic of League opponents. In

October 1919, former state auditor Carl Kositzky began publishing incendiary attacks on the League under the guise of a monthly journal titled the *Red Flame*. The inflammatory charges it hurled against the NPL—Sovietism, free love, revolution, and tyranny—defined the outer boundaries of Red-baiting. Sensational and seditious, the *Red Flame* pegged the NPL as an antidemocratic and un-American threat to the nation. In 1920, the NPL-controlled legislature tried to protect the League by passing a law prohibiting the display of red flags.[81]

The *Red Flame*'s effort to associate the NPL with Bolshevism did more to smear the League outside North Dakota than it did inside the state. Opponents elsewhere picked up and reprinted its columns and claims. The journal's unidentified publishers touted that "loyalist organizations all over the country are making use of its material." But seditious headlines and controversial charges did not spur the growth of the anti-League forces in North Dakota. Careful organizing by the Independent Voters Association did. Comparatively tame, the IVA fed on the growing dissent within the NPL and cobbled together a broad coalition of League opponents in North Dakota. In the state that seemed to embody League success, the rapid organization of otherwise scattered opposition seized on the ongoing fragmentation of the NPL.[82]

✳

Red-baiting and mob violence did not define direct responses to the emerging Nonpartisan League in Saskatchewan and Alberta. Instead, just as Canadian farmers were the first to import the NPL idea from North Dakota, those north of the border led the way into the League's future—failure or fusion. Torn apart by controversy over democracy within the organization, the Saskatchewan League fell apart in 1917. A brief rebirth of the NPL there in 1920 did not take hold. But in Alberta, the League pushed the much larger United Farmers of Alberta (UFA) to embrace electoral politics. As a result, the Alberta NPL dissolved into the UFA. The comparative absence of anti-Red politics pointed at the League in western Canada made success more likely. That success, however, came through influence rather than dramatic membership growth.[83]

Electoral failure in June 1917 shattered the young Saskatchewan NPL. Disheartened agrarians vented their anger on each other. The organization's July 1917 convention included competing charges of disloyalty and division. After raucous debate, the body expelled a number of members—including a prominent leader of the large Saskatchewan Grain Growers' Association. Then questions about internal governance emerged. Vesting most power in the executive—following the North Dakota model—left the organization open to criticism. One Leaguer declared that "though the league was supposed to be democratic . . . its constitution was strongly undemocratic in almost every particular." The convention quickly agreed. It scrapped the old structure.[84]

A new executive committee took power and rewrote the organization's constitution. But a renewal of democracy did little to ease the group's woes. The damage had been done. The Saskatchewan League had only $392 on hand, besides its thirteen automobiles, and the "scarcity of funds" forced the fledgling movement's newspaper to be "greatly reduced in size." Without allies in bigger farm organizations, it soon nearly disappeared altogether. The infighting destroyed the group's remaining credibility.[85]

In late 1919, agrarians in southern Saskatchewan called League members to Swift Current for a convention. They hoped to jump-start the organization. Soliciting memberships, holding meetings, and publishing at least one pamphlet, they again mimicked efforts south of the border. Organizers focused their work on Ukrainian communities. The leader of the province's Liberal Party, Premier William Martin, kept close tabs on the emerging group. Yet within months, the League again fell apart. The expansion effort came up against jaded farmers involved in the previous, failed incarnation of the NPL. In Saskatchewan, the dream of organized agrarians in a nonpartisan movement died.[86]

Alberta's Nonpartisan League worked through 1918 to show its relevance and to avoid alienating the much larger United Farmers of Alberta. The wartime Union government in Canada—which brought the major parties together in a patriotic coalition—sunk the League in late 1917's federal elections but created an opportunity for the province's farmers to justify their position. Firmly committed to the elimination of party politics, Alberta's League pointed to the new national coalition as a "plutocratic organization" that nonetheless "levelled the old party fortifications to a considerable degree." This meant "the way has been cleared for the advance of the League," especially "for the attainment of our fundamental economic reforms."[87]

"Well-known to nearly all the farmers in the province," the NPL attracted new members and raised money by democratizing itself. At its March 1918 annual convention—timed to coincide with the UFA's annual convention—it created local committees that coordinated their work with an elected executive board. This gave the membership more power. It also helped to align the League alongside the apolitical but similarly organized UFA. As the membership of both the two overlapped, this proved crucial.[88]

The Alberta NPL's membership soon began calling on the UFA to enter electoral politics. Anxious to avoid a fragmented farmers movement, John Glambeck, an ardent Leaguer and prominent UFA member from Milo,

Alberta, believed that "alongside our splendid economic Union"—the UFA—
"we must have political power." That meant electing "men from our own
ranks." Even as these pleas grew in number, the League earned a reputation
for thoughtful and responsive political work. Much-admired provincial legislators Louise McKinney and James Weir (both prominent UFA members)
cemented the Alberta League's good standing with the province's farmers.
Though it disagreed with the League's politics, the *Calgary News Telegram*, for
instance, editorialized that the "much-derided organization" did a better job
of serving democracy and the people than the established parties.[89]

All this made the League more appealing to the province's agrarians. As
the League expanded throughout 1918, it pushed the United Farmers of Alberta to participate in electoral politics. That summer, the UFA firmly reminded the *Alberta Non-Partisan*'s editor that the group refused to endorse
the Alberta NPL or any political party. Meanwhile, the NPL committed itself
to moderation. It praised the UFA in editorials and continued to draw in
prominent UFA leaders. Even so, the League gently insisted that "the sooner
these movements learn to practice between themselves the cooperation they
preach, the better it will be for both."[90]

The Alberta NPL also avoided alienating the UFA with perceived radicalism. It tried to distance itself from the NPL in the United States, awash
in controversy. The *Alberta Non-Partisan*'s editor, John Ford, privately noted
significant differences between his paper and the *Nonpartisan Leader*. "To any
student of politics and economics," Ford believed his paper was better. The
main organ of the League south of the border remained "much more blatant
in regard to 'Big Biz' and not so very consistent." The Albertans also decided
that any invitation to Arthur Townley to speak in Alberta would spark nothing but trouble.[91]

Townley's interest in such an invitation remains unclear. In the United
States, the Nonpartisan League resolutely ignored parallel efforts in Canada.
Even after John Ford wrote from Calgary to the Saint Paul office in April 1918
asking for literature and pamphlets, the American NPL seemed to pay little
attention. The Saint Paul office sent a variety of materials but did not follow
up with a query of their own. As late as 1921, the NPL in the United States assured a Canadian investigator that it had "no organization in Canada, never
had any in Canada, is not now trying to promote any organization in Canada, and has no plans for invading the Dominion." Not even the *Nonpartisan
Leader*'s "several thousand subscribers in Western Canada" could pique the
American movement's interest.[92]

Alberta's Leaguers continued their work regardless. The now nearly six thousand farmers belonging to both the UFA and the NPL—one-third of the larger organization—loudly demanded that the former enter politics. Immediately before the UFA's annual convention in January 1919, William Irvine, the lecturer and secretary of the Alberta NPL, called on the UFA to formally endorse the League. Henry Wise Wood, the UFA's leader, distrusted Irvine and the NPL and hoped to keep his organization out of politics. Nonetheless, he also needed to avoid looming fissures among Alberta's farmers.[93]

At the convention, Wood and his allies adroitly changed election rules to ensure that the NPL minority could not seize a sizable number of seats on the executive board. Then, calling for "pure democracy in action," they committed the United Farmers and Farm Women of Alberta to federal politics. Instead of forming a party or embracing the League, Wood insisted that the UFA work from the grassroots. Each UFA local would independently decide how to enter politics. James Weir, a League legislator, embraced the plan and threw the NPL's support behind the proposal. He addressed the assembled, believing that the plan was "big enough and wide enough and good enough for any democrat in this or any other country." Though it fell short of NPL demands, the UFA's embrace of direct political action gave the League the opportunity to utilize its newly organized members at the grassroots level and tout its own authentic commitment to democracy.[94]

Still worried that the UFA's "planless plan which spells defeat" might harm the province's farm movement, in the months that followed, Irvine pushed for the creation of a joint committee made up of representatives from both groups. The rank and file of both organizations agreed. At a meeting in early May 1919, Irvine and two other NPL representatives brought five propositions to UFA leaders. They represented "the fundamental principles of the Nonpartisan movement." At the direction of Wood, the UFA agreed to support a "business government" centered on nonpartisanship, provincial-level campaigning, "keeping the political direction separate from the commercial" work of the UFA, platforms devised by the members, and a farmer-funded "political department" that would keep outside influences from determining policy. The populist orientation of the NPL dovetailed with Wood's insistence that "all democratic organization begins at the bottom and works upward." It remained only for each UFA local to decide whether or not to adopt them.[95]

Believing "its great important work was to bring the farmers' organizations to the point of taking direct political action," the League excitedly celebrated its triumph. The NPL not only pushed the UFA into politics but also defined how it would become political. In the wake of this success, the

League's leaders called on its members to "amalgamate" with the UFA and to push for the five propositions—along with a sixth, making the *Alberta Non-Partisan* the official newspaper for farmer politics in the province—in the UFA local deliberations over the summer.[96]

At the first meeting in Medicine Hat, over one hundred farmer delegates stood alongside Henry Wise Wood and NPL president J. C. Buckley and agreed to the propositions. The NPL members reveled in the results. In a resolution that even Saint Paul's *Nonpartisan Leader* took note of, the UFA local formally recognized that "the Nonpartisan league has been very success-ful in maturing farmers' demand for independent action" and invited "our Nonpartisan members to throw in all their resources and influences with this body so as to preserve unity of action."[97]

When UFA delegates met in Macleod—the heart of the NPL insurgency—Louise McKinney and William Irvine backed Henry Wise Wood's call for democracy. The assembled approved every principle proposed by the joint committee. Only the effort to make the *Alberta Non-Partisan* the official or-gan of the movement failed. In Lethbridge, similar results led Irvine, a former minister, to pledge that "the Nonpartisan League was willing to sacrifice its own life and its identity in order that it might win eternal life." Noted suffrag-ist Irene Parlby called for the end of "machine politics" to hearty applause. At the meeting in Calgary, tensions emerged. League legislator James Weir put an end to sporadic infighting by pointing out the full-throated approval of the merger in the other districts. At the meeting in Red Deer, delegates con-fronted local members of the Liberal Party who proposed joint action with derisive laughter. Then they approved the propositions. A similar proposal at the UFA local in Wetaskiwin fell short as the delegates agreed to the NPL-UFA merger. Delegates meeting in Canrose, Wainright, and Edmonton also approved the union.[98]

In July 1919, UFA delegates from all the locals came together to build the proposed political department of the farmers movement. Irvine, satisfied with the results, declared it to be "as democratic as democracy knows how to make it." In a November 1919 special election in Cochrane, the UFA elected its first representative to the provincial legislature. It was just a taste of things to come. In the 1921 elections, the UFA seized control of the province.[99]

All told, League experiences in Saskatchewan and Alberta suggested the significance of democracy and flexibility for the broader NPL's future. Au-tocracy sowed suspicion, while democratic commitments provided a firm foundation for coalitions with others. Fusion—carefully handled—could trump factionalization. During the 1920 election season, the American NPL confronted similar issues. But League leaders in the United States ignored the Alberta NPL's achievements and chose different paths.

✳

Despite the troubles facing the NPL, thousands of farmers remained committed to fomenting political change. In January 1920, Charles Wooster, a former Populist state legislator and elderly Leaguer in Nebraska, reminded his neighbors why the NPL mattered. Wooster firmly believed "in a government of the people, by the people and for the people; and not in a government of the Garys, the Rockefellers, the Armours and the Morgans by the Garys, the Rockefellers, the Armours and the Morgans and for the Garys, the Rockefellers, the Armours and the Morgans." He called on fellow agrarians to "stand solidly together" in order to convince "wage earners, the people of small salaries, small-property holders and small business men of the cities and towns" to join together and create a "truly democratic government representing the great overwhelming mass of the people and not the mere shell of a democracy without the substance."[100]

Shared by many farmers, such sentiments powered the NPL as it approached the 1920 elections. One commentator suggested that despite growing complications and "bitterness against the League," members retained a "genuinely religious fervor." S. A. Olsness, a League official in North Dakota, admitted to a friend that "these principles and their success have become almost a religion with me." The intensification of this thinking among the rank and file kept discussions about democracy in the NPL at the fore. It also—finally—led to a reorganization of the League that many thought would give it a better chance at victory.[101]

In order to better tap into member energy for the election season, Townley finally agreed to substantive changes in the organization in December 1919. "A more inclusive form of organization" emerged from a national convention of League leaders. It created a structure that included permanent county committees. At that level, local farmers would make decisions. Equally important, the money collected from membership dues would be apportioned to fund the county, state, and national organizations alike. For the first time ever, money—the League's life blood—would be distributed and used democratically. Finally, Townley declared his commitment to "convert a following into a self-governing, self-actuating organization." He offered his resignation. This gesture resulted in a unanimous vote of confidence in Townley's leadership.[102]

A need to tailor NPL efforts to specific states—more than any newfound commitment to democracy on Townley's part—drove this shift. Farmers believed that "the League has grown to a point where control can no longer be so centralized." Moving power within the League from the national office to local committees and state managers allowed those with a more complete understanding of local conditions to make key decisions. Local variations mattered. The changes in primary laws ensured different electoral machineries. Efforts to ally with labor unions varied from state to state. League opponents utilized a bewildering array of strategies. The economic questions that drove members to join depended on the region of the country they lived in. As NPL leaders looked forward to the 1920 elections, they hoped that democratization would not only empower members but also provide a foundation for more political victories.[103]

Townley and others recognized that nothing less than the future of the NPL as a national force was at stake. Transcending its roots in the northern plains wheat economy and becoming a major player in agricultural politics depended on securing power in a state beyond North Dakota. Failing to do so would ensure further fragmentation. In July 1920, Townley assured members that "we're going into the hardest political battle in the history of America." Anxious to ensure that the NPL would not "fight a losing battle anymore," he called on Leaguers to throw everything they had into the 1920 elections.[104]

His call came amid rapidly falling prices for farm commodities. The end of World War I brought the products of European farmers back into global agricultural markets. Simultaneously, worldwide demand dropped. The falling value of crops directly affected the NPL and its members. Even as Townley called on members to focus their energies, the national office announced cutbacks. They noted that "membership fees will not sustain" the organization, "let alone leave anything for campaigns." The national edition of the *Nonpartisan Leader*, as well as the state newspapers, became biweeklies. With the elections just a few months away, the financial shortfall came at the worst time.[105]

Nonetheless, Townley and his advisers directed state managers to pursue three strategies for success. First, they tamped down all aspirations for a presence in national electoral politics. Second, they called for alliances with organized labor. Third, they insisted on the significance of the women's vote in the upcoming elections.

In January 1920, Leaguers in South Dakota nominated North Dakota governor Lynn Frazier to be on the state ballot as a candidate for the presidency. Frazier rejected the suggestion. Nonetheless, the idea gained traction. A few farmers even distributed lapel pins that said "Lynn J. Frazier For President." When the rank and file refused to let the idea die, League leaders flatly declared that the "time is not yet ripe" for an NPL-backed presidential campaign.[106]

Committed to nonpartisanship, the League also distanced itself from the organization of a national third party. When a broad coalition of dissatisfied politicians, farmers, progressive reformers, and laborers came together to create a Farmer-Labor Party in Chicago in July 1920, the League sent representatives but refused to join. It insisted that "the League has never and never will merge, amalgamate, or affiliate with any other organization or party." Even so, the South Dakota NPL—a formal political party because of that state's new primary law—broke with the national organization and embraced the effort. Tensions caused by increased local control began to emerge.[107]

Even so, building on tentative connections made during the 1918 election season in Minnesota, North Dakota, and Montana, the NPL reached out to organized labor. The national office strongly encouraged alliances with workers. Outside of North Dakota, agrarian voters alone simply could not secure electoral victories. In states where farmers did not make up a majority, electoral success hinged on finding like-minded allies. Most members agreed. "In the Nonpartisan League strongholds of the Northwest," one reporter found "farmers eager to unite with organized labor." Conveniently, the American Federation of Labor (AFL) instructed its members to engage in primary endorsements. This freed up trade unions to ally themselves with the League. One way or another, NPL-led farmer-labor coalitions emerged in North Dakota, Minnesota, Idaho, South Dakota, Washington, Montana, Kansas, Colorado, Nebraska, Wisconsin, and Texas.[108]

In North Dakota and Minnesota, organized labor had already created corollaries to the NPL by organizing Working People's Nonpartisan Leagues (WPNPLs). In other states, new models for collaboration emerged. The changes to primary laws in Idaho and South Dakota pushed laborers to ally themselves with the NPL in third parties in order to effect change. The sizable number of Leaguers in both states meant that the NPL drove the effort and did much to shape the results. Careful negotiations with the state federations of labor in both states ensured the stability of these new third parties.[109]

More tortured routes to a farmer-labor party emerged in Washington and Nebraska. In the wake of the failed general strike in Seattle in 1919, divisions emerged there about how to move forward. One combination of laborers and agrarians—largely made up of AFL and NPL members—wanted to pursue nonpartisan politics along the lines of the League model. Others—including the Washington State Grange master William Bouck, former Socialists, and Seattle's large shipyard unions—hoped to forge a new and independent party. Agreeing to disagree, the factions squabbled over the best route to pursue but put off any final decision until early 1920.[110]

In July 1920, farmers and workers finally met in separate but simultaneous conventions in Yakima to determine how best to proceed. Talks soon broke down. The NPL and one of the railway unions "refused to join . . . in forming a new party." Instead, sticking to the League's original tactics, they planned to run candidates in the Republican primaries that fall. Furthermore, the two organizations challenged their opponents by sponsoring a new Workers' Nonpartisan League to bring together laborers of every stripe who preferred nonpartisan action.[111]

That effort failed. Just a week later, the NPL reversed itself. Agreeing to support an independent party in statewide races and labor-backed contests in Puget Sound—as well as nonpartisan candidates for Republican primaries in eastern counties where the League felt strong—the farmers and workers patched together a compromise. To complete the confusion, Leaguers on the Cascade Slope led by Grange leader and NPL member William Bouck then broke with the state organization and decided to focus all their attention on the independent party. In the end, most of the rest of the NPL reluctantly followed them. However awkwardly, farmers and laborers in Washington State came together for the first time to nominate a shared slate of candidates in September 1920.[112]

Meanwhile, in Nebraska, the NPL suffered a serious reversal. Soon after the League endorsed Elmer Young for governor, a bout of influenza forced him to withdraw from the primaries. Left without a candidate for the state's top office, the League's chances in the contest collapsed. Days later, farmers decided to cast their lot with organized labor in a third-party campaign. Calling for a convention, the group joined with railway and other unions. Leading progressives also received invitations. A new platform and new candidates sprang from the May 1920 meeting. As part of the coalition, the NPL devoted itself to a unified independent platform in which farmer concerns proved secondary to labor issues.[113]

In Montana, Kansas, and Colorado, alliances emerged when laborers directly rejected the existing parties. As trade unionists in Montana formally condemned the AFL's Samuel Gompers "for reactionary policies," they pledged their support for the NPL. The Kansas State Federation of Labor invited NPL representatives to join in crafting a shared platform for the primaries. In Colorado, a special meeting in June 1920 saw the AFL, railway unions, and the League all agree to coordinate a nonpartisan effort targeting the state's Democratic Party. The combination proved powerful enough to spark a new opposition effort. Led by Denver lawyer Tyson Dines, businessmen joined together to form the United Americans. As a testament to the strength of the new farmer-labor coalition, they immediately began publishing anti-NPL pamphlets.[114]

Where the League struggled, laborers swept up the farmer's organization into new alliances. A fierce fight between the NPL and Wisconsin's Society of Equity wrecked efforts to grow the organization there. Furthermore, Victor Berger's powerful Socialist machine centered in Milwaukee made no secret of its distrust of the League. For years Berger hoped to bring farmers into the Socialist fold. He soon got his chance.[115]

By 1919, the League's national office saw little choice but to work with Berger. William Zumach, Berger's close associate, took over the Wisconsin NPL. Zumach told one confidant that "the Wisconsin Socialist organization is too well organized and too powerful to be swallowed up" by the NPL. Taking the Socialist Party's lead, the League joined with the Wisconsin Federation of Labor, the railroad unions, and progressives led by Robert La Follette, to form a new coalition in the Republican primaries.[116]

In Texas, the small group of farmers—largely former Socialists—who survived wartime repression reached out to local trade unions, especially in Houston. Finding a voice in the *Southland Farmer* (an agrarian paper based in the city), the alliance raised the visibility of the NPL in the Lone Star State and led to the League's first state convention there. Taking advantage of the state Federation of Labor's unhappiness with the Democratic Party, the two organizations coordinated their efforts. The fusion of unions and agrarians came up with "straight farmer labor candidates," but they did poorly in the July 1920 primaries.[117]

Working with labor unions (largely based in cities) helped the League to neuter the threat of a strong anti-NPL vote among women in cities throughout the West. As a new force in the electorate, women posed a special challenge to the League. Fears that "a large feminine vote will be cast in the towns, where opposition to the farmers' program is strong," even as "women of the country will . . . find it too inconvenient, to go to the polls," loomed large. Townley and others were right to be worried. In North Dakota, for instance, the Women's Christian Temperance Union, concerned that statewide prohibition might be overturned by initiative (a tactic made legal by the NPL), offered its public support to the Independent Voters Association.[118]

Readying for the 1920 electoral contests, League precinct meetings not only included women but also solicited their input. The *Colorado Leader* reminded farmers to "take your wives to the convention" because "they too can sign the petitions." In Nebraska, women at local precinct meetings were "cordially invited to take part in the deliberation." Furthermore, they could now be "elected as delegates to the county or state convention." Fully participating in the electoral aspects of civic work of rural communities for the first time, women proved crucial.[119]

Such a change did not come naturally. The inclusion of women in the political process, for instance, required frequent reminders that "we cannot win without the woman vote!" and that unless every farmer saw "that his wife votes . . . he will lose all chance of fair representation in our government." Nebraska's state NPL committee commanded farmers to "take the women to the polls. All the city women are going to vote. The country women must vote or else the city will have twice as many votes as the country." In Colorado, the NPL announced that "the women . . . know what the profiteer has done to the family pocketbook." In fact, they admitted, "We cannot win without the woman vote!" To help women who had never before cast a ballot, NPL auxiliaries in North Dakota devised "a little cardboard guide card" with "the League candidates names in capital letters." One farm wife remembered that "many were the women that reached down into their bosoms to pull out their card when they got behind the curtain at the polling place." All told, League-affiliated "women had a powerful pencil at the polls."[120]

✶

Entering the primary season, the NPL believed that its focus on state politics, farmer-labor alliances, and women voters would pay off. As returns rolled in from state after state, the results were mixed. Put to the test, NPL leaders began to worry about November.

Without a gubernatorial candidate in Nebraska, the League held low expectations for the primary there. In the wake of a three-day May blizzard followed by heavy rains, rural voters struggled to turn out. The NPL candidate for attorney general suffered a narrow loss. Nonetheless, farmers nominated six state senators and eighteen state representatives. In June, the League believed the Republican Party primary in Minnesota portended better things to come. But then the League's gubernatorial candidate—Henrik Shipstead, a Norwegian American dentist—fell eight thousand votes short, despite carrying twenty-four more counties than his predecessor in the 1918 primary. With the exception of one state supreme court justice, its other candidates for statewide office fell short as well. Only two of the five men the League nominated for the US House of Representatives won. Democrats voted in the Republican primary in droves. The defeat stung.[121]

The news from North Dakota later that month also left much to be desired. At first glance, the NPL retained its firm grip on the state. In the face of intense IVA opposition in the Republican primaries, Governor Lynn Frazier defeated the fiery former Leaguer William Langer. Most other NPL candidates held their seats. The IVA took only the secretary of state and state treasurer spots on the November ballot. But incumbent US Congressman John

Baer—the League's cartoonist—lost. Equally important, the NPL margins of victory shrunk dramatically. Though it seemed clear that the League would hold on to North Dakota, controversies over NPL-affiliated businesses and Townley's firm hand on the organization cut into the League's constituency.[122]

Montana proved to be another matter. There, on August 27, the NPL and its labor allies swept the Democratic Party primary. Burton Wheeler, a Butte-based lawyer and former US district attorney, beat out the incumbent lieutenant governor for the right to run for the top spot. The League candidates won every other statewide office by at least ten thousand votes. At the local level, "twenty-eight of the 51 counties in the state qualified as 100-per-cent League . . . nominating every League candidate for the legislature and for county office." A similar formula in Colorado tamed the establishment candidates in the Democratic Party primary there. James Collins, longtime Farmers Union president and NPL state executive member, took the governor's spot. Union activist Ed Anderson took the lieutenant governor spot. The farmer-labor alliance won three other statewide nominations. Furthermore, "League candidates for the Legislature were nominated in every county in which the farmers entered the contest." Given the Democratic Party's dominance in both states, this boded well for the League.[123]

In September, the League secured another victory, this time in Wisconsin. Its nominees won the governor's race, the lieutenant governor's race, the secretary of state's race, and three US Congress nominations. Yet in many ways, the wins proved hollow. Despite the League's claims of victory, more powerful groups in Wisconsin—including US Senator La Follette's allies and the Socialist Party—ultimately did more to ensure victory than the NPL.[124]

In three other states—Idaho, Washington, and South Dakota—the NPL did not enter the primaries. This gave the harried national staff time to make adjustments before the general elections in November. Opponents and adherents alike knew that the future of the organization depended on it.

Most significantly, the NPL decided to continue the fight for Minnesota, home of the NPL's headquarters. Seizure of the state would give the organization a huge boost. Working alongside organized labor, Townley and others formed an independent League-labor ticket to run against the Republican and Democratic primary winners. To further the effort, a long-planned daily newspaper—the *Minnesota Daily Star*—finally began publishing in August 1920. It emanated from a new building in the staunchly antilabor city, one that also housed the national and state NPL headquarters, the WPNPL offices, reporters for the *Minneapolis Labor Review*, and the *Nonpartisan Leader* staff. Putting farm and labor leaders together under one roof, the new headquarters firmed up the coalition for the coming fight.[125]

The NPL also became more sophisticated in its election planning. Instead of merely mobilizing as many citizens as possible, it engaged in a more systematic effort. For instance, as campaign preparations mounted, staffers worked to pinpoint the geographic location of the League's strength in states across the rural West. The detailed maps identified individual members and made it clear which counties could be counted on. The diagrams showed exactly where the NPL needed to do more to bring allied voters to the polls. In North Dakota, the League even distributed cards to collect demographic information—township, post office, NPL or IVA membership, religion, children, and nationality—on the state's voters.[126]

At the national level, both major parties feared the League. The Democrats tolerated the League takeover of the Colorado and Montana parties. The Republicans, however, still smarted from NPL inroads in North Dakota and Minnesota. Courting voters of all stripes, James M. Cox, the Democratic nominee, visited North Dakota and attacked "big business." A few days later, he noted only that "if the state government" of North Dakota "had been honest and square with the farmers," the NPL would not exist. In contrast, when Republican presidential candidate Warren G. Harding spoke on agriculture at the Minnesota State Fair, he blandly endorsed farm cooperatives but remained silent on the NPL.[127]

Former president William Howard Taft took a different tack. Touring North Dakota that summer, the staunch Republican backed establishment politicians in both parties who supported the IVA's attempt to rid the state of the NPL. Taft then condemned the League in three widely circulated columns in the *Philadelphia Public Ledger*. "Townleyism is socialism," he proclaimed, and it could only be defeated by "non-leaguers, without regard to party." This galvanized politicos across the West to work across party lines to defeat the NPL. Privately, Taft deemed the League's use of the open primary "unfair and fraudulent."[128]

Others engaged in more direct attacks. In Dillon, Montana, gubernatorial candidate Burton Wheeler remembered being forced to speak at a ranch outside of town after the town fathers banned any political meetings not approved by local Republican or Democratic party officials. Before he could begin, Wheeler saw "a phalanx of men, apparently white collared professional fellows" approach and cry "get a rope." Retreating to a boxcar on a nearby railroad siding north of town, Wheeler and a sympathetic farmhand faced down yet another mob. "When the posse drove up and started to open the door of the boxcar," Wheeler's "protector cocked his gun" and called out: "I'll shoot anyone full of lead who opens that door!" The crowd parted. As word

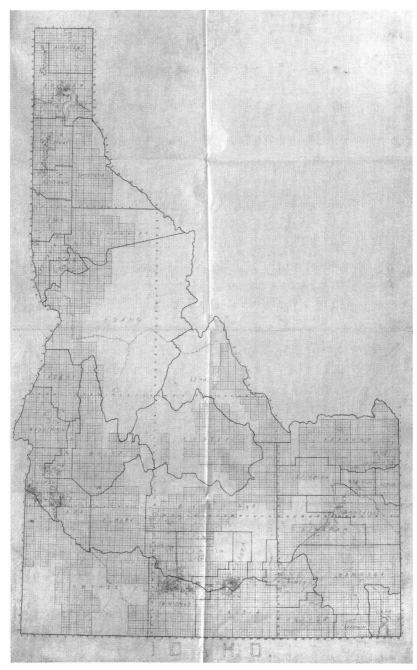

FIGURE 23. Planning for the 1920 election season, NPL staff began carefully locating their current members on detailed state maps. In Idaho, the Snake River valley and the Panhandle proved to be league strongholds. (Minnesota Historical Society, 146.D.6.2)

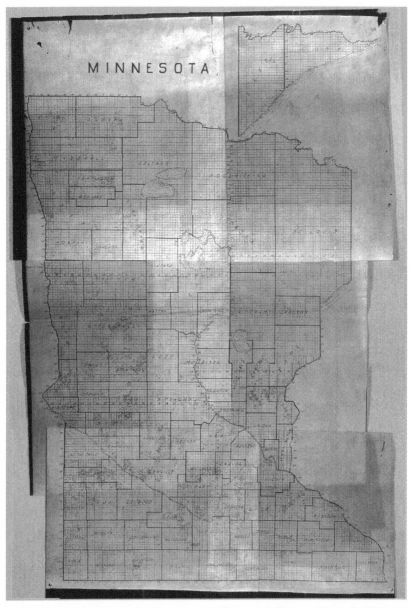

FIGURE 24. Another state membership map prepared by league staff in 1920, this one for Minnesota. Note the sizable presence of NPL members in the Red River and Minnesota River valleys. (Minnesota Historical Society, 146.D.6.2)

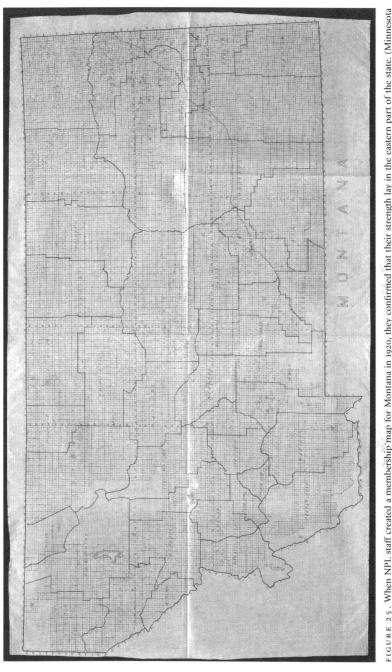

FIGURE 25. When NPL staff created a membership map for Montana in 1920, they confirmed that their strength lay in the eastern part of the state. (Minnesota Historical Society, 146.D.6.2)

of the incident spread, opposition newspapermen ridiculed Wheeler. They began calling him "Boxcar Burt."[129]

Finally, motivating the League's members proved an uphill battle. The gradual fragmentation of the NPL itself disheartened some. For others—especially in North Dakota—the League seemed to have already achieved its primary goals. Against all odds, the people finally received their state-owned mill, elevator, and bank. The IVA continued to assault the League but quietly admitted its support of the NPL's original program. Even Townley, at the dedication of the new mill and elevator in Grand Forks in January 1920, admitted to the crowd that seeing the cornerstone of the League's program come to fruition made him "feel as if my work was nearly done." More than a few farmers agreed.[130]

*

The November 1920 election results both confirmed the gradual fragmentation of the NPL and accelerated the process. Comparing the outcome to the 1918 elections for an Associated Press reporter, League staffer Oliver Morris bravely insisted that "we feel that the League has made wonderful progress in two years." Facts on the ground trumped this hopeful attitude. Indeed, a reporter for the *Nation* pointed out that "the disaster was more sweeping than either friend or foe had anticipated." In fact, "the Nonpartisan League, on the whole the most hopeful democratic movement in the Northwest for a generation, has gone down to defeat."[131]

Even though the NPL turned out enormous numbers at the polls—receiving 1,180,000 votes for its gubernatorial candidates, compared to 230,000 in 1918—the League's strategies for victory largely fell short. The refusal to engage in national politics gave both major parties an opportunity to oppose the League as they wished, state by state. Alliances with unions failed to give the NPL winning margins. The difference turned on the "failure of certain groups of women"—those in rural areas and working-class neighborhoods—"to register in the primaries and to vote." The organization's losses paralleled those of their agrarian members, who faced commodity prices that continued to spiral downward that fall.[132]

In Minnesota the defeat was complete. All but one candidate for major office—Oscar Keller, reelected to represent Saint Paul in the US House of Representatives—fell short. His election depended almost entirely on the union vote. To ensure a League loss, Democrats voted in huge numbers for J. A. O. Preus, the Republican candidate for governor. The NPL also failed to increase its representation in the statehouse. Nebraska saw similar results.

League candidates ran well behind major party candidates for statewide office. In the state legislature, the NPL remained a distinct minority, with only sixteen representatives in both houses.[133]

Montana and Colorado offered little more. Theodore Roosevelt's 1912 campaign manager and former US senator Joseph Dixon, like Wheeler, ran a campaign that drew on long-standing frustration over the Anaconda Company's control of Montana. Seemingly less radical than Wheeler—whom Anaconda-funded publicists painted as a friend of Bolshevism—the Republican drew enough fearful Democrats to his side to trump the League candidate by more than thirty-five thousand votes. When the legislature met again, fewer Leaguers would be seen in Helena. Farther south, the Republican ticket swept Colorado. Only six NPL representatives won election to the state legislature.[134]

Third-party efforts in Idaho, South Dakota, and Washington State offered only dead ends. In Idaho, the NPL farmer-labor coalition endorsed a number of Democratic candidates. All of them lost. Frank Gooding—former governor and Leaguer Ray McKaig's archenemy—won the US Senate contest. Longtime NPL opponent Peter Norbeck crushed League leader Tom Ayres in the battle for a US Senate seat in South Dakota. At the state level, all the NPL candidates went down to defeat, as only a handful of representatives persisted in Pierre. In Washington State, the awkward alliance between agrarians and workers posted better returns than the Democratic Party. But becoming the second-most powerful party in the state soothed few laborers or Leaguers. Just three NPL-endorsed candidates joined the state legislature.[135]

In the days following the election, it became clear that only Wisconsin and North Dakota could be counted as League victories. The NPL trumpeted its wins in those two states. Yet because the NPL did not even direct the effort of the League in Wisconsin, its endorsed candidates owed the NPL little. Trying to claim newly elected governor John Blaine as its own, the *Nonpartisan Leader* touted his farm ownership as well as his support of the initiative and referendum. Yet Blaine followed the lead of Robert La Follette, not Arthur Townley.

Voters in North Dakota approved five IVA-backed initiatives. Leaguers breathed easier only after the rural vote rolled in. The NPL not only narrowly held on to the governor's office and most state offices but also elected North Dakota State Agricultural College professor Edwin F. Ladd to the US Senate. Ladd, popular because of his studies that proved the large-scale cheating in grain grading in North Dakota, became the first NPL politician to reach that body. League leader William Lemke took over the state's attorney general's office. Losses in the legislature, however, loomed large. The League held on to the state senate by the slimmest of margins but lost the house of representatives to the IVA.[136]

Blaming poor results on the national Republican landslide, bad weather in the Upper Midwest, and the distance rural women had to travel to reach the polls, NPL leaders denied the depth of these defeats. But outsiders recognized that they further weakened Arthur Townley's position. Even before the election, the *Nation* reported that Townley seemed more about "putting over the Nonpartisan League . . . for the sake of winning than for the good results that victory may achieve." The inability to deliver elections further diminished his stature. For the most Leaguers, the reverses intensified the stakes—especially in North Dakota. Bitter personal attacks between the IVA and the NPL became the order of the day as state politics there stalled out in gridlock. Frustrated by fragmentation, the League faltered.[137]

✳

Within weeks, the League's leadership called on farmers to look for better things to come. "It all depends on the spirit that the League members show in the next two years," wrote the editor of the *Nonpartisan Leader*. That spirit seemed unbowed. Carl Thurow, an elderly Nebraskan, claimed that he had "lost everything through the animal will power of the special interests." In response, he wrote, he "joined the army of the Nonpartisan League" and used "my pen, my vote, and my suggestions" as weapons. He did not plan on putting them down anytime soon. Across multiple states, members of the Nonpartisan League redoubled their efforts to effect political change.[138]

Their commitment, however, could not overshadow growing concerns about the broader meaning of the 1920 electoral failure. Economic challenges—which only deepened as the weeks dragged on—provoked new fears about the future of farming. William Alexander, a Leaguer from Mona, Montana, described "the deadly work" done by "high interest and small crops" in his district. The neighbors that persisted did anything they could to "stave off starvation and mortgage foreclosure." Hard times made the NPL more appealing. But they also made it more difficult for farmers to financially support the League with paid memberships.[139]

Anxiety increased when data from the 1920 census suggested that more Americans lived in cities than in the country. Agrarians saw this as a demographic crisis. As former US congressman and League cartoonist John Baer put it, "Farming must be made a more profitable undertaking if our nation is to progress." A corollary issue—that the falling number of farmers reduced their effectiveness as a voting bloc—remained implicit.[140]

As an organization, the NPL responded to the election losses and to these growing crises with two initiatives. First, members demanded further institutional change, reducing the reach of the national office and putting more power in the hands of the rank and file. Second, with decreasing numbers of agrarians, recruiting and involving rural women in the movement took on a new significance. Every voter in country districts counted more than ever. Democratization of the organization, in this context, largely meant renewing the effort to enroll women.

Taking advantage of the lull between the 1920 and 1922 election seasons, the national NPL convention in Minneapolis focused on the League itself. Admittedly "warm discussions" there in March 1921 reoriented the organization. Reeling from election losses, the assembled charted a new path. Despite defeats, they publicly affirmed the leadership provided by Townley, Francis Wood, and William Lemke. They also excitedly pointed to the election of North Dakota's new US senator Edwin F. Ladd. Anxious to expand their power in the nation's capital—unlike Townley and his lieutenants—the delegates believed that Ladd's election meant that their "program [would] become, not merely a state but a national issue."[141]

More substantive changes involved decentralizing the League's internal structure. No longer would the NPL be, as one reporter described it, "an institution in which the few decide and the many follow." The farmers moved "the work of keeping the books and accounts" to the state NPL offices. Most of the membership dues followed. Delegates reduced the national NPL staff by two-thirds. They directed the *Nonpartisan Leader* to solicit subscriptions apart from memberships. Each state would hold a convention and elect "new state committees of farmers." Those new committees would "be in complete charge of the affairs of the League in the states."[142]

Arthur Townley lost control of the organization's money. The delegates also removed him from his role as chief decision maker and political strategist. Townley remained the titular head of the national committee, but now the real power rested with farmers in individual states. Even the *Nonpartisan Leader*, once the flagship publication of the organization, would need to fend for itself. Agrarians believed that state-level NPL and local farmer-owned papers offered better outlets for confronting state and local issues. The gathered Leaguers altogether rejected the notion of centralized, hierarchical power. For the first time ever, the NPL would be as democratic as its ideals and rhetoric.

The intense debates at this meeting represented a turning point in the NPL's trajectory. They indicate that the members understood the significance of the 1920 election defeats. The attempt to seize a state government outside of North Dakota—an achievement that just months before seemed to be within the League's grasp—fell short. To many, the failure highlighted the limits of Townley's tightly centralized leadership. Unable to coordinate a series of complicated campaigns and negotiate widely varying local contexts, he no longer held farmers spellbound. State-level leaders felt free to challenge business as usual. The long-standing push for democratization within the NPL broached the final obstacle. The assembly broke Townley's grip on the League once and for all.

The convention did take one last piece of advice from Townley. He and his advisors insisted that the local NPL women's clubs receive more financial support. Everyone agreed that the future of the NPL depended on their success. Indeed, the meeting itself included "several women delegates," indicating the growing power they held in the organization. Just two months later, the committee running North Dakota's NPL gave the woman representing the clubs a right to vote on the state organization's affairs. They only retracted that right when NPL club women protested that they had not elected her. The farm women believed that she could not represent the thousands of North Dakota women anxious for League success without democratic backing. Clearly, the men recognized the women's clubs—and their commitment to democracy—as a force to be respected.[143]

The need to harness the energy of farm women soon led to calls to create a thousand women's clubs by fall 1921. To do so, existing NPL women's clubs turned to the time-tested canvass. "That little Ford that darts from farmhouse to farmhouse," national NPL Women's Club manager Kate Gregg claimed, "has in it . . . three club members who are out signing up the members for a club in the adjoining township." Mimicking the tried and true tactics of paid League employees, in just a few months these volunteers rapidly increased the number of women's clubs in state after state. As a women's club organizer in Nebraska put it, "The result of team work made possible by the league will astonish us."[144]

The campaign extended the NPL women's clubs into the most remote rural districts. For instance, Renville County, Minnesota, soon sported twenty-five of them. On July 4, 1921, they worked together to sponsor a holiday parade and picnic. More than five thousand people attended. The League's 1920 gubernatorial candidate, Henrik Shipstead, addressed the crowd. A float showing "'the League goat' with 'Big Biz' up a tree" proved a crowd favorite. Baseball games faced local towns off against each other. The day closed with live music and a dance. Organized entirely by NPL women, the event signaled the ways in which rural women embraced civic action. Activist and author Helen Keller noted the growing involvement of women in the organization and believed that the shift helped to make the League "the most hopeful thing in America."[145]

✶

In 1921, a reporter from the *Christian Science Monitor* nonetheless argued that the League had entered "a new phase of the struggle for its existence." The rural economic crisis continued to threaten the NPL's cash flow. In North Dakota, twenty-six banks failed that year. As NPL memberships expired, farmers proved unable to renew. Without money, agrarians struggled to stay on their farms and stay firm in their politics. Not even the extension of credit to cover membership fees could keep the League from shrinking rapidly.[146]

Dependent on those fees, the NPL suffered greatly. "Hard times have forced the League to cut down its working force," wrote one League employee. Throughout 1921, circulation numbers for the *Nonpartisan Leader* fell. Then the *Leader* cut the subscription cost. In Minneapolis, NPL leaders hoped that farmers unable to pay for a full membership might at least be able to afford a $1.50 annual subscription to the *Leader*. Many could not. Eventually, the NPL reduced its membership fee, by then $18 in some states, to $6.50. But cash-strapped farmers had no money to spare.[147]

New agrarian organizations began to compete with the NPL for farmers' limited funds. Across Wisconsin, Minnesota, and North Dakota, the Farm Bureau began a membership drive. Born from United States Department of Agriculture extension service work, the anti-League group began making inroads among agrarians as a less divisive option for cooperation in politics. The bureau claimed that the NPL had failed to deliver what it promised. In response, Leaguers in Morton County, North Dakota, called on their fellow agrarians to "sound a warning . . . against the Farm Bureau" because the organization's dues would compete for the farmers' limited funds. Already struggling to retain members, the NPL could do little to combat the new option.[148]

League enemies took notice of these challenges. Nicolay Grevstad—a high-ranking Republican Party operative—told one US senator that "Mr. Townley is pretty well done up in so far as his personal influence and leadership is concerned." Indeed, he "and the others of the 'Inner circle' at headquarters are getting short of power and funds." Nonetheless, "the sentiment of unrest and dreams created by the League remains." Grevstad rightly believed that empowered Leaguers would "remain a distinct and potent force throughout the entire west."[149]

This fear encouraged further efforts to suppress the NPL. In March 1921, a mob in Great Bend, Kansas, confronted former US senator Ralph Burton and forced him to abandon plans to lecture to Leaguers in nearby Ellinwood.

Later the same day, a crowd seized the state organizer and secretary. Beating them both badly, the men then tarred them and, lacking feathers, rolled them in a nearby field of grass. The two Leaguers fled, hiding in a nearby straw stack before being taken in by a sympathetic farmer. Governor Henry Allen's newspaper, the *Wichita Beacon*, defended the mob.[150]

League enemies also looked to the future. Anxious to memorialize the NPL as a Socialist plot, journalist Asher Howard (a recent member of the Minneapolis Chamber of Commerce's Board of Directors) not only produced a widely distributed and damning portrayal of the NPL but also deposited the evidence he gathered to critique the League at the Minnesota Historical Society. He did so "to block any attempt in the future to confuse the facts in regard to this feature of the organization."[151]

Adversaries even attacked the heart of the League program—North Dakota's state-owned industries. Legitimated by a US Supreme Court ruling in 1920, the Bank of North Dakota nonetheless struggled. Banks across North Dakota, weakened by the agricultural depression, teetered on the edge of failure. Ongoing boycotts of the state bank by the major financial institutions of the Midwest and even Wall Street suggested ongoing opposition from the nation's financiers. Then, using an initiative procedure in the 1920 election, the Independent Voters Association successfully passed a new law eliminating the requirement that all of the state's public funds be deposited in the bank.

Undercapitalized and unable to secure any credit, the institution charged with building the state mill and elevator simply ran out of money. The League reached out to its archenemies—Minneapolis bankers—and offered to scale back its program in North Dakota in return for $ 2 million in state bank bonds. The attempts to negotiate for funding broke down when the IVA insisted that the NPL give up control of the Industrial Commission altogether. As the *Nation* put it, the failure of these negotiations highlighted "the frank, direct, and unabashed control of the politics of a sovereign State exercised by the financial interests in and about it." In the end, Leaguers and their supporters needed to raise the money to institute the constitutionally mandated state industries on their own.[152]

Taking out advertisements in national publications and appealing to labor unions and farm organizations to buy bonds, NPL leaders desperately tried to keep the state bank alive. North Dakota governor Lynn Frazier visited progressive leaders up and down the East Coast, trying to convince them to back the bond sale. The League insisted—rightly—that "the great banking and public utility interests" were "making a final effort to ruin it, if possible." The *Nation* editorialized in favor of buying North Dakota state bank bonds, telling readers to "strike a direct blow at the financial control which would dictate to and coerce the will of a sovereign people."[153]

The impressive response—from allies, skeptics, and even those who knew little about farm troubles or the League itself—stunned many opponents. The notoriously conservative *Chicago Tribune* agreed to advertise the bond sale. A church in Cincinnati, Ohio, promised to sell $100,000 worth of bonds. Unions across the country, in particular, rose to the challenge. The American Federation of Labor agreed to endorse the state bank, encouraging members to make deposits and buy bonds. As a result, in Chicago, the AFL organized a $2 million drive. In Missouri, a convention of labor leaders unanimously called on workers to purchase bank bonds. Supporting the Bank of North Dakota directly, the Railroad Telegraphers' Union deposited $50,000 in the institution. The nationwide sales of bonds kept the bank from collapsing altogether. By the end of the year, infused with new capital, the state bank prepared to recommence construction on the half-finished state mill and elevator.[154]

The campaign pointed to wide-ranging sympathy for the farmers' struggle. It ensured that the League's broader program persisted. Yet the effort consumed much of the money and energy that might otherwise have been spent reenrolling farmers in the NPL to bolster its organizational infrastructure and prepare for the future.

*

Before it could do either, opponents began organizing a recall election in North Dakota. Recalls gave citizens the right to replace officials before term completion. Alongside referendums and initiatives, the recall became a cornerstone of efforts to institute direct democracy in early twentieth-century America. During the 1919 legislative session, the NPL drafted and passed a constitutional amendment to institute the recall in North Dakota. A petition signed by at last 30 percent of the voters in a district would trigger a special election for the designated offices. With the League's commitment to popular politics, the legislation sailed through the statehouse. Citizens overwhelmingly approved the proposed amendment in March 1920.[155]

Emboldened by their seizure of the state's house of representatives and talk among frustrated financiers in Minneapolis, Theodore Nelson and other IVA leaders considered distributing petitions to recall Governor Lynn Frazier, newly elected attorney general William Lemke, and Commissioner of Agriculture John Hagan. Nelson, firmly committed to the notion that "action in the I.V.A. on fundamental matters must come from the bottom . . . and not from the top as in the Nonpartisan League" sent questionnaires to the group's organizers across the state. Most rejected the proposed recall. Many IVA farmers found the idea distasteful.[156]

The staff and volunteers polled thought success more likely if the recall targeted not only League officials but also NPL programs such as the state bank, the issuing of bonds to support the bank, rural credits, and the Industrial Commission. This would provide two avenues for disassembling the state institutions—one by political control, the other by referendum. IVA leaders then turned to "a committee of bankers"—their other constituency—to gauge their support for a recall. This group, unlike the struggling agrarians that made up the IVA rank and file, kept the organization solvent. They liked the idea.

Announced in the IVA's paper, the *Independent*, the top-down decision to pursue a recall proved controversial. To unify the IVA behind the effort, Nelson called a state convention. In March 1921, after intense debates in which members from the eastern part of the state pushed hard for a recall, a divided IVA narrowly agreed to pursue one.[157]

Leaguers opposed the recall outright. They pointed out that the recall election itself would cost thousands of dollars and do little more than widen rifts in already deeply divided communities. "The people," announced the *Nonpartisan Leader*, "wanted to rest from political strife." More directly, one banker sympathetic to the NPL wrote to Theodore Nelson, insisting that "we should get down to business and leave the politics alone at these hard times."[158]

Many IVA supporters agreed. Some refused to sign the petition. A representative of the prohibition movement—which firmly backed the association—recognized faltering League support but also "found a great many I.V.A. that are strongly opposed to a recall election." The small businessmen who made up a sizable portion of the membership also felt conflicted. One not only was "tired of keeping things stirred up all the time" but also worried that "if there is a Recall and the League winn [sic] they will never be gotten out."[159]

Other potential IVA allies refused to back the recall. Though the *Bismarck Tribune* had consistently combated the League, the newspaper accused the IVA of duplicity. If elected, the opposition claimed it would uphold the NPL program even as it offered up initiatives that gutted it. The *Tribune* much preferred a "straight out-and-out fight against the League and its program, without any trimming or bait to League voters" to supporting an effort created by the "outlaw Leaguers" in charge of the IVA. Usher Burdick, a staunch Republican and former candidate for governor, noted that he "found no interest among farmers in the proposed recall." Even the remnants of Alexander McKenzie's political machine chose not to directly support the IVA's effort.[160]

Confident that its careful counting of voters—which showed that Lynn Frazier's election majorities in North Dakota shrank dramatically between 1916 and 1920, especially in eastern counties—pointed to a potential recall victory, IVA leaders stuck to their plan. Ignoring rank-and-file fatigue, they pushed ahead. Nelson and the IVA candidates insisted on their commitment to the League's original 1915 program while claiming to end NPL corruption and overreach. If this approach could motivate enough North Dakotans to sign the petitions, the men counted on voters opposed to the recall to nonetheless vote against the NPL in the recall itself.[161]

Under immense pressure, Theodore Nelson confided in a friend that "this petition business, and without funds, not only makes my head swim, but makes me gray." Rallying the IVA faithful and raising money proved difficult. Entreaties to the Minneapolis Chamber of Commerce for financial support garnered little. The campaign worked with only half the funds available to the IVA during the previous year's elections. The stakes were high. If the IVA failed for a fourth time to defeat the League, the organization would collapse.[162]

Even so, the petition circulation campaign seemed to bear out a shift in voter opinion. As IVA organizers took the forms from farm to farm, the numbers of signatures grew. Attacks on the League for its "misuse of public funds" and its leaders propensity to "abuse public confidence" gained traction. In turn, the IVA's candidates for governor, attorney general, and commissioner of agriculture and labor all committed to supporting the original 1915 NPL program. Minnesota governor J. A. O. Preus and South Dakota governor Peter Norbeck supported the effort. Both drew hundreds to IVA-sponsored events across the state. Comparing the numbers to those who had voted against League candidates just months before, Nelson noted that "there have been substantial changes in favor of the Independents in practically every rural precinct." The election would be won or lost in the countryside.[163]

The IVA leader was right. By early 1921, thousands of North Dakota's agrarians had drifted from the NPL orbit. Yet their feelings about the recall remained complex. George Howard, for instance, grew wheat on 240 acres in Walsh County. Excited by the League program and the prospect of a farmer governor from nearby Hoople, he and his neighbors strongly backed the NPL in 1916. But by 1919, the longtime Republican's concerns about the movement grew. His neighbor, state representative Nels Hedalen, bolted from the NPL in early 1919 and helped organize the IVA. Howard soon started receiving free copies of the IVA's *Independent*. When his neighbor circulated an IVA-sponsored recall petition in 1921, Howard sympathized but chose not to sign it. Simultaneously, he believed the NPL had served its purpose and overstepped its original program. He would not support the recall but would vote against the NPL if given the chance.[164]

Agrarians such as Howard made it difficult for the newly democratized NPL to respond to the recall effectively. Farmers in Griggs County, North Dakota, resolved that "shame and dishonor be forever placed on the shoulders of these corrupt class of men . . . at this time circulating recall petitions against . . . men placed in trusty positions by the farmers." Convincing dissatisfied neighbors proved difficult. "Now in the hands of the members themselves," as chairman A. A. Liederbach put it, the organization depended on county- and precinct-level organizing to defeat the recall. This meant more responsibility fell on individual NPL members to convince their neighbors to reject the IVA's bid for power. Liederbach put it bluntly: "Don't be a slacker." Yet farmers weary of NPL overreach drifted from the organization's orbit and chose not to engage in the necessary work.[165]

Internal tensions also surfaced during the recall campaign. Desperate to raise funds to fight the recall, the national NPL office wanted to sell its North Dakota-based daily, the *Fargo Courier-News*. The newly autonomous North Dakota state NPL committee refused to do so. Weeks later, the frustrated *Courier-News* editorial staff—aligned with Arthur Townley and the national office—accused North Dakota's NPL leadership of misusing recall campaign funds raised by already hard-pressed League members. Recognizing that such controversy could undercut the NPL's fight against the recall, North Dakota attorney general and League leader William Lemke stepped in. In July 1921, the staff retracted its charges.[166]

The stakes were high for the NPL too. A successful recall in North Dakota meant, as one reporter suggested, the "doom of the movement elsewhere." NPL leaders and members, recognizing the recall's import, did everything in their power to defeat it. Desperate to stem the IVA tide, Leaguers and their allies even turned to underhanded personal attacks. Echoing the previous election's talk of free love and immorality, the Fargo Trades and Labor Assembly condemned Theodore Nelson in an incendiary brochure. The publication accused the IVA leader of conducting an affair and abandoning his family (Nelson divorced his ailing wife in 1916 and married his stenographer that same year). It also reproached prominent IVA leader E. W. Everson for visiting prostitutes on trips to Saint Paul. Clearly marked as "NOT INTENDED FOR GENERAL DISTRIBUTION," the booklet suggested that "care must be taken that it does note reach the hands of children of immature minds." The especially forward attempt to smear the IVA only hardened the battle lines.[167]

As the recall approached, speakers on both sides "found much apathy among the voters." It would be a close contest. Then, just five days before the recall, the US Supreme Court announced it would not hear Arthur Townley's appeal of his disloyalty conviction in Minnesota. In light of newly appointed chief justice William Howard Taft's public animosity toward the NPL, Townley anticipated the decision. Earlier in 1921, he told an NPL member concerned that he might land in jail "to send me some cake when I am in." "It will not be so bad," he told her. "I need a rest and I can do a lot of writing and I think it will be best." The NPL leader believed that news of his failed appeal would boost League turnout.[168]

NPL supporters and opponents alike understood the significance of the week to come. The pending incarceration of Arthur Townley and the looming recall vote in North Dakota would reveal the extent to which the Nonpartisan League had succumbed to internal and external pressures. Controversies fueled by poor decision making and inadvisable business ventures produced political consequences. The alienation of other agrarian groups drove erstwhile allies away. Townley's tight hold on the reins of power turned friends into enemies. It also weakened the hand of the more democratically minded once they finally began to govern the NPL. Politicians' effort to reverse democratic processes such as the primary produced procedural roadblocks. Red-baiting opponents burnished the League's reputation for Bolshevism. Hard times in the countryside left farmers unable to financially support their own organization. In the wake of the 1920 electoral defeats, a commentator suggested that the NPL was "guilty of many political sins . . . which have offended the popular conscience." He predicted that "the League will . . . become chastened by two lean years." That chastening came much sooner than expected.[169]

Legacies

As an organization, the League had to cease functioning. But although the visible machinery largely melted away, a sentiment and a point of view had been established in the minds of hundreds of thousands of farmers and ranchers.

THE NATION, August 1, 1923

Early in the morning on October 29, 1921, jubilant members of the Independent Voters Association (IVA) in Dickinson, North Dakota, telegrammed President Warren Harding. The state's voters had weighed in. It looked as though IVA candidates had defeated the Nonpartisan League's governor Lynn Frazier, attorney general William Lemke, and commissioner of agriculture John Hagan. For the first time in American history, citizens successfully recalled statewide office holders. Flush with victory, the IVA adherents half-jokingly asked the president for "permission 'for the readmission of North Dakota into the United States.'"[1]

Front pages all over the country thrilled at the never before seen spectacle of a recall. The initial numbers—as expected—suggested a strong anti-League slant. Yet the decision depended on turnout in rural districts. The outcome was soon settled. With 80 percent of precincts reporting, the NPL-owned *Fargo Courier-News* conceded defeat. By a narrow margin—just a few thousand votes—all of the IVA candidates had beaten their League opponents. Later that night, thousands of IVA supporters in Fargo jammed the streets as car horns blared and fireworks exploded. One automobile sported a new bumper sticker: "Back in the U.S.A." Poking fun at disheartened Leaguers in Minot, the winners claimed their victory over the NPL by parading a goat through the streets with a red flag tied to its tail.[2]

Four days later, the "deposed dictator of North Dakota"—Arthur Townley—arrived in Jackson, Minnesota, to serve his ninety-day jail sentence for fostering disloyalty during World War I. After three years of failed court appeals, the head of the League seemed resigned to his fate. While Townley presented himself to the county sheriff, the young *Minneapolis Tribune* reporter Lorena Hickok (later a close confidant of Eleanor Roosevelt) watched more than one hundred Leaguers gather in the street to "bid their leader good-bye." One man "pushed through the crowd and handed Townley a bible." The well-wisher then "exploded into a long string of oaths" targeting the League's enemies. When the jail door closed, George Brewer, an NPL organizer, winced. "I wish I were God for just one minute," he muttered. In response, an opponent yelled out: "We've got your man where we want him now!"[3]

In the space of one week, the League's fortunes had taken a tremendous turn for the worse. Combined with Townley's incarceration, the recall eliminated most of the NPL's remaining momentum. *Nonpartisan Leader* editor Oliver Morris suggested that failure resulted from North Dakota's NPL leaders' rejection of "President Townley's advice and leadership." In the *Nation*, Morris went further, , claiming that "the cry of 'more democracy in the League' led to defeat." Silenced, the organization's now-sullied leader—jailed for "saying things about the war that no one will now question"—nonetheless accepted his sentence "without bitterness." In tandem, the two events represented the nadir of the NPL.[4]

John Hagan privately admitted that he and other Leaguers "were probably a little overconfident" about the recall. But he also blamed the organization's "very lame" propaganda campaign. Furthermore, "even though it should pass out of existence, the things for which we fought, the people will continue to fight for." Hagan was right. Even as its candidates won, the IVA's initiatives gutting the state's industrial program fell short. Not a single one passed. Though farmers in eastern precincts deserted the NPL candidates in sizable numbers, enough stuck to the League program to protect the state bank, mill, and elevator. Unwilling to undo the platform that so many voted for in 1916, agrarians troubled by the NPL's shortcomings nonetheless defended the organization's achievements.[5]

Even as the League began to fade from the national political scene, former members persevered in their commitments. Manifested in a wide range of movements and politics, the political economy of the NPL lived on. The League's loss of the first statewide recall in history proved to be only one of many legacies. Most remain overlooked. The few who do take the NPL's legacies seriously usually cite the brilliance of Townley's mobilizing, the radicalism of the farmer members, the aberrant eruption of people power, the movement's narrow regional reach, and the role played by the League as a foundation for midcentury midwestern liberalism. At best, these orthodox understandings fall short. At worst, they are flat wrong.[6]

Furthermore, such perspectives turn the NPL into a fascinating and even bizarre relic, one with little import in the years that followed. Largely defunct by the middle of 1923, the insurgency embodied by the Nonpartisan League nonetheless offered compelling legacies. It showed the significance of citizen-centered democracy in the Progressive Era. It empowered rural women not only to participate in but also to affect electoral politics. It fused farmer and labor interests into powerful third parties in both the United States and Canada. It launched the careers of insurgent US senators who wielded influence in that body into the late 1930s. It established the legality and promise of state-owned enterprises. It successfully extended older antimonopolist popular politics into an urbanizing and industrializing America. It applied principles of economic cooperation to politics. It deployed novel electoral tactics to make politicians more responsive to voters. It combined contemporary corporate efficiencies with a critique of corporations. It articulated an agrarian moral economy that envisioned an alternate future for American capitalism. Ultimately, the League offered the biggest challenge to party politics as usual in twentieth-century US history.

The struggle over how to remember the NPL began even before its demise. Between the 1921 recall and the passing of the League as a national force in the mid-1920s, proponents and opponents engaged in pitched political battles even as they worked to define the marks left by the NPL. Put to use by politicians and the people alike, the League example points us toward a complicated history of memory that opens up larger questions about American politics and the potential for change. If we remember the NPL in conventional ways, it offers us little. But if we bring more nuanced visions to bear, the legacies of the League point us toward an innovative approach to democracy and the economy. The NPL, as experienced by thousands of farmers across the North American West, offered an alternative future. Its decline did not destroy their hopes and dreams. Indeed, the attitudes and inclinations of the League's rank and file persisted in American politics. By defining a potential democratic response to the challenges of the future, the NPL continues to resonate in the twenty-first century as a distinctive manifestation of a lost tradition in American politics.

*

In the wake of the recall, divisions in the NPL grew. Allegations of legal incongruities in the recall petitions led a few farmers to instigate a lawsuit. Some agrarians backed the plan. At a meeting in Ransom County, for instance, NPL farmers insisted that "scheming corrupt politicians, and capatilistic [sic] interests" brought on the "fraudulent and unlawful" recall. Neither the NPL nor its recalled officials wanted anything to do with litigation. They rightly worried about bad publicity. Other upset North Dakota Leaguers insisted on putting Townley back in charge. The movement to do so went nowhere. Meanwhile, the IVA officials entering office tried "to smooth the bitterness caused by five years of terrific political strife."[7]

With the 1922 primaries just a few months off, Townley used his time in jail to write a column in the shrunken, soon-to-be monthly *National Leader*. He hoped to reinvigorate his supporters and regain control in North Dakota and Minnesota. In the former, farmer leaders refused to relinquish power and readied for yet another primary fight. In the latter, farmer-labor leaders, forced to adapt to changes in the state's primary laws, envisioned an outright third party. Townley opposed both plans. He argued that instead of participating in primaries or nominating its own candidates, the NPL should serve as a voting bloc that would deliver votes to the most sympathetic candidates from the major parties. This way, the organization might continue to influence elections without becoming an easily defeated third party.[8]

Freed in late January 1922, Townley spoke to supporters at an American Civil Liberties Union–sponsored event in Minneapolis. Addressing the mixed farmer and labor audience, he asserted that "I am the only fellow in the country who made you stick together long enough to get any real results." Townley then argued that party politicians always succeeded because agrarians and workers failed to stick with the NPL. They "had damn few more brains" than a barnyard animal, he said. That said, he admitted that in North Dakota, the League "jumped into the sea before we learned how to swim in a puddle." In fact, he believed, "When it comes to running a government, you and I are dubs." After angering the constituency he needed most, Townley meekly retracted his critique a few days later.[9]

Even before the NPL in both states met in conventions, the democratic awakening of League members led most to reject Townley and his controversial plan for the future. In January 1922, labor leaders in Minnesota scheduled farmer-labor caucuses. Voters' willing participation represented the new Farmer-Labor Party's rejection of Townley's idea. In North Dakota, most farmers wished to continue to participate in the Republican Party primary. Some worried that adapting Townley's plan would mean the reassertion of his control over the League. Others believed the original NPL idea still made sense. One empowered League farmer thanked Townley "'for leading us to the green pastures and clear waters" but said that "if he thinks we will follow him to the slaughter house he will find we are not sheep." Despite support from Lynn Frazier and John Hagan, the "balance of power" plan advocated by Townley went down to defeat.[10]

Thoroughly beaten in his bid to regain control, Townley announced his resignation as president of the Nonpartisan League. He blamed his inattention to falling membership numbers. But something deeper gnawed at Townley. Oliver Morris, his ally at the *National Leader*, blamed his departure on the internal push for more democracy. "Constantly increasing demand for more and more decentralization and more and more and more local autonomy" forced the otherwise successful Townley from his post. Morris was right. The growing empowerment of League members led them to see their famous leader as extraneous. They recognized that Townley's unchecked overreach—especially with League-related commercial enterprises—damaged their movement. Fully convinced of their own power, they no longer tolerated an NPL with an autocratic leader. The very qualities he worked to inculcate among farmers—democracy, discipline, and action—succeeded in driving him out of the NPL. Irony pervaded the second of the League's many legacies.[11]

*

Townley's departure did not, however, stifle the growing room for women in the NPL. More began to run for office. As early as 1918, Ina Williams served in the Washington state legislature as a Granger endorsed by the Nonpartisan League. The better-known Jeannette Rankin—the first woman to serve in the US Congress—received support from the Montana NPL that same year. But few women stood for office, and even fewer earned the support of organized farmers. Persistent beliefs that women should focus their energies on families and enter into politics only to address domestic issues kept many from running in campaigns. Rural women often internalized these limits.[12]

As the NPL's commitment to rural women expanded, so did their political presence. With a solid voting bloc of rural women behind them, women found more traction for candidacy. Ella Thilquist, secretary of an NPL auxiliary in Minnesota, reported in 1922 that her group "had the pleasure of having with us a lady from one of the neighboring clubs who gave an interesting talk. She is candidate for county superintendent of schools from our county an[d] we are going to do all we can to help elect her as we know that office will be capably taken care of if she is elected." Organizing themselves, NPL women began imagining each other as politicians as well as voters.[13]

Marie Weekes, editor of the *Norfolk Press*, ran for US Congress in Nebraska in 1920 with the League's backing. Minnesota Leaguers endorsed Lilly Anderson for state treasurer in 1920. They touted her as a "real farm girl." But as state secretary of the women's auxiliary of the state's American Society of

Equity chapter, Anderson offered much more. The *Minnesota Leader* crowed that "the action of the farmers and workers" in this case proved "more than honeyed words their belief in the justice of votes for women and the right of women to share in the responsibilities of government." One prominent local suffragist declared that "it is the first time any political organization has offered something tangible to the women."[14]

That same year, the NPL in Montana followed suit. When Ella Dorothy Lord came in second in the balloting for the League's endorsement as state auditor, a clamor arose as men and women alike pointed to her longtime support of laborers and farmers. The convention then nominated Lord for state treasurer. Apparently the "conference had made up its mind to have a woman on the ticket." Then the crowd backed Margaret A. Hannah, wife of a former state senator and a leader of Montana's NPL women's clubs, for superintendent of public instruction.[15]

In Colorado's 1920 elections, the NPL supported Mary Bradford for superintendent of public instruction and May Peake for the state senate. In Washington State, Alfa Ventzke—described by the *Nonpartisan Leader* as "a practical farm woman"—not only ran for the superintendent of public instruction's job in 1920 but also served as the state NPL's secretary. Somehow, she found time to retain her standing in the state Grange and also run her sheep ranch in the Methow valley.[16]

The expansion of women's power in the NPL in 1921 led to even more women running for higher office. Minnie T. Craig, the first president of the North Dakota NPL women's clubs, won a seat in that state's legislature in 1922. She campaigned for her seat alongside NPL employee Ina Brickner. They traveled from township to township and every night after meeting with voters "pitched [their] tent, made [their] bed on the ground, and spent the night in the school-yard." In Minnesota, the NPL backed Myrtle Cain, a Minnesota labor organizer and suffragist, in her run for the state legislature in 1922. She won the race and became one of the first four women legislators in state history. In Washington State, the League endorsed suffragist Frances Axtell in the Republican Party primary for US senator.[17]

A teacher from South Dakota, however, became the most prominent woman politician in the Nonpartisan League. Alice Lorraine Daly grew up in Saint Paul, earning a bachelor's and a master's degree from the University of Minnesota. As a young woman, she earned a spot on the front page of the *New York Times* in 1913 by calling off an engagement with US Senator James Brady of Idaho. She became the head of public speaking at the State Normal School in Madison, South Dakota, in 1915. Nicknamed the "Joan of Arc of

South Dakota," Daly worked as a suffragist and pacifist, even becoming Jane Addams's choice to chair the Women's Peace Party in that state.[18]

Outraged by wartime repression focused on the NPL, Daly came to believe that she "was teaching . . . farmers' sons and daughters not living truth but falsehood" when teaching them about "the first amendment of the Constitution." Ostracized for her support of the League and local labor unions, she resigned her post in 1920. But with the newfound empowerment of farm women providing real clout, Daly emerged as the League's candidate for South Dakota governor in 1922.[19]

As the first American woman nominated by a significant party for a governor's race, Daly ran a vigorous campaign, reaching out to South Dakota's union workers as well as NPL farmers. She declared to voters that her "campaign will not be a campaign against anyone. It will be a campaign against conditions, against the weaknesses of our present, out worn economic system, and for a more modern system of finance, credit, public control and ownership, a system that will lift from the masses of our people the burdens of mortgages, taxes and ever-recurring payments of interest that keep them in a state of servitude." Coming in third place that November, Daly continued to support farmer-labor coalitions in South Dakota for the rest of her life. More than any other Leaguer, she proved especially popular among members of the NPL women's auxiliaries across the West.[20]

The League's support of women candidates gave farm women long-desired access to electoral power. Desperate to transcend imposing challenges, agrarians adjusted the League's structure and focus in the hope such changes would spark new momentum for the 1922 elections. Equally important, the changes bequeathed to rural women a new sense of their political significance as well as a growing sense of civic agency. By April 1922, for instance, women once largely ignored by the NPL spent months debating the pros and cons of birth control in the organization's *National Leader*. This too-often ignored legacy created the foundation for rural women's political activism in the decades to come.[21]

*

Townley's departure also did not detract from the appeal of the NPL model. Though newspaper editors across Oregon pledged in 1920 to keep the NPL out of the state, Grange leader Charles Spence finally realized his longtime dream. Though he failed to convince the state's agrarians of the worthiness of direct political action through the Nonpartisan League, he brought the Farmers Union and the Grange together around shared concerns over taxation in late 1921. In 1922, Spence wound up as the farmers' gubernatorial candidate.[22]

Loosed from the national office, Walter Thomas Mills moved to California. After a mass meeting in San Francisco in January 1922, he began recruiting farm men and women in the Central Valley. The effort soon fizzled out. South Dakota's NPL made a few promising inroads into Wyoming in 1921. Following up, a Nebraska-born North Dakota farmer named Harry Lux moved into Wyoming in early 1922 and began signing up farmers. Though he soon left to work for the NPL in Nebraska, a small group persisted. They joined with laborers and progressives to influence races for the governor and the US Senate in 1922 but did not survive the election season.[23]

Local control defined these efforts, as it did in states with existing League organizations. The nominal leadership of the national office gave way to little more than cheerleading. Truly on their own, state NPL leaders reoriented their organizations to better face on-the-ground realities. In Wisconsin, the League effectively merged into a progressive Republican coalition guided by Governor John Blaine. In Nebraska, the NPL adopted Townley's balance-of-power plan. Narrowly rejecting a bid to join a progressive third party, Leaguers there cut against the existing momentum. The decision angered emerging allies, drove the NPL's leadership out of the organization, and left farmers looking elsewhere. The Montana NPL also chose to throw its weight behind major-party candidates. As a body that did little more than endorse nominees, it dissipated as a force in the Democratic Party after deciding whom to back in 1922. A tiny remnant soldiered on as an explicitly farmer-labor coalition sympathetic to the newborn Communist movement.[24]

In every state, NPL members anxiously looked toward Washington, DC, as both the source of and answer to their considerable problems. Both shifts led to a new direction in League politics. In convention after convention, agrarians broke with long-standing rules limiting them to state politics. Finally in charge of their own organization, many chose to focus on nominating and electing officials to fight for farmers in Congress. Because successful congressional elections depended on state-oriented campaigns, these efforts emerged as a logical extension of long-standing League policies. At the same time, they kept Leaguers from creating a comprehensive national agricultural plan. As the rural depression deepened, the NPL's response remained rooted in the states, even as farmers began to desire a national answer to the failing agrarian economy.

In North Dakota, just months after the recall, the League put forward former governor Lynn Frazier for the US Senate. He defeated his IVA-sponsored opponent in the Republican primary that June and then triumphed in the 1922 general election. Frazier joined his state's E. F. Ladd and Idaho's long-time progressive William Borah as NPL-backed officials in the nation's highest legislative body. The same year, farmers and workers in Minnesota successfully supported Henrik Shipstead—the League's failed 1920 gubernatorial candidate—in his bid against Senator Frank Kellogg (later a winner of the Nobel Peace Prize). According to one commentator, unlike most politicians, the victorious Shipstead saw "the people behind the ballot box."[25]

"League-minded candidates" ran for the US Senate in other states as well. Burton Wheeler, the NPL candidate for governor of Montana in 1920, won with the League's support and allied himself with Ladd, Frazier, and Shipstead. Nebraska's Ralph B. Howell earned the NPL's official endorsement and, drawing on a broad coalition of progressives, won. In Washington State, Clarence Dill narrowly defeated his Republican challenger with the backing of the NPL and its labor allies.[26]

Suddenly, there were seven US senators representing NPL interests. Joining other western senators that one opponent later called "sons of the wild jackass," they formed the core of a bipartisan farm bloc soon to sway federal policy. The League's influence in this group grew further. In 1923, after a special election called when Minnesota's Knute Nelson died, longtime equity leader and Leaguer Magnus Johnson entered the US Senate under the banner of the Farmer-Labor Party. Upon Senator Edwin F. Ladd's death in 1925, League newspaperman Gerald Nye, from Fryburg, North Dakota, got himself appointed to replace Ladd. Nye was elected to the seat the following year.[27]

This new national presence made a huge impact. After all, the *National Leader* crowed, "ninety-six men in the United States senate really govern the country." Farmers now turned to senators with "real power" to influence the nation's agricultural policy. The sudden ascendance of candidates backed by the NPL to the US Senate belied the group's fading power. However obliquely, the NPL idea—to nominate and elect politicians who followed the League program regardless of party—persisted. With more local control than ever, the NPL reached the heights of electoral politics even as the national organization bottomed out.[28]

In May 1923, the NPL sold off its North Dakota daily, the *Fargo Courier-News*. The League- and union-owned *Minnesota Daily Star* fell into bankruptcy later that year. A former NPL journalist writing in the *New York Times* described the newspaper, the "world's greatest experiment in co-operative journalism," as a failure. The national NPL office—flat broke—finally closed in July 1923. It still held $2 million in postdated checks and a sizable debt.[29]

The critique of internal democracy from the pro-Townley leadership circle continued to the end. In its final issue, the *National Leader* proclaimed that "state management of campaigns seemed necessary and useful" despite

FIGURE 26. NPL members of the Farmer-Labor Party backed Magnus Johnson, an Equity Cooperative Exchange board member, in his bid for the US Senate in the 1923 special election in Minnesota. The
former state legislator cultivated his agrarian background to appeal to rural voters. (Minnesota Historical
Society, Por 5679 p8)

admitting that the trend proved "contrary" to the leadership's "best judgment." In the end, "state management . . . cut down the income of the national
office, crippled its activities and decreased the means of the League to expand
into new states." Conveniently ignoring the agrarian movements that birthed
the NPL, the paper went on to describe Townley as the man who "almost over
night" forged "an organization . . . built from nothing. " In fact, farmers had
already organized themselves in cooperative movements, and democracy did
not demand diffusion. The inability of depression-stricken farmers to pay
even a greatly reduced membership fee stymied national organizing. Structured to be dependent on credit, the organization found little succor in tough
economic times.[30]

"Notwithstanding all these handicaps," another editorial suggested, "the
League won its greatest victories in 1922." It had. "No organization of modern
times," it went on, "has had so far reaching an effect on national politics as the

League." It was true. The presence of seven NPL-aligned US senators proved it. As another crucial legacy of a brief movement, their rise to power had few parallels.[31]

*

Through the 1920s, the most successful state NPL organizations formed lasting farmer-labor coalitions in third parties. Everywhere except in Alberta and North Dakota, this tack proved the most productive. Casting aside the Nonpartisan League name, the movement became more diffuse and less oriented to agricultural concerns, even as it took up the banner of broader reforms. North of the border, the Alberta NPL disappeared, but its legacy deeply influenced the United Farmers of Alberta. Meanwhile, in its home state, the League refused to break up and soldiered on as an organization struggling for control of the Republican Party.

Even before the League's national office closed, sympathetic editors at the *Nation* analyzed the results of the 1922 US Senate races and declared that "the impossible has happened. The farmers and labor have got together, and a new party is born." Though affiliated senators held different party designations—Democratic, Republican, independent, or Farmer-Labor— they stood "for the same thing and they [knew] it." "The same forces elected them regardless of the label," and they would work together to counter the influence of both major parties.[32]

Without a doubt, the *Nation* averred, "this we owe largely to the Nonpartisan League." Though the organization no longer controlled a single state, "its soul goes marching on." In an oblique reference to Minneapolis milling and banking interests, the editors noted that "the same powers were arrayed against the Dakota program of State grain elevators and State rural credits as were stoutly fighting union labor." Furthermore, the farmers' movement "discovered the effective method of creating a third party." By pointing out "the fatuity of party labels" in a way "that Western voters understand and approve," it became "an authentic movement of the masses."[33]

"Borne by a great underground swell," this way of approaching electoral politics was "not the product of any one or two, or three, powerful personalities." And it proved fiercely independent: "It rejoices in Senator Borah's growing national leadership, but it is still a little skeptical of his economic thinking." These voters "remained proud of [Senator Robert] La Follette" but refused to tie their "fortunes to his personal career." Indeed, "it is fortunate . . . in being a movement growing from the bottom up." All told, the emerging farmer-labor coalition seemed to be "the most hopeful political movement since the birth of the Republican Party in anti-slavery days."[34]

The seeds for this shift grew directly from NPL planning. As Arthur Townley became less and less central to the movement, League leaders in both state and national offices firmed up alliances with their union allies. Earlier attempts to connect the movement to organized labor in 1918 and 1920 bore fruit. Unions across the West knew from previous elections that the League earnestly sought their support. NPL secretary Henry Teigan, a former North Dakota Socialist Party member, urged the building of farmer-labor parties wherever the League organized. His friend Albert E. Bowen—the longtime NPL lecturer and the originator of the League idea—also backed the creation of farmer-labor alliances. When Knud Wefald won his race for the US House of Representatives from Minnesota's Ninth Congressional District on the Farmer-Labor ticket in 1922, Bowen sent him a celebratory telegram as "the first Farm Labor [*sic*] member ever elected to congress in America."[35]

Unlike his closest colleagues, Townley consistently warned against the creation of third parties. In the wake of the 1922 US Senate victories, he used his monthly column in the NPL's *National Leader* to point out the long history of failed third parties. To be sure, "special interests" controlled both major parties. But "machine politicians" ached for the League to convert its organization into easily defeated third parties—as their attempted manipulations of the direct primary suggested. These enemies of the people counted on voters "not carrying their protest further than from one party to the other" and thus they could simply "prepare in advance for the vote of protest." In this, Townley noted, the established parties "have not been wrong."[36]

As the state organizations threw themselves into third-party coalitions with unions, Townley continued to rail against the trend. Using a "no-party or nonpartisan organization" allowed for broad coalitions among voters who could agree on "certain principles" that represented shared interests. This— not party affiliation—"moved the man who gives little or no attention to politics." It allowed voters to avoid being divided by party sentiment and instead "direct the united strength of all toward reforms" and "strike toward immediate political success." Only when farmers called "off this party monkey business and all stand together" in their own interest by endorsing suitable candidates would victories follow.[37]

Discredited by his own stand against internal democracy, Townley found few adherents. Outside of North Dakota, statewide offices seemed almost impossible to win with the pure nonpartisan approach. More important, changes in primary laws in many states made his insights largely moot. The evidence and the context seemed to be against the League's most famous leader. So did broader trends. Aborted efforts to establish a national farmer-labor movement nonetheless suggested the broader support for such a move.

State-level insurgencies—driven by local concerns and more easily organized than a national movement—continued to appeal as a path to power.[38]

The Idaho NPL led the way toward a full-blown third party. Established politicians angered farmers and workers alike by eliminating the direct primary and reinstituted party conventions in 1919. As the 1920 elections approach, Leaguers found few allies. Progressives proved suspicious. Mormons in southeastern Idaho turned away from the organization. Labor unions worried about a crushing defeat such as the one suffered two years before rejected all League overtures. Unable to find friends, the Idaho NPL ran a short ticket of independents on the November ballot. Though these candidates lost, they polled enough support to suggest that a farmer-labor party offered a way out of their quandary.[39]

Overcoming internal opposition, Ray McKaig forged ahead. He urged the creation of a separate Idaho Farmers Political Association to ensure continuity in agrarian protest. He hoped to modify the Idaho NPL's platform to appeal to the broadest range of allies. Results soon suggested the wisdom of McKaig's strategy. Pointing to the major parties' denial of direct democracy, he recruited the support of Republican US senator William Borah. Committed to inclusion, Borah railed against his own party's support of the change. Insurgent Democrats and Republicans joined in, even as trade unions signed on. In order to ensure their comfort, the Idaho NPL dissolved. A new Progressive Party took its place.[40]

Almost immediately, it eclipsed Idaho's Democratic Party. The 1922 elections saw gains in the state legislature and in county governments. Agrarians allied with the movement hoped for more victories. In 1923, irrigation farmer Annie Pike Greenwood touted the newfound civic agency of "farmers in their old mackinaws or sheepskin coats and overalls" who led and provided the backbone for this broader coalition. William Borah, running for reelection to the US Senate, took notice. In 1924, he depended on the state's Progressive Party for victory.[41]

In Minnesota, the Democratic Party's interest in fusing with the emerging Farmer-Labor Party in 1922 showed the potential of a powerful third-party plan. Labor leaders fended off the merger, which appealed to many farmers. Overcoming these internal differences, the electoral victories forged by the farmer-labor alliance in 1922 and 1923 reinforced the desire for a third party. Moving from federated action to unified action took time.[42]

In July 1923, Minnesota farmer-labor leaders—including NPL secretary Henry Teigan and trade unionist William Mahoney—called a meeting of the like-minded in hopes of creating a national farmer-labor movement. Their successful election of two US senators and a congressman gave them real clout. They drew on NPL connections from across the West—including Idaho's Ray McKaig—to show the potential geographic reach of a farmer-labor coalition. Attendees decided not to propose a permanent party but instead hoped to unite around a single presidential candidate, Wisconsin US senator Robert La Follette. Outmaneuvered by the AFL and the railway unions, the farmer-labor effort ultimately failed. La Follette rejected the Minnesota-led initiative for its tolerance of communism, scuttling the emerging coalition for a third party. He then ran with Montanan Burton Wheeler as his vice presidential candidate at the head of an independent ticket. The pair lost the 1924 presidential election, but the La Follette-Wheeler ticket still clearly appealed to Leaguers. The insurgents came in second in Idaho, Minnesota, Montana, North Dakota, and South Dakota, all former hotbeds of NPL organizing.[43]

As the Minnesota NPL gradually shrunk, Mahoney and his unions became the more powerful partner. Preparing for the 1924 elections, workers saw few reasons to retain separate organizations. Though leading Leaguers such as Magnus Johnson and Lynn Frazier advised the NPL to retain an independent organization, many rank-and-file farmers came to see the utility of unity. The crucial moment came in March 1924. The annual convention settled the League's future in Minnesota. During the debate, one "good Lutheran farmer" dramatically rose to speak. He insisted that "next in importance to spreading the gospel is the union of the farmers and industrial workers." One observer described the outburst as an expression of "a sentiment which [had] become a religion with tens of thousands." By meeting's end, the Minnesota NPL narrowly agreed to formally dissolve in order to merge with the state's trade unions in the Farmer-Labor Party. The losers, suspicious of the influence of communism on Mahoney and his unions, openly wept as the Minnesota NPL passed into history.[44]

The merger proved rocky. "The League farmers do not subscribe to communist doctrines," wrote one former Leaguer. As Mahoney began to back out of his plan to quietly include Communists in the emerging coalition, bad press and internal struggle weakened the Farmer-Labor Party's appeal to the general public. When local Communists responded to Mahoney's hamhanded attempt to distance the party from them by officially endorsing the Farmer-Labor ticket, many voters turned away. Thoroughly Red-baited, even the moderate Farmer-Labor candidates such as Magnus Johnson (the sitting US senator) and Floyd Olson (the party's gubernatorial candidate) fell short in November.[45]

Even so, the Farmer-Labor Party built on its earlier successes and replaced the Democratic Party as the "other" major party in Minnesota. It survived the late 1920s by depending on Henrik Shipstead's reputation and popularity, which earned him reelection to the US Senate in 1928. The Great Depression improved Farmer-Labor fortunes. Skillfully proposing a response to the emerging economic catastrophe, in 1930, the party successfully elected Floyd Olson governor. More victories in 1932 established the party as a major player in Minnesota through the rest of the decade.[46]

Many former members of the NPL persisted in advocating farm concerns and moderation in Farmer-Labor policy. Still more, finding the earlier charges of radicalism false, rejoined the effort. Former Leaguer Henry Teigan even served a term (1936–1938) in the US House of Representatives under the Farmer-Labor banner. Governor Elmer Benson (1936–1938) joined the Nonpartisan League in 1918 when he returned home from World War I service. True to his roots, when his political career ended, Benson returned to rural Minnesota.[47]

The NPL's insistence on partnership with unions even influenced farmers in places where the League never proved especially powerful. In Oklahoma, attempts to jump-start the NPL in 1919 and 1920 largely failed. But in 1921, when labor leaders tied to the Farmers Union proposed a Farmer-Labor Reconstruction League made up of workers and agrarians, the remnants of the NPL joined in. Modeled on nonpartisan strategies pioneered by the NPL, the Reconstruction League also included calls for a state bank, state-owned mills, and tax exemptions on farm improvements. Accused by opponents of "Townleyism," the movement nonetheless gathered strength. African American farmers joined the coalition despite the racism of their fellow agrarians. In 1922, it triumphed first in the Democratic Party primary and then in the general election. Agricultural reforms followed. A fearsome response by small-town businessmen—in the form of the Ku Klux Klan—drove the Reconstruction League's governor, Jack Walton, to overstep his authority a year later. He soon found himself impeached. Nonetheless, farmer-labor cooperation modeled on the NPL example firmly changed the state's trajectory.[48]

Despite election reverses in Texas, the two-thousand-strong NPL there stuck by its union allies. In fact, it became the driving force behind a new partnership in 1922. Drawing together the state's Farm Labor Union of America (FLUA), the railroad brotherhoods, and the Texas State Federation of Labor, the League—and its women's clubs—formed the Non-Partisan Political Conference, which endorsed candidates for office in the November elections. Their candidate for governor, Fred Rodgers, earned almost two hundred thousand votes. Further encouraged by this combined farmer-labor action, the state NPL decided to merge with the FLUA and propel the coalition forward as the Farm-Labor Political Conference. The group readied itself to put forward a full farmer-labor slate for the 1924 election season. Though it succeeded in multiple counties, opponents used its association with prominent Texas NPL members to smear the group. Internal discord followed. The FLUA in Texas began to decline.[49]

The NPL's insistence on close ties to organized labor even persisted north of the border, as Leaguers in Alberta also drove farm organizations to reach out to labor. Pushed by the NPL to enter politics, the United Farmers of Alberta (UFA) took over that province's parliament in 1921. Initially, the newly politicized farm organization was wary of unions. Former Alberta NPL leaders such as William Irvine pushed the UFA to pursue fruitful alliances with laborers in Calgary and Edmonton. Irvine and other former Leaguers believed that democracy—and electoral success—depended on farmer-labor cooperation. Hewing to his own philosophy, he ran as a UFA-backed labor candidate for the House of Commons from Calgary in 1921 and won. Though UFA and labor representatives frustrated other reformers by insisting on nonpartisanship, at the federal level, the so-called Ginger Group (in which Irvine emerged as a central figure) came to include UFA and union backers alike. It served as Canada's farmer-labor coalition for nearly a decade.[50]

In 1924, a reporter for the *Nation* rightly argued that all these farmer-labor collaborations stemmed from the Nonpartisan League's "contagious educational political influence" on workers and agrarians alike. The political awakening the NPL gave farmers "not merely in the Northwest but also in the Middle and Southwest" the wherewithal to envision new electoral alliances. The "populist education" generated by the League and the "cooperative movement" it grew from pushed agrarians to see direct connections between their deteriorating situation and those of urban wage workers in economic hard times. Politically, their vision of unity paid off. At the moment when laborers started to outnumber farmers across North America, the Nonpartisan League's insistence on uniting workers and farmers in a movement for change proved yet another powerful legacy.[51]

✷

Broader political and economic shifts in the 1920s bore out the NPL's fears of a bleak future for farmers. The corporate power—embodied by Minnesota's milling and transportation interests—that animated the NPL insurgency in its earliest days seemed to triumph. Intellectuals imagined America's citizens as untrustworthy in politics. Agriculture itself gradually became more large scale and industrial. The rural economy never fully recovered from its early 1920s downturn. More farm families lost their farms. Only the ongoing struggles of the loosely allied members of the US Senate's farm bloc offered hope.

By the mid-1920s, corporations moved into the front rank of American life. Exerting influence throughout the culture, these companies consolidated their power. Concentrated capital in the hands of a few—the very thing League farmers feared—became the norm. Many called for a new investors' democracy to replace the outmoded model of republican small-property owners. The widespread ownership of corporate shares, they argued—not regulation—offered the best way to make corporations accountable to the broader polity. Most Americans acquiesced. With the exception of the depressed countryside, unending prosperity seemed at hand. Productivity increased, though industrial wages failed to match their growth. Touting the attractions of consumption and antiunion stances, businesses backed a new emphasis on management and expertise. A broad consensus backed the new order.[52]

The celebration of management as a triumph of rational growth seeped into thinking about democracy, citizenship, and politics. Bureaucracies offered safe havens for trained experts to make decisions for the public. The multiplicity of perspectives and commitments in a diverse nation demanded management. Centralized power exercised by experts offered the brightest prospect for a stable and prosperous future. However gentle and thoughtful, state coercion needed to replace untrustworthy and unpredictable citizen-centered action.[53]

A few intellectuals fought this trend. Inspired in no small part by his months in Kansas as an NPL newspaper editor, Waldo Frank declared that "man in modern America is in a state of danger." In 1929, he called for a unified nation in which commitment to civic society might trump social, economic, and ethnic difference. Instead of a "uniform" and "unconscious" machine-age society in which corporations served as "an idol which combines the values of success and the machine," he envisioned a "conscious people, a varied and integral people." Ridiculed as silly and sentimental, Frank's vision gained little traction.[54]

In this context, the moral economy of lower-middle-class farmers who in previous years backed the NPL began to decline. Decades-old conceptions of productive work, economic equity, active citizenship, and self-sufficiency became outdated. Those who held fast to such perspectives seemed backward in the emerging mass culture. The gleaming attractions of consumer culture and cities defined progress. New aspirations emerged. Middle managers in large companies filling their urban homes with the latest appliances pointed the way to the future. Consumption, leisure, house ownership, and professionalization offered a new middle-class ethic better matched to the triumphant industrial society.[55]

New visions also altered agricultural practices. Industrial values propagated by government bureaucrats and corporate innovators promised a bright new future for a depressed rural America. Small-scale family farming needed to give way to mechanized and rationalized agriculture, demanding technical expertise and larger farms. Businessmen, bankers, agricultural agents, and scientists promised that careful planning and quantification offered a way out of the economic disaster that defined farm life in the 1920s. Even the simple act of giving up horses and purchasing a tractor swept rural families up in a dense web of relationships that provided momentum for this change.[56]

Resisting these trends proved difficult. Agrarians appreciated much about the new promise of technology and progress but hoped to embrace it on their own terms. They anxiously looked to their elected representatives to articulate their complicated response. Farmers who hewed to the NPL's vision—which itself mixed a commitment to innovation and change as well as long-standing agrarian values—found little succor. In the changing context of the 1920s and 1930s, the League's successful extension of antimonopolist popular politics into an urbanizing and industrializing America struck many as antiquated. Its use of principles of economic cooperation in politics seemed old fashioned. Its work to establish the legality and promise of state-owned enterprises flew in the face of an economy dominated by corporations.[57]

Nonetheless, the Nonpartisan League label persisted in its home state. North Dakota's organization soldiered on as a voting bloc working inside the state's deeply divided but still dominant Republican Party. Notably absent from the farmer-labor turn, the League suffered from exhaustion. Stalemate characterized North Dakota politics in the 1920s. The IVA put state industries on a solid foundation even as they reined in what they envisioned as the excesses of the League legislatures. Rural women provided the difference for the League's return to power in the late 1920s. But the NPL then fell prey to internal struggles.

Nostalgia for the original insurgency set in among beaten-down farmers. Longtime Leaguer John Miklethun pointed out that "we were fighting for principles them days, there were no fighting for Jobs nor individuals at all." "It would be for the best interest of the *common good*," he told a friend, "if we trace our steps back and renew" the people-driven NPL. Instead, the former apostate William Langer seized on the opportunity presented by the organization's strife. In 1928, he refashioned the League into his personal political machine. Langer's reanimation of NPL rhetoric in the midst of the Great Depression helped him win the governorship in 1932. Overcoming controversy, he parlayed his support into a US Senate victory over his old comrade and then foe Lynn Frazier in 1940.[58]

Other western agrarians, frustrated by the League's dissolution, sought avenues outside of politics. Many retreated to strictly economic cooperative organizing. Multiple options emerged. The Equity Cooperative Exchange opened its own terminal elevator in Saint Paul in 1917 but slowly stumbled into bankruptcy. By 1925, the Farmers Union took over the equity elevator and its regional network. Inspired by the successful example of Canadian farmers, others sought to create wheat pools. A few thousand followed Arthur Townley out of the Nonpartisan League and into his next venture—the National Producers' Alliance. It brought farmers together to pool their crops in order to control the prices received for commodities. Cash-poor agrarians hoped to protect themselves with the one asset they still possessed. Townley left the group after a year, and the organization lasted only two more. In 1926, it too folded into the Farmers Union.[59]

The latter rose from the ashes of these movements as the strongest cooperative organization of the 1920s and 1930s. Braced by the passage of the Capper-Volstead Act of 1922, which protected cooperatives from federal antitrust regulation, the Farmers Union worked through local chapters to build up the commitment to co-ops in rural communities. It demanded economic justice, called for farmer unity, and railed against university agricultural extension agents. The Farmers Union added consumer as well as producer enterprises to the agrarian repertoire by sponsoring farmer-owned gas and oil providers as well as insurance companies. Trained organizers reinvigorated "saddle sore farmers" while a thorough educational program increased their ranks. In North Dakota, former Leaguer Charles Talbott led the crusade. Former NPL accountant M. W. Thatcher kept individual cooperative elevators connected through the union.[60]

A handful of farmers turned to radical politics. Longtime Washington State agrarian leader William Bouck tried to capitalize on the thousands of NPL-educated farmers in the Northwest by creating a new left-leaning agrarian organization in 1926. He traveled from his home state to Minnesota and back holding rallies and raising hopes, but the effort soon fizzled out. NPL organizer Alfred Knutson joined and helped lead the Communist-affiliated United Farmers Educational League. It retained a tiny but dedicated core of members across the northern tier into the 1930s. Many had been active members of the NPL in its heyday.[61]

Most former Leaguers had put their political trust in the agrarian bloc that they had elevated to the US Senate earlier in the decade. Far from radical, the loosely organized senators—including a few who had opposed the

NPL—came together only as shared interests aligned. From 1924 on, their energy coalesced around policies tied together in the McNary-Haugen Act. Seeking federal intervention to help agrarians facing the deepest depression yet seen in rural America, senators pushed for tariff protection and price supports. They hoped to extend the benefits of government policies supporting various industries to farmers. Ironically, opponents in the business community claimed that the search for structural equity would undercut the self-sufficiency of American farmers. They chose to ignore the ways in which federal policies helped businesses. Instead, they pushed farmers to create more independent cooperatives. The NPL experience had convinced many farmers that cooperatives alone were not enough.[62]

Agrarians demanded that McNary-Haugen be passed. Their representatives in Congress worked hard to find allies. Proponents of the legislation—tarred by one opponent as "only a few fanatics . . . who more recently were wrecking the Northwest with Townley's Non-Partisan League"—did everything they could. In the end, both houses passed the legislation twice. President Calvin Coolidge vetoed the bill both times.[63]

Between 1921 and 1933, senators originally elected with NPL support, such as Burton Wheeler (D-MT), Clarence Dill (D-WA), and Gerald Nye (R-ND), proved the most consistent supporters of McNary-Haugen and other pro-farmer legislation in the US Congress. Henrik Shipstead (FL-MN), and Lynn Frazier (R-ND) joined them by backing such measures at least 85 percent of the time. These men represented three different parties—Democratic, Republican, and Farmer-Labor—and kept the long-departed League's demands for government intervention to bring equity to the marketplace alive.[64]

They also formed the core of a bloc of western progressives in the US Senate. But without a democratic movement of farmers that cut across party lines to consistently back them, they offered few on-the-ground successes for struggling farmers. Furthermore, H. L. Mencken's *American Mercury*—a guiding light of the new urban middle class—belittled them as "a sorry band of weaklings and time servers" with "little intellectual or moral fibre." According to the magazine's contemptuous correspondent, the fortunes of "their pet cure-all, public ownership" fell prey to their own tendency to "pother, trim, and hedge."[65]

Meanwhile, the growing power of the United States Department of Agriculture (USDA) in agrarian life pointed toward a new future for the rural Midwest and West. Those who fought the McNary-Haugen Act turned to the USDA to foster programs that encouraged efficiency, industrialization, and diversification as alternatives to structural changes in commodity markets. As the agency became more powerful, bureaucrats gained more leverage

FIGURE 27. On February 25, 1924 (*left to right*), US Senators Magnus Johnson (MN), Lynn Frazier (ND), Charles McNary (OR), Henrik Shipstead (MN), and Arthur Capper (KS) called on President Calvin Coolidge to lobby for the passage of the McNary-Haugen Act. Despite repeated attempts, the proposed legislation was never signed into law. (Library of Congress Prints and Photographs Division, LC-H234-A-7899)

in agricultural production. Farmers' flight from rural America and the loss of their economic independence—two clear trends of the 1920s—became officially accepted costs of integrating agrarians into industrial society. Using cooperative models in politics and insisting that politicians be responsive to voters—key goals of the NPL—seemed more distant than ever.

When economic disaster expanded from the countryside into the city in the early 1930s, democratically spawned efforts for economic equity struck many politicians and officials as quaint. Expertise exercised by trained bureaucrats offered a much brighter future for a rapidly complicating society. This growing commitment to technocratic intervention and top-down decision making provided the foundation for confronting the agricultural crisis during the Great Depression.[66]

⋆

The election of Franklin Delano Roosevelt to the presidency in 1932 signaled serious change. That year, agrarians voted for the New Yorker in droves. They were not alone. Facing down the Great Depression demanded inventiveness. Roosevelt's electoral coalition eventually remade the Democratic Party and launched the nation in a new direction. In the 1930s, a new national political consensus began to emerge around the New Deal. The policies Roosevelt pursued became the wellspring of midcentury liberalism. But many agrarians ultimately felt left out. Liberalism's widely accepted reliance on technocratic expertise and interest-group politics replaced older visions of democracy like those articulated by the NPL.

Most farmers primed for insurgency supported Roosevelt in the 1932 election. Herbert Hoover's refusal to support a more ambitious farm policy drove farmers from the Republican Party. Prominent former Leaguers such as Burton Wheeler, Lynn Frazier, Henrik Shipstead, and Gerald Nye backed Roosevelt too. In the months before the new president took office, agriculture experts proposed a new scheme for farm relief. Devised by M. L. Wilson, an agricultural economist and USDA agent from Montana who opposed the NPL when it spread across that state in the late 1910s, it was intended to reduce production. Farmers would be paid a subsidy not to grow their usual crops. This would limit the market supply and drive up prices for commodities.[67]

Other options abounded. As early as 1925, Lynn Frazier argued for a farmer-run government corporation to fix the price of wheat. The idea went nowhere. In 1933, Frazier proposed refinancing farm mortgages to help agrarians stay on the land. His plan earned the endorsement of eight state legislatures. But Roosevelt's careful strategizing ensured that momentum for production allotments won out. In early 1933, the Agricultural Adjustment Act became federal law. Concerned about the expansion of centralized power and the top-down planning it represented, Frazier, Nye, and Borah voted against it. Ruled unconstitutional in 1936, a new version passed in 1938. It established the precedent of sizable federal price supports, a central feature of the nation's agricultural policy today. The law also paved the way for a host of other federal farm programs.[68]

Many politicians whose careers began in the NPL held a different perspective on farm relief. They demanded equity and empowerment that would preserve agrarian independence rather than spur dependence on bureaucratic experts and government programs. These men did not stand alone. The latent ideology that drove the League insurgency—a lower-middle-class vision of using government to spur an equitable economic playing field, encourage democratic cooperation, and rein in corporate power—still counted important adherents.

Across the rural Midwest, farmers joined together in a new insurgency to articulate this perspective. The Farm Holiday Association called on agrarians to control their own production in order to dictate market prices. Refusing to sell their commodities, farmers in eight states hoped to protect their property and way of life. The association included many former Leaguers in its ranks, including Alice Lorraine Daly, who used her *Dakota Farm Press* (Aberdeen, South Dakota) to outline agrarian demands. A farm strike in October 1932 gave way to foreclosure fights and political support from North Dakota's NPL governor William Langer in 1933 and 1934.[69]

In New York City, a like-minded community of urban intellectuals and anticommunist radicals gathered around a new magazine called *Common Sense*. Founded in 1932 by Alfred Bingham, its first issues depended on the investigative reporting of former NPL journalist Walter W. Liggett. Establishing itself as a non-Marxist alternative to liberalism, *Common Sense* attracted luminaries such as John Dewey, John Dos Passos, Upton Sinclair, and Lewis Mumford. Dewey's nonpartisan political organization (itself an application of the NPL's example)—the League for Independent Political Action—attached itself to *Common Sense*.

Recognizing the rural discontent embodied in the Farm Holiday Association, the group looked to farmers for support. It reached out to Thomas R. Amlie, a former US representative from Wisconsin, to create a coalition with farmer-labor parties in the Midwest. Amlie—a North Dakotan who spent 1919 as an NPL organizer in the Badger State—courted the rambunctious and popular Floyd Olson, Farmer-Labor governor of Minnesota, to lead a national movement. But the effort, dependent on an opportunity-oriented Olson, fizzled out. Meanwhile, the Farm Holiday Association dissipated as farmer-labor parties and federal monies drew down discontent.[70]

Despite the failure to create a new political coalition to oppose the New Deal, the agrarian perspective found expression in at least one piece of federal legislation that affected farmers. In 1934, Frazier and William Lemke—a former Townley lieutenant now representing North Dakota's NPL in the US House of Representatives—jointly sponsored legislation to give farmers the same rights in bankruptcy proceedings as corporations. It helped to empower agrarians and keep creditors from seizing their property. Henrik Shipstead backed the proposed law. So did many Farmers Union members.[71]

Opposed by bankers, insurance companies, and agricultural economists as "socialist," the act narrowly made it through Congress. Presidential advisors resisted, but FDR grudgingly signed the bill into law in 1934. Struck down by the Supreme Court in 1935, a modified version again became law that same year. But because local creditors fiercely opposed the policy, district courts tended to dismiss petitioners trying to use the new provisions. Only a few thousand farmers ever benefited from the legislation. Lemke's later efforts to pass farm mortgage refinancing reforms and to establish a national bank not unlike North Dakota's state bank produced nothing but opposition from FDR and his allies.[72]

The differences between farmer politicians and the New Dealers implementing Roosevelt's program grew as the 1930s dragged on. In 1936, no less an authority than journalist Walter Lippmann outlined those differences. To be sure, Roosevelt and the aging Idaho senator William Borah agreed "that large corporate wealth has exercised too much power. But they are radically different in their general feeling as to how to deal with the problem." Unlike the president, Borah—along with a handful of senators like him, most with roots in the NPL—stood out as the ultimate example of an older and long-standing rural opposition to "all concentration of power, political or economic," someone "who [was] against private privilege and private monopoly, against political bureaucracy and centralized government." Instead, Borah and his colleagues believed "in widely distributed private property, . . . competition," and "a government of limited powers."[73]

Because the USDA administered farm relief, the new policies greatly expanded the government's reach. Dependent on trained experts working in a hierarchy bringing scientific planning to bear on relief, Roosevelt's New Deal programs generally embedded technocrats as the wellspring of power and action. This included the agricultural programs. Farmer-run groups such as the Farm Bureau and the Farmers Union no longer held the most sway in agrarian life. The emerging agricultural bureaucracy of policy makers and scholars (urban and rural alike) did.[74]

Further liberal emphasis on consumption's central role in the American economy also alienated farmers once associated with the NPL. Building on trends established in the 1920s, New Dealers worked hard to firm up the formation of a new middle class, one premised on industrial employment, professionalization, purchasing power, and single-family home ownership. Intentionally replacing an old middle class of small-property-owning farmers, businesspeople, and artisans, Democratic programs largely succeeded in making liberal visions of the future a reality.[75]

Farmers met this shift with mixed feelings. Many benefited from New Deal programs such as rural electrification. Others expressed their discontent with government subsidies and price supports. A 1938 poll showed the deep divides among American agrarians on the issue. They struggled to adhere to older ways as they aspired to and adjusted to the new. The power of productive work, the strength of cooperative bonds, a commitment to economic independence, and the centrality of family persisted despite the New Deal. Resisting the pull of relief measures sometimes proved a matter of pride.

Constantly critiquing the distant urban origins of agricultural reform, agrarians often refused to be reformed.[76]

Even so, the rural clamor for government relief grew every year. The USDA's insistence that the Agricultural Adjustment Act be administered locally by county agriculture agents—a faint nod to the democratic ethos that farmers forced bureaucrats to respect—sometimes amplified neighborly differences. At the least, it made the program more palatable. Innovation always appealed to agrarians anxious to improve their lives, as long as it readily fit into their daily routines. Consumer culture, in particular, held many allures—especially for farm women anxious to reduce their daily labors.

Ultimately, economic want trumped other concerns. The moral imperative of the old middle class—that small-property-owning farmers held a central place in the social and cultural fabric of the nation by producing the source of all wealth—gave way to the demand that rural prosperity mattered as much as urban prosperity. Crushed by the combination of drought and depression, agrarians desperate for aid warily accepted the dictates of bureaucrats. Giving government the right to determine how much to grow, they compromised in order to stay on their farms. Preserving their way of life demanded it. Simultaneously proud to persist and embarrassed by relief, farm families lived out contradictions that gradually undercut the world they knew.[77]

In other parts of the United States, resistance to this technocratic turn emerged in popular politics. Embodied by the machine politics of Louisiana's flamboyant Huey Long, the right-wing reaction of radio's Father Charles Coughlin, and the old-age pension plan pitched by physician Francis Townsend, various movements responded to the New Deal. The old middle class's long-standing suspicion of centralized power—whether it was political or economic—endured. Concerned about the novel interventions of the New Deal, these crusades fueled the rise of the Union Party, which promised a new president in 1936—the former NPL leader William Lemke.[78]

Lemke, angered by Roosevelt's resistance to his proposed farm policies, took up the third party's standard. Coughlin backed the candidacy of this largely unknown New Deal dissident in the hope that unhappy agrarians might join the movement to defeat Roosevelt. Attacked by established parties as well as politicians, the Union Party garnered little support. Coughlin's anti-Semitism alienated many—including Lemke. Subordinating his distaste for the movement's racist streak to his intense dedication to monetary reform for farmers, Lemke soldiered on despite the dissonance. Frazier backed his old friend and fellow Leaguer, haranguing the two major parties for accepting corporate money and influence.[79]

The Union Party never seriously challenged FDR. Western farmers who once found their voice in the NPL—which Lemke helped to create and lead—generally rejected his alliance with reactionaries. Lemke earned only 12.8 percent of the presidential vote in his home state of North Dakota. He was, however, reelected to the US House, where he continued to advocate for farmers and rail against the New Deal.[80]

Farmers' rejection of the Union Party did not signal acceptance of the New Deal. Agrarian concerns over the centralization of power in Washington, DC, persisted. Two incidents involving former NPL politicians and senior US senators—Burton Wheeler (MT), Gerald Nye (ND), Lynn Frazier (ND), and Henrik Shipstead (MN)—illustrated farmers' unease with liberal aims to expand the scope of government in daily life as well as the Democratic Party's willingness to stick by corporate leaders.[81]

In 1937, Roosevelt, frustrated by a conservative Supreme Court that consistently struck down New Deal programs as unconstitutional, resolved to change the court itself. He proposed legislation granting the president the power to expand the number of justices. As chief executive, he would do the nominating. This would create a pro–New Deal majority on the court. Proponents demanded the change. Critics called it a "court-packing" scheme. Chief among them was Burton Wheeler. Backed by Republicans such as Frazier and Nye as well as Shipstead, the dissident Democrat created a coalition to defeat the bill. Though the former Leaguers believed in judicial reform, they saw Roosevelt's effort as a bald grab for executive power, one that frightened their constituents. They also ended their longtime but already fraying alliance with organized labor, which continued to support Roosevelt.[82]

Simultaneously, war clouds gathered in Europe and Asia. The Spanish Civil War (1936–1939) pointed to the competing ambitions of fascism and communism, while the Japanese invasion of China (1937) signaled the former's aspirations. In 1934, Gerald Nye led a Senate subcommittee investigation into war profiteering by the munitions industry during World War I. The evidence confirmed the undue influence of businesses in the march up to and prosecution of the war. It justified the NPL's initial stance against America's entry in that conflict. The findings convinced Nye that "the removal of the element of profit from war would materially remove the danger of more war." World War I taught liberals to distrust the capacity of citizens to act democratically. Nye, Frazier, and the handful of farm-bloc senators that remained learned a different lesson—a distrust of corporations partnering with government.[83]

The subcommittee's findings not only made Nye a highly visible broker on foreign affairs in Washington, DC, but also firmed up his commitment to neutrality. Senate colleagues such as Frazier, Wheeler, and Shipstead shared

FIGURE 28. North Dakota's US Senator Gerald Nye in his office in Washington, DC, 1936. (Library of Congress Prints and Photographs Division, LC-H22-D-460)

it for similar reasons. At first the stance drew widespread support. But when Germany invaded Poland in 1939, many—including Franklin Roosevelt—began pushing for US intervention. Nye railed against the president and led the charge to keep America out of the war. He and Wheeler became prominent spokesmen on the America First Committee and virulently opposed the Lend-Lease Program for Great Britain. Wheeler famously declared that Roosevelt's "New Deal triple-A foreign policy" (a direct reference to the Agricultural Adjustment Act) would "plow under every fourth American boy." Much like William Lemke, the men uneasily worked alongside disturbing anti-Semitic strains in affiliated movements because they needed allies. Constituents back home—including many former Leaguers—backed their stance.[84]

Pearl Harbor changed everything. Though the event forced every farm-bloc politician to vote for war, the decision to do so gave them little hope or comfort. When Roosevelt took the New Deal coalition into the global conflict in 1941, Nye and the others made their final break with urban laborers. To them, the once solid partners in the agrarian fight against corporate power now seemed to be in big business's pocket. Contemporaries saw the dissolution of the farm-labor alliance as a radical rural lurch to the right. Instead, it represented a consistent focus on agrarian concerns around which the broader political spectrum changed. An emerging liberal consensus, fearful of both fascism and communism, tried to bridge the gap with an emerging welfare state buoyed by a commitment to corporate capitalism. They took organized labor with them. Desperate for allies, the abandoned farm-bloc senators made unusual and unfortunate alliances.[85]

Unfairly characterized as backward, anti-Semitic, short-sighted, and even antipatriotic, Nye's firmly held stance on neutrality—one that he shared with others with roots in the NPL insurgency—anticipated the necessary mix of government and corporate power needed to successfully prosecute a modern war. World War II begat a new partnership between business and the state that amplified, enriched, and expanded the reach of both. The postwar emergence of a national security state backed by a far-reaching military-industrial complex represented the very thing Nye and his colleagues feared most.

The effect of modern liberal politics—one rarely acknowledged today—was centralized governmental power mutually supported by centralized corporate power. It confirmed the prescience of Nye and many others. Longstanding fears of centralized economic and political power loomed larger as the two began to merge. Historically, American popular politics tended to reject concentrated power of any sort. When the New Deal went to war, it refigured mass citizen participation and dramatically expanded governmental authority. It also combined the reach of large-scale corporate entities with growing state power. That combination redefined democracy as the just distribution of consumer goods and citizenship as the act of voting in formal elections. It also cemented the rightward turn in American popular politics that we live with today.[86]

The Senate farm bloc's resistance to court packing and America's entry into World War II represented the last stand of the Nonpartisan League's way of thinking in American politics. Derided in Washington, the aging men fell into decline. Lynn Frazier lost his seat to William Langer—another Nonpartisan Leaguer and fellow isolationist—in 1940. Gerald Nye lost his in 1944. In the 1946 elections, opponents defeated Burton Wheeler and Henrik Shipstead. The Democratic Party successfully forged a new consensus, one that defined American politics for the next thirty years. Liberals soon added insult to injury by claiming the NPL tradition as its own.[87]

*

Midcentury liberals rejected older visions of civic agency in favor of centralized, technical, bureaucratic power. They hoped to help interest groups navigate a corporate-state partnership. Even so, triumphant midwestern liberals claimed the NPL as the touchstone for their own politics. This memorialization drew away attention from alternative legacies, especially those that emerged in Canada. When the consensus forged by the Democratic Party began to fray in the 1960s and 1970s, a younger generation reclaimed the League as radical, not liberal. This reclamation of the NPL as a radical group persists. It befuddles today's commentators, leaving them unable to explain the presence of state-owned industries in conservative North Dakota.

As World War II neared its end, New Deal liberalism surged in the Midwest. In 1944, a merger between Minnesota's relatively weak Democratic Party and the once-powerful Farmer-Labor Party created the Democratic Farmer-Labor Party (DFL). After four more years of internal struggle, the dynamic young mayor of Minneapolis, Hubert H. Humphrey, brought the loosely knit coalition of New Dealers, Popular Front labor leaders, reformers, Communists, trade unions, and farmers under his firm control. He and his small circle of advisers propagated national Democratic Party policies. They ejected Communists and their sympathizers while drawing in the state's small African American community to undo the national party's local reputation for racism. The break with radicals mirrored a trend in national politics in the 1940s. Leftists and liberals no longer worked in coalitions. Furthermore, sharpened distinctions left little room for more complicated political perspectives and stances.[88]

Humphrey launched his national career in 1948 with a fierce attack on segregation at the Democratic National Convention in order to strengthen his antiracist credentials at home. The DFL went on to play a central role in the national party during the 1950s, 1960s, and 1970s. To Americans everywhere, Humphrey, Eugene McCarthy, and Walter Mondale symbolized the core values of American liberalism. Alongside Humphrey's other lieutenants— Orville Freeman, Eugenie Anderson, and Arthur Naftalin—they pushed the broader party away from urban machine politics and toward a consensus on civil rights and the balancing of group interests even as they remained firmly anticommunist.[89]

As they did so, they deployed the unique farmer-labor history of the DFL to assure voters that a populist streak lay at the heart of their enterprise. Publicly, they identified the continuities between their politics and earlier forms of agrarian protest. In so doing, Humphrey and the others laid claim to the NPL's insurgent spirit. Anxious to avoid any real insurgency, they used this discourse to soften their clear commitment to the two-party system and the bureaucratic solutions of the Cold War state.

This effort started in the early 1940s. Fargo-born Arthur Naftalin— Uni-

versity of Minnesota political scientist, DFL staffer, close confidant of Hubert
Humphrey, and eventually, mayor of Minneapolis—began his dissertation
on the history of Minnesota's Farmer-Labor Party. Much of the work turned
on the NPL's role in the creation of the Farmer-Labor Party and on Arthur
Townley. Naftalin not only envisioned the top-down orientation of Townley as
critical to the League's success but also found his rejection of third parties en-
thralling. Townley's NPL, in Naftalin's reading, served not as a citizen-centered
movement for political and economic democracy but as a nascent interest
group for a specific economic segment of society. Ignoring the League's effort to bring the democratic principles of cooperative organizing into politics,
Naftalin believed the NPL prefigured post–World War II liberalism's desire
to manage competing interests in American society. His dissertation further
argued that the insurgent tradition claimed by farmer-labor advocates within
the DFL really belonged to Humphrey and his associates. Naftalin completed
the dissertation in 1948, just as the DFL fell into line with the national party.[90]

As Naftalin finished his degree, a younger colleague in the university's political science department slaved away on his own dissertation. Robert Morlan
focused his work entirely on the Nonpartisan League. Like Naftalin, he also
actively worked in the DFL to make Humphrey's vision of liberalism a reality.
Doggedly tracking down League luminaries such as Arthur Townley, William
Lemke, John Baer, Joseph Gilbert, and Howard R. Wood, Morlan used inter-
views with participants to supplement his study of NPL publications and opposition pamphlets and newspapers. The revised work, published in 1955 as
Political Prairie Fire: The Nonpartisan League, 1915–1922, became the definitive
book on the movement.[91]

Morlan's deft prose and compelling plot drew in readers. He stressed "the
personality and organizing genius" of Arthur Townley, "one of the great natural leaders of protest movements which this country has produced." Morlan
also firmly insisted that the League "laid much of the foundation of modern
midwestern liberalism." Historian Larry Remele, the author of the introduction to the 1985 reprint edition agreed, suggesting that Morlan's book showed
how "a vibrant insurgent tradition" in the DFL "links the modern party to its
agrarian roots."[92]

Compelling and richly detailed, *Political Prairie Fire* even came to influence
the memories of participants. In the 1960s and 1970s, old men and women
swept up in the first years of the NPL often told League tales in language
that eerily echoed those described in Morlan's text. Only a few dissented. IVA
leader Theodore Nelson, penning his privately published memoirs in Oregon, found Morlan's lionization of Townley disgusting. "Instead of eulogiz-
ing Arthur C. Townley as a progressive minded political leader," the former

NPL opponent believed Morlan should have called him "Pilot of Political Plunderers."[93]

Inspired in no small part by the reclamation of the League by Minnesota's youthful DFL activists, young liberals in North Dakota's Democratic Party began an effort to realign politics in that state. Transforming the Nonpartisan League from a faction of the Republican Party into a new partner in the mid-century liberal consensus proved difficult. Powered by the Farmers Union and guided by Quentin Burdick—eventually a US senator—it took them nine years to bring two-party politics to North Dakota. In 1956, they finally succeeded. The liberal Democratic-NPL party that emerged tried to balance a commitment to federal farm programs that encouraged industrial agriculture with echoes of the League's demand for equity for family farmers.[94]

The League's founders paid little heed to these renovations of the NPL by liberals. Albert E. Bowen lived out a quiet retirement in Oklahoma City. Arthur Townley—who became a rabid, right-wing anticommunist—died alone in an automobile accident outside Makoti, North Dakota, in 1959. Midwestern liberals conveniently ignored these men's original intent to build an insurgency. Instead, they transformed the Nonpartisan League's status as the biggest challenge to party politics as usual in twentieth-century US history into a historical justification of the reigning liberal consensus.[95]

North of the border, the NPL's legacy followed a different trajectory. In Canada, the League's commitment to democracy over technocracy persisted. In 1931, new leadership in the United Farmers of Alberta took the province's reigning power in a novel direction. Hit hard by the depression, farmers in the prairie provinces needed allies. Their new leader, Robert Gardiner, a leading UFA MP, used his time in Ottawa as a member of the "Ginger Group" to build relationships with like-minded colleagues from across the dominion. Under his guidance, in 1932 the UFA issued a call for a meeting gathering together all those in Canada interested in establishing a new politics to confront the economic crisis.

Former Alberta NPL leader (and UFA MP) William Irvine worked closely with his colleague J. S. Woodsworth to encourage attendance. Irvine had served one term as a labor MP. Woodsworth, a prominent Winnipeg-based labor leader, garnered support from agrarians by pointing to his brief 1918 stint as an NPL organizer in Alberta. Together, they possessed the connections to create a farmer-labor coalition. Meeting in Irvine's office in Ottawa, they gathered together agrarian and union leaders from across the nation and prepared for the Calgary conference.[96]

Irvine and Woodsworth, meeting with reformers from across the country in Calgary, showed their commitment to nonpartisanship and insisted on a movement, not a party. The labor leaders present ensured a Socialist presence. The large contingent of representatives from the UFA and Ontario farm groups insisted on muting the socialist language. The resulting Cooperative Commonwealth Federation (CCF) proudly claimed a farmer-labor coalition with a socialist twist. Rooted in the popular politics of the prairie provinces, Woodsworth, Irvine, and others hoped to create a distinctive Canadian variant of socialism, one that reflected agrarian cooperative experiences and perspectives. Others in the coalition advocated a labor-inflected and urban-centered socialist party. The power of the former slowly eroded, even as the movement rose to power in the 1940s.[97]

Meanwhile, many in Alberta's provincial UFA leadership rejected the turn to socialism. Some constituents drifted toward a new political option in Alberta. The Social Credit Party, founded by an evangelical minister, focused on reforming Canada's monetary policy. It rejected the socialist leanings of the CCF even as it drew in agrarian voters frustrated with the UFA and the emerging urban working class in Calgary and Edmonton. In 1935, Social Credit swept away the UFA. Social Credit then took a sharp rightward turn after winning reelection in 1940. After World War II, it built a power base on the province's newfound oil wealth. Rigidity replaced reform.[98]

In Saskatchewan, the CCF's clout grew. As a party, it ran the largely agricultural province from 1944 to 1961. Though it gradually drifted from an insistence on equitability in a cooperative commonwealth, it created numerous public-owned industries during its first term. In 1961, it joined with mainline labor unions to found the federal New Democratic Party. The Saskatchewan CCF's best-known achievement—a single-payer universal health care system that became the model for all of Canada—continues to be recognized today. In 2004, Saskatchewan's CCF premier Tommy Douglas (1944–1960) won the title of Greatest Canadian.[99]

As a direct descendent of the Nonpartisan League (and many other movements), the CCF transformed all of Canada. Though it confronted—and was shaped by—Canada's postwar liberal turn, its work for democracy and equity persisted. In contrast, except in North Dakota, in the United States the NPL lingered as little more than a resonant nomenclature. Subsumed into the dominant liberalism of the era, the League seemed part of a distant past.

*

In the 1970s, a generation influenced by the New Left and its excesses discovered the League anew. Two young members of a Marxist film collective in San Francisco—John Hanson and Rob Nilsson—returned to their North Dakota roots. They hoped to create a film about the rise of the Nonpartisan

League. Excited by the young men's interest in the state's past, North Dakotans welcomed them with open arms. Convinced that they could tell the story of the "radical roots of a conservative state," the unknown filmmakers depended largely on support offered by the North Dakota Humanities Council. Between 1975 and 1977, public backing for the film that became *Northern Lights* (1978)—expressed in the willingness of locals to donate spaces, serve as extras, and even suggest appropriate scenes—provided a grassroots flavor to the entire enterprise. Meanwhile, colleagues in the Marxist film collective back in San Francisco knew little of what they were doing.[100]

A miniscule budget required enrolling local farmers to work alongside a handful of underpaid professional actors. With their talented colleague, cinematographer Judy Irola, the men finished most shots with only one camera in one or two takes. Hanson and Nilsson often fine-tuned the script just minutes before they filmed specific scenes. Shuttling between the West Coast and Crosby, North Dakota, the filmmakers relied heavily on the recollections of old-timers such as former North Dakota socialist and NPL organizer Henry Martinson. In fact, more than half a century after the NPL's rise and fall, Hanson and Nilsson used the film *Northern Lights* to turn League organizer Henry Martinson into an unlikely movie star.[101]

More than anyone the filmmakers encountered, Martinson spoke with real authority. As a homesteader in western North Dakota, he knew the difficulties of wheat farming firsthand. Martinson's participation in the state's Socialist Party meant that he personally knew leading figures of the League such as Albert Bowen, Arthur Townley, and Arthur LeSueur. As an NPL employee in 1918 and 1919, Martinson enrolled farmers across Minnesota and North Dakota. He went on to serve North Dakota as deputy commissioner in the state's Department of Agriculture and Labor from 1937 to 1965. In 1979, Martinson even became North Dakota's poet laureate. Recognizing the authenticity in the elderly radical's sparkling personality, the young filmmakers used Martinson at the beginning and end of *Northern Lights* to play the role of an old man recalling the movement's glory days.[102]

Long after most Americans had forgotten the NPL, Martinson's compelling performance in *Northern Lights* helped bring it new notoriety. After its premier in Crosby, North Dakota, in July 1978, the plucky little film made its way to theaters across the Upper Midwest. Then, in May 1979, the movie won the Camera d'Or prize for the best first feature at the Cannes Film Festival. Feted alongside fellow prize winners such as Terrence Malick, Sally Field, and Jack Lemmon, the feisty filmmakers suddenly became darlings of a nascent independent film movement.[103]

Critics across the country took notice. Alongside the spectacular cinematography, Martinson's performance received special attention. Vincent Canby, the *New York Times* film columnist, suggested that Martinson was the reason to see the movie. As the centerpiece of "the stunningly photographed, fictionalized story," he heralded the old man's memory as "our periscope on the past." Indeed, more than the compelling prairie landscapes or hardscrabble tale, Martinson, the ninety-four-year-old who "still does push-ups," was "what one remembers best of all." *Chicago Sun-Times* movie reviewer Roger Ebert also noticed Martinson, observing that the "old man" telling the story of the "now nearly forgotten" NPL "is real, and so are the events he will recall."[104]

Hanson and Nilsson chose their narrator wisely. Martinson's presence provided the authority of the story unfolding on the screen. Moviegoers thrilled at the stark realism. The old man's rendering of the League as a radical movement that put the "powers that be on the run" appealed to many Americans aching for social change amid the economic and political dislocations of the 1970s. Martinson embodied—and perpetuated—a romantic vision of the past. Even the filmmakers, anxious to avoid doctrinaire preaching, fell sway to Martinson's compelling vision: ideologically driven agrarians embracing political revolution in the most unlikely of places.[105]

Yet Martinson's own biography belied this vision of the NPL. As the last secretary of North Dakota's Socialist Party, he saw comrade after comrade leave the party to join the League's staff. But he resisted. Martinson held fast to his conviction that the NPL lacked a revolutionary spirit. Not until 1918, when the Socialist Party completely fell apart, did he finally sign on as a League organizer. Even then, he only worked for the NPL in Minnesota and North Dakota. He did not experience the movement as one that extended geographically across the North American West.[106]

Like so many others, Martinson remembered what he wanted to when ruminating on the Nonpartisan League experience. In the wake of the Great Depression, World War II, the Cold War, and the triumph of liberalism in American politics, his memory seized on the novelty of a moment when socialism existed as an ideological option. Rendering the League's rise and fall through this lens, the elderly activist hoped to remind later generations that pre–Cold War socialism constituted an indigenous radical tradition in America. He chose to emphasize this as the NPL's legacy.

Martinson's vision—and the physical separation from heady Bay Area sectionalism—encouraged Hanson and Nilsson to reflect on how they might create their "own language" and "forms" free from any New Left party or

FIGURE 29. The State Bank of North Dakota building in Bismarck, completed in 2008, illustrates the institution's success in its mission to promote "agriculture, commerce, and industry" across the state. (Photograph by author)

sectarian line. The popular democracy inherent in the League story inspired them. Soon after the film's production, the filmmakers' collective broke up. But they could not escape the reach of the New Left. As the movie gained more and more critical acclaim, frustrated second-wave feminists rightly argued that the absence of powerful women in the film distorted the historical experience. Largely ignoring this critique, young radicals and left-leaning liberals thrilled by *Northern Lights* rushed to learn more. Most found their way to Morlan's *Political Prairie Fire*. By 1985, demand for the thirty-year-old book proved strong enough for it to find its way back into print.[107]

It took two more decades for another wave of interest in the NPL to emerge. The economic downturn of 2008—and North Dakota's continued prosperity despite it—piqued national interest in the League's state-owned industries. Filmmaker Michael Moore, frustrated by the neoliberal ascendency, profiled North Dakota's state bank in his *Capitalism: A Love Story* (2009). Soon thereafter, he called on left-leaning activists to work to establish similar banks across the country. Media outlets such as *Newsweek*, the *Huffington Post*, *Mother Jones*, *Salon*, and the *New York Times* followed up with stories about rising interest in the Bank of North Dakota. Its successful promotion of businesses, low-interest student loans for residents, and profitability for the state itself grabbed attention. They also noted the anomaly of its support in a generally conservative place. Ignoring what closely informed observers call the "potent mix of egalitarianism and populism" that characterizes contemporary North Dakota politics, boosters and detractors alike understood the bank as the outcome of a socialist movement rather than a result of a cooperative capitalist call for equity in the marketplace. Tied tightly to socialism in most writers' imaginations, a real movement for state banks failed to materialize.[108]

✶

One hundred years after its founding, the NPL refuses to completely disappear. Shaped by ever-changing political contexts, the League's legacies morph to meet the needs of new generations. This makes understanding the complicated mixture of popular politics and agrarian moral economy at the heart of the Nonpartisan League experience more difficult. Despite dangers and limitations, our current predicament demands cutting through these memories and creating the past anew for our time.

FIGURE 30. North Dakota's State Mill and Elevator produces a variety of flour-based products that can be found across the Upper Midwest, like these seen on grocery store shelves in Grand Forks, North Dakota, in 2014. (Photograph by author)

The intense commitment to citizen action in electoral politics distinguishes the NPL insurgency from political movements and parties today. Making democracy a daily habit meant reimagining one's relationship to the political process as well as one's fellow citizens. Politics was not something that only politicians practiced. For many Leaguers, politics took place not just on election day but in rural schoolhouses, farmer-led precinct caucuses, conversations with neighbors, community picnics, churches, and women's club meetings—wherever and whenever farmers and their families gathered. Prodded by an NPL organizer, the *Nonpartisan Leader,* or even electoral success, farm men and women gradually built their capacity to work together across difference toward a common goal. Fostered by experience and example, this capacity changed lives as well as policy. The League's rapid success showed that neither parties nor longtime officeholders determined the destiny of desired change. In many cases, citizen-politicians, elevated by their neighbors, took the place of professional politicians.

In fact, the NPL transformed voters' relationship to party politics. Instead of political campaigns funded by the rich, its membership model made for political financing from the bottom up. Each purchase of a membership represented a financial as well as an ideological investment in an organization that represented and engaged self-interest. The work to create and then build consensus around a platform—before selecting candidates—shifted voters' attention from a nominee's personality and popularity to the policies that mattered. In turn, citizens then held elected officials accountable, regardless of party affiliation. Equally important, that platform rejected ideological litmus tests and neutered potentially divisive issues through a fixation on shared interests. The creative use of the direct primary brought the platform to bear on party politicians who did not welcome it.

Newly elected Leaguers, like their constituents, demanded democratic intervention in markets, not bureaucratic oversight of the same. The NPL's insistence on government competition in particular economic sectors transcends current debates over the size and influence of government or the scope of government regulation. Instead of focusing on big or small government, Leaguers concentrated their energies on how public policy might better reflect the will of the citizenry. Local and state government would be the institutions through which citizens cooperated to mutual individual benefit.

Driven by a commitment to cooperation, the NPL railed against corporations but not capitalism. Creating an equal chance to succeed or fail in a marketplace—not the equal distribution of wealth or consumer goods—defined most Leaguers' vision. In an age of transnational capital and globalization, curbing corporate power has never seemed further from reach. Yet few know that citizens deeply vested in small-scale capitalism have consistently offered the most powerful criticism of corporations.[109] The NPL hoped to ensure that individuals enjoyed the same economic rights as concentrated capital in any given market. Public ownership of competitive productive and financial institutions in the marketplace offered the vehicle for exactly that. In an age of profound economic inequity, this fix deserves more attention.

Farmers in the NPL also drew on long experience in cooperation to integrate economic needs and political power. Their understanding of cooperatives proved more expansive than the mere creation and support of producer or consumer co-ops. It represented an alternative vision of capitalism itself, one deeply rooted in the American experience and tightly tied to democracy. In 1953, the former NPL leader Joseph Gilbert described members' ideal relationship between politics and economics this way: "Cooperation is democracy applied to making a living." He and others harbored a full-fledged ambition to create citizen-centered democracy in both governance and the economy. They understood economic democracy—equity—and political democracy as interdependent. As a modern reclamation of a long-standing strain of American political economy, the NPL illustrates the capacity of this tradition to be reenvisioned for our time. Any such work should keep these forebears in mind even as it, like the League, introduces innovation.[110]

Finally, the League illustrates the limits of our contemporary political imagination. Its defiance of easy categorizations today reveals a great deal about contemporary America. Whenever we envision politics along a simplified left-right spectrum, we do ourselves a disservice. Our everyday political language puts limits on what we can see and imagine. It does not allow for richer conversations about democracy and its relation to the economy—past or present. Any effort to reanimate our decrepit civic institutions will require not only the reclamation of less familiar traditions but also a discourse that challenges settled assumptions. The health of our democracy depends on it.

Recovering the Nonpartisan League story for the twenty-first century matters because memory resists acquiescence. It is important to remember the League as a antimonopolist popular force with national and even transnational reach, as a powerful movement for citizen-centered democracy during the Progressive Era, as a tool for rural women's empowerment, as a bridge between farmers and laborers in industrializing America, as a pioneer in bringing cooperative thinking to politics, as inventors of an innovative approach to elections, as a thoroughly modern combination of corporate efficiencies and corporate critique, as proffering an alternative form of capitalism, and as a retort to entrenched party politics, because these insights help us better understand the relevance of the NPL to today. The men and women who created and fought for the League—as well as those who fiercely opposed it—deserve nothing less.

Epilogue

The history of the League is, therefore, an outstanding example of the power which an organized minority can wield in American politics if it uses the proper tactics.

SAMUEL P. HUNTINGTON, "The Election Tactics
of the Nonpartisan League," 1950

Today, many Americans find any form of popular politics unsettling. Enthusiastic and engaged citizens questioning settled assumptions threaten those who know better. These deviations from the norms of party politics strike the establishment as aberrations. The nation, they claim, is best governed by those who pay attention, those with training at the best colleges and universities, those deeply steeped in statistical analysis, memo writing, and *realpolitik*. Democracy—whether on the right or the left—demands limits.

Yet scandals continue to emanate from America's city halls, statehouses, and most notably, Washington, DC. Billions in federal spending and corporate lobbying make that metropolitan area the wealthiest in the nation. Beltway denizens strive for celebrity alongside power, conspicuously consuming. Bureaucratic struggles, predicated on procedural nuance, define daily life. Idealism dies. Money pervades. Cronyism triumphs. Seamy tell-alls titillate the ruling class. Settled assumptions seem frail, at best.

Freedom—more of it than in any generation in human history, we are told—will allow us to determine our futures as we please in the midst of ecological (and thus collective) catastrophe. Saving America's largest banks while forcing millions out of their mortgages and homes comes to seem like common sense. Promises to improve daily living with access to cheap credit and consumer goods (a project most American politicians dedicate themselves to, despite other differences) shrink the political imagination. The public's sense of possibility in politics atrophies.

The average person realizes and accepts this. Long cynical about the concentration of power—economic or political, or in our corporatized and privatizing society, both—citizens turn away from politics. Distant bureaucracies dictate. Injustices in daily life deflate. Equality (fictional as it may still be for some) without equity offers little. Solutions seem far off. Unmoored from sureties, citizens stop voting. They feel disconnected. They fulfill the stereotype outlined for them by elites. "We, the people" just don't seem to care much.

All this makes any eruption of popular sentiment unusual. On the rare occasions one begins to threaten an established party, concern mounts. Reason, not emotion, it is hoped, will prevail. Training, the ruling elite argues, must trump passion. Liberals and conservatives alike often agree: easily manipulated by media as corrupt as our politics, the public cannot be trusted. Amazingly, come election time, politicians of every stripe then turn to populist themes and rhetoric to connect with the few citizens still swept up in political life. In this way, populism—a twisted and hollow ghost of the real thing—still drives politics today.

This take on popular politics even extends to our past. Memory and history—contested, flawed, and slippery—play tricks on the living. The rich variety of popular movements in American political history tend to be remembered as xenophobic, irrational, and antimodern. At best, they embody a nostalgic, long-gone past. At worst, they represented a nadir in America's otherwise persistent trajectory of improvement. Those who disagree often lift up equally problematic parables of grassroots gods.

Without lapsing into romantic portrayals, we must better understand not only the context in which the NPL evolved but also the ways in which it illuminates our lives today. The present-day context for popular politics—a consumer society, the corporate capture of capitalism, money in politics, and the entrenchment of institutionalized bureaucracies—have distorted the content and form of popular politics as well as our understanding of them. By fashioning the League's story for our time—just as past generations sought to do so in theirs—this book hopes to contribute to the restoration of a broader political imaginary.

In an age of corporate power, dissatisfaction with government, and economic hardship not unlike that faced by western farmers in the 1910s, the tactics and ideology of insurgent democracy embodied by the NPL remains a powerful model for creating citizen-centered politics. Antimonopoly politics rooted in the experiences of middling folk provided an antidote to corporate control. The dangerous defiance of internal democracy outlined the limits of League tactics and organization.

Though an astute contemporary called the Nonpartisan League "one of the century's outstanding political events in America," the movement garners little attention today.[1] To be sure, we seem far removed from a world where agrarians innovated, where over 50 percent of the nation's population lived on a farm, and where citizens believed they could change their political and economic situation. But League farmers envisioned democracy—the capacity of people working together despite their differences to solve common problems—not as a naïve ideal but as a realistic aspiration. They imagined interdependence as a precondition of independence. They embraced responsibilities alongside rights. They prized productive work—not conspicuous consumption—as the pinnacle of a life lived well. They envisioned economic equity to be as crucial as legal and social equality. The men and women transformed by their experience in the Nonpartisan League have much to teach our society, one where citizens often rebuff the right to rule and where economic divides grow deeper every year.

In its rejection of party-based ideologies, loyalties, and structures, the NPL shows us that two-party political systems can be transformed without constitutional transformations or revolutionary tumult. Its emphasis on cooperative organization and economic equity suggests ways to undercut the concentration of wealth in the hands of a few. Finally, its nonpartisan approach to primaries and caucuses, organized around a clear platform and a committed citizen-based membership, suggests an alternative way into electoral politics for those without money and clout. We would do well to heed its example.

Acknowledgments

Writing this book was hard work. My debts are many. It's as simple as that.

The process always begins with archivists and librarians. Many thanks to Doris Peterson, University of South Dakota; Linda Hein, Nebraska State Historical Society; Carolyn Bowler, Idaho State Historical Society; John Bye and Michael Robinson, North Dakota Institute for Regional Studies; Donna McCrea, University of Montana; Zoe Anne Stoltz, Montana State Historical Society; Jim Davis, State Historical Society of North Dakota; Jennifer Jones, South Dakota Oral History Center; Carl Hallberg, Wyoming State Archives; Jim Bowman and Susan M. Kooyman, Glenbow Archives; and the staffs at the Kansas State Historical Society, the Saskatchewan Archives Board (Saskatoon), and the Special Collections at the University of North Dakota. At the Minnesota Historical Society, Patrick Coleman—himself an NPL scholar—tracked down long-lost sources and helped me identify especially telling photographs.

A Dean's Summer Scholarship Grant from Augsburg College helped me start the project. Monies from the institution's Mildred Joel Fund for Canadian Studies helped me finish it. At Augsburg, fellow historians Phil Adamo, Jacqueline deVries, Bill Green, Don Gustafson, and Maheen Zaman graciously put up with my absences, distractions, and obsessions. Other colleagues, including Lars Christiansen, Sarah Combellick-Bidney, Robert Cowgill, Darcey Engen, Christina Erickson, Nancy Fischer, Doug Green, Elise Marubbio, Kathleen McBride, Sarah Meyers, Tim Pippert, Mohamed Sallam, Mary Laurel True, Marty Stortz, and Joe Underhill provided insights in conversation and solace in commiseration. Harry Boyte, Dennis Donovan, and Elaine Eschenbacher taught me much about democracy in practice as well as theory. From start to finish, Bill Wittenbreer proved to be an invaluable partner. At the end, Karen Kaivola saved me from myself. Finally, Matt Charboneau did excellent work as an undergraduate research assistant. As for the other alumni of HIS 480, consider this book your final assignment. Let me know what you think.

I spent the 2011–2012 academic year as the Lloyd Lewis Fellow in American History at the Newberry Library in Chicago. It made all the difference. Dick Brown, Chris Cantwell, Heath Carter, Diane Dillon, Leon Fink, Michael Goode, Danny Greene, Ben Johnson, Liesl Olson, Susan Sleeper Smith, Scott Stevens, and Daniel Wasserman-Soler as well as my fellow fellows Randy Head, Judith Miller, Monica Prasad, Ben Robinson, Helen Thompson, and Sarah Rivett challenged me to do better and then showed me how.

Others provided sources, insights, and encouragement. Tom Alter helped me understand agrarian politics in Texas and shared his research finds. Robert Caulkins was the first to use declassified Bureau of Information files related to the NPL and generously alerted me to them. Thomas Contois shared his as-yet unpublished manuscript, a gem that needs to find its way into print. Jason McCollom helped me understand transnational connections among agrarians in new ways. Marcia Anderson, Annette Atkins, Jane Carroll, Jeff Kolnick, Debbie Miller, and Shannon Smith helped me figure out what I was trying to say. Erik and Susan Holland always offered hearty support. Winston Chrislock and Odd Lovoll cheered me on as I embarked on an unanticipated journey down a path pioneered by Winston's father. That path—also marked by the work of Larry Remele, Scott Ellsworth, and William Pratt—made this book possible.

A long time ago, David L. Holmes, Fredrika Teute, and James Axtell ensured that I learned how to write, read, think, and make sense out of the sources. Later on, Anne M. Butler (whom I dearly miss), Sara M. Evans, and a host of others helped me to sharpen those skills. I am grateful to all of them for it.

Lawrence Culver got me going by challenging me to prove the NPL's significance. When I convinced him, I knew I was on to something. Cynthia Prescott graciously invited me to speak at the University of North Dakota, where I started to work out the big picture. Lori Lahlum perused drafts, shared sources, and kept the encouragement coming. Kirsten Delegard's sage advice and close reading propelled me forward at a critical moment. Mark Fiege has believed in me for a long time. His faith matters greatly. David Rich Lewis persuaded others that this was the book I was born to write. I hope I didn't fail him. Robert Johnston read every word—more than once—and made this book possible. Jeff Kolnick introduced me to the literature, critiqued draft chapters, and offered unwavering support. I couldn't have done it without him. Don't let him tell you otherwise.

I'm grateful that Matt Klingle insisted I send a proposal to the University of Chicago Press. There, Robert Devens believed in my work as much as I did. After Robert left, Timothy Mennel patiently and skillfully guided me through a foreign world. He enrolled anonymous referees who consistently challenged me to get things right. Though I have fallen short of their aspirations, I thank them for making the book much better. The manuscript also benefited from George Roupe's careful and precise copyediting.

For years, Mark Fiege, Matt Klingle, David Rich Lewis, Gregory Smoak, Jay Taylor, and Bob Wilson—better known as the VBL—offered up insanity that kept me sane. Every June, the guys at the Marino asked, "How's the book coming along?" Well, here it is. Save your questions for Shazam's Lounge. Scott Carlson, Curtis Christoff, Mark and Laura Dahl, Cathy Fitch, Kasha Foster, A. Scott Harris, Matthew Lungerhausen, Michael Noble, and Evan Roberts, never read a word but contributed more than they know.

My own ancestors supported and then opposed the NPL. That discovery helped me better understand them as well as the League. I wrote a fair bit of this book sitting at my grandfather Erling Vinje's desk. As a North Dakota dirt farmer with an eighth-grade education, he insisted on the value of hard work. His daughter Gail and Arlen Johnson kept me well fed, sheltered, and in good spirits whenever I was in Bismarck. Cary and Tina Clark and Gunnel Clark continue to lavish undeserved attention on me. Robert Johannes, Lisa Clark, Brian Quinn, Annika Quinn, and Duncan Quinn are the best. Jay and Jean Lansing rarely asked about the book, which was usually what I needed. Kevin Lucken joined us just in time to have some real fun. I love Jill, David, and Chloe Edsen fiercely. Here's to the good times.

My parents, George and Darlene Lansing, made it all possible for me. I will never be able to repay them. The values I hold dear come from them. I hope this makes them both proud. As for Nina Clark, all I can say is that she is everything, and everything is for her.

Notes

Prologue

1. Thomas L. Haskell, "Objectivity Is Not Neutrality: Rhetoric vs. Practice in Peter Novick's *That Noble Dream*," *History and Theory* 29, no. 2 (May 1990): 129–157. Various traditions inform my inclination for a usable past: feminist scholarship, social critics of American democracy and life such as Christopher Lasch, and historians working in the midwestern vein, identified by David S. Brown in *Beyond the Frontier: The Midwestern Voice in American Historical Writing* (Chicago: University of Chicago Press, 2009).

Chapter One

1. *Bismarck (ND) Tribune*, June 28, 1916.

2. "The Political Storm Becomes a Hurricane," *Nonpartisan Leader* (Fargo, ND), March 9, 1916.

3. *Bismarck Tribune*, June 30, 1916; *Nonpartisan Leader*, July 13, 1916.

4. *Nonpartisan Leader*, July 13, 1916.

5. David Danbom, *Born in the Country: A History of Rural America* (Baltimore: Johns Hopkins University Press, 1995), 161–167.

6. Robert F. Zeidel, "Peopling the Empire: The Great Northern Railroad and the Recruitment of Immigrant Settlers to North Dakota," *North Dakota History* 60, no. 2 (Spring 1993): 14–23; Eric Foner, *Give Me Liberty: An American History*, 3rd ed., vol. 2, *From 1865* (New York: W. W. Norton, 2012), 598.

7. Catherine McNichol Stock, *Main Street in Crisis: The Great Depression and the Old Middle Class on the Northern Plains* (Chapel Hill: University of North Carolina Press, 1992), 13–14; Elwyn B. Robinson, *History of North Dakota* (Lincoln: University of Nebraska Press, 1966), 235–247; *Fourteenth Census of the United States*, 1920 Bulletin: Agriculture, North Dakota (Washington, DC: Government Printing Office, 1921), 3.

8. Robinson, *History of North Dakota*, 247–248.

9. Barbara Handy Marchello, *Women of the Northern Plains: Gender and Settlement on the Homestead Frontier, 1870–1930* (Saint Paul: Minnesota Historical Society Press, 2005), 85–115.

10. Stock, *Main Street in Crisis*, 41–62.

11. Ibid., 63–85; John C. Hudson, *Plains Country Towns* (Minneapolis: University of Minnesota Press, 1985), 104–130. Robert D. Johnston's *The Radical Middle Class: Populist Democracy and the Question of Capitalism in Progressive Era Portland, Oregon* (Princeton, NJ: Princeton University Press, 2003) elucidates the complicated and contingent nature of the middle class in early twentieth-century America.

12. Mary W. M. Hargreaves, *Dry Farming in the Northern Great Plains, 1900–1925* (Cambridge, MA: Harvard University Press, 1957), 8–17; Theodore Saloutos, "The Spring Wheat Farmer in a Maturing Economy, 1870–1920," *Journal of Economic History* 6, no. 2 (November 1946): 173–190.

13. Robinson, *History of North Dakota*, 248–249.

14. William Cronon, *Nature's Metropolis: Chicago and the Great West* (New York: W. W. Norton, 1991), 97–147, 376–377.

15. Lucile M. Kane, *The Waterfall That Built a City: The Falls of St. Anthony in Minneapolis* (Saint Paul: Minnesota Historical Society Press, 1966).

16. Barbara Levorsen, *The Quiet Conquest: A History of the Lives and Times of the First Settlers of Central North Dakota* (Hawley, MN: Hawley Herald, 1974), 47.

17. "A Farmer's Testimony," *Nation* 112, no. 2921 (June 29, 1921): 919.

18. Levorsen, *The Quiet Conquest*, 34.

19. James H. Stock, "Real Estate Mortgages, Foreclosures, and Midwestern Agrarian Unrest, 1865–1920," *Journal of Economic History* 44, no. 1 (March 1983): 89–105; Allen G. Bogue, "Foreclosure Tenancy on the Northern Plains," *Agricultural History* 39, no. 1 (January 1965): 3–16; *Fourteenth Census of the United States*, 1920 Bulletin: Agriculture, North Dakota, 4.; E. F. Ladd, *A Revelation of Facts Not Generally Known: Senator E. F. Ladd of North Dakota Makes Timely Reply to Former President Taft's Malicious Attack on the Farmers of the Country* (Washington, DC, 1921).

20. Alvin S. Tostlebee, "The Bank of North Dakota: An Experiment in Agrarian Banking," *Studies in History, Economic, and Public Law* 64, no. 1 (1924): 32–35; Charles Edward Russell, "The Non-Partisan League," in *Papers and Proceedings, Eleventh Annual Meeting of the American Sociological Society* (Chicago: University of Chicago Press, 1917), 32–33.

21. Larry Remele, "'Things as They Should Be': Jeffersonian Idealism and Rural Rebellion in Minnesota and North Dakota, 1910–1920," *Minnesota History* 51, no. 1 (Spring 1988): 15–22; Adam Ward Rome, "American Farmers as Entrepreneurs, 1870–1900," *Agricultural History* 56, no. 1 (January 1982): 37–49; Hudson, *Plains County Towns*, 128. By 1920 in Walsh County, North Dakota, "a typical farmer would have 22 indebted 'neighbors.'" Stock, "Real Estate Mortgages," 95.

22. Charles Postel, *The Populist Vision* (New York: Oxford University Press, 2007); Larry Remele, "'God Helps Those Who Help Themselves': The Farmer's Alliance and Dakota Statehood," *Montana: The Magazine of Western History* 37, no. 4 (Autumn 1987): 22–33; Glenn V. Brudvig, "The Farmer's Alliance and Populist Movement in North Dakota, 1884–1896," (master's thesis, University of North Dakota, 1956); Robinson, *History of North Dakota*, 223–224; John D. Hicks, *The Populist Revolt: A History of the Farmers' Alliance and the People's Party* (Minneapolis: University of Minnesota Press, 1931), 287–288.

23. Robert P. Wilkins, "Alexander McKenzie and the Politics of Bossism," in *The North Dakota Political Tradition*, ed. Thomas Howard (Ames: Iowa State University Press, 1981), 3–39; Charles N. Glabb, "The Revolution of 1906: N.D. vs. McKenzie," *North Dakota Quarterly* 24, no. 4 (Fall 1956): 101–109.

24. Glabb, "The Revolution of 1906," and Glabb, "John Burke and the Progressive Revolt," in

The North Dakota Political Tradition, ed. Thomas Howard (Ames: Iowa State University Press, 1981), 40–65.

25. Theodore Saloutos and John D. Hicks, *Twentieth-Century Populism: Agricultural Discontent in the Middle West, 1900–1939* (Lincoln: University of Nebraska Press, 1951), 111–140; Theodore G. Nelson, *Scrapbook Memoirs* (Salem, OR: Your Town, 1957), 34–48; Scott Ellsworth, "The Origins of the Nonpartisan League," (PhD diss., Duke University, 1982), 126–183.

26. Jason McCollom, "Political Harvests: Transnational Farmers' Movements in North Dakota and Saskatchewan" (PhD diss., University of Arkansas, 2014); Louis Aubrey Wood, *A History of Farmers' Movements in Canada* (1924; repr., Toronto: University of Toronto Press, 1975), 183–222; Saloutos and Hicks, *Twentieth-Century Populism*, 134.

27. For more on the cooperative mind-set, see Brett Fairbairn, "The Meaning of Rochdale: The Rochdale Pioneers and the Co-operative Principles," Occasional Paper Series, Centre for the Study of Co-operatives, University of Saskatchewan, 1994; Steven J. Keillor, *Cooperative Commonwealth: Co-ops in Rural Minnesota, 1859–1939* (Saint Paul: Minnesota Historical Society Press, 2000).

28. Theodore Saloutos, "The Rise of the Equity Cooperative Exchange," *Mississippi Valley Historical Review* 32, no. 1 (June 1945): 31–62; Usher L. Burdick, *The Life of George Sperry Loftus: Militant Farm Leader of the Northwest* (Baltimore: Wirth Brothers, 1939), 40; James Manahan, *Trials of a Lawyer* (Minneapolis: Farnham, 1933), 208–212.

29. Laura O'Day, "Buffalo as a Flour Milling Center," *Economic Geography* 8, no. 1 (January 1932): 81–93.

30. Quoted in Ellsworth, "Origins of the Nonpartisan League," 102–103; Ole Olson, interview by unknown, October 10, 1951, New Rockford, North Dakota, North Dakota Institute for Regional Studies, North Dakota State University, Fargo, ND (hereafter NDIRS).

31. Larry Remele, "North Dakota's Forgotten Farmers Union, 1913–1920," *North Dakota History* 45, no. 1 (Spring 1978): 4–21; Walter S. Losk, "The Nonpartisan League, the Farmers Union, and the Press of North Dakota" (master's thesis, University of Minnesota, 1951), 39–40; *Bismarck Tribune*, June 23, 1916.

32. Robert H. Bahmer, "The Economic and Political Background of the Nonpartisan League" (PhD diss., University of Minnesota, 1941), 162–260.

33. Danbom, *"Our Purpose Is to Serve": The First Century of the North Dakota Agricultural Experiment Station* (Fargo: North Dakota Institute for Regional Studies, 1990), 60–72; Hargreaves, *Dry Farming in the Northern Great Plains*, 187–195.

34. Paul R. Fossum, "The Agrarian Movement in North Dakota," *Johns Hopkins University Studies in Historical and Political Science* 43 (1925): 71–75.

35. "The A.C. Mill," no date, no author, folder 18, box 2, E. F. Ladd Papers, NDIRS; E. F. Ladd, "Is the Present System of Grading Wheat Equitable?" *Special Bulletin No. 14*, 3 (Fargo: North Dakota Agricultural College Experiment Station, January 1915); Ladd, "Chemical and Physical Constraints for Wheat Products," *Bulletin No. 114* (Fargo: North Dakota Agricultural College Experiment Station, January 1916).

36. Robinson, *History of North Dakota*, 275; William Langer, *The Nonpartisan League: Its Birth, Activities and Leaders* (Mandan, ND: Morton County Farmers Press, 1920), 12.

37. *Iconoclast* (Minot, ND), February 20, 1914; *Twelfth Census of the United States*, James River Valley Township, Dickey County, North Dakota, 1900; *Thirteenth Census of the United States*, Martin Township, Walsh County, North Dakota, 1910; Ellsworth, "Origins of the Nonpartisan League," 218; *Bowman (ND) Citizen*, March 12, 1914; *Bowman Citizen*, November 19, 1914.

38. Nick Salvatore, *Eugene V. Debs: Citizen and Socialist* (Urbana: University of Illinois Press, 1982). The great exception was Oklahoma, where socialism thrived. See James R. Green, *Grass-Roots Socialism: Radical Movements in the Southwest, 1895–1943* (Baton Rouge: Louisiana State University Press, 1978); and Jim Bissett, *Agrarian Socialism in America: Marx, Jefferson, and Jesus in the Oklahoma Countryside, 1904–1920* (Norman: University of Oklahoma Press, 2002).

39. William C. Pratt, "Socialism on the Northern Plains, 1900–1924," *South Dakota History* 18 (Spring–Summer 1988): 1–35; Jackson Putnam, "The Role of the North Dakota Socialist Party in North Dakota History," *North Dakota Quarterly* 24, no. 4 (Fall 1958): 114–122; Perry Joel Hornbacher, "The North Dakota Socialist Party" (master's thesis, North Dakota State University, 1982); Donald B. Marti, "Answering the Agrarian Question: Socialists, Farmers, and Algie Martin Simons," *Agricultural History* 65, no. 3 (Summer 1991): 53–69; Henry P. Richardson, "Scientific Organizing and the Farmer," *International Socialist Review* 15, no. 9 (March 1915): 554–558; Charles Johnston, "Socialism and the American Farmer," *North American Review* 682 (September 1912): 307.

40. "North Dakota State Platform, 1911," folder 4, box 1, Arthur LeSueur Papers, Minnesota Historical Society, Saint Paul, MN (hereafter MNHS); Ellsworth, "Origins of the Nonpartisan League," 184–193; William A. Henke, *Prairie Politics: Parties and Platforms in North Dakota, 1889–1914* (Bismarck: State Historical Society of North Dakota, 1974), 112–121, 133–137; *Iconoclast*, March 6, 1914; A. Halvorson to Arthur LeSueur, March 6, 1914, folder 6, box 1, Arthur LeSueur Papers, MNHS.

41. Joseph Gilbert, interview by Kathryn Johnson, January 27, 1954, Minneapolis, MN, Joseph Gilbert Papers, MNHS.

42. *Iconoclast*, January 20, 1914; *Iconoclast*, May 29, 1914; *Iconoclast*, March 20, 1914; Arthur LeSueur to Charles H. Kerr, June 18, 1914, folder 6, box 1, Arthur LeSueur Papers, MNHS; *Bowman Citizen*, September 3, 1914.

43. Larry Remele, *The Lost Years of A. C. Townley (after the Nonpartisan League)* (Bismarck: North Dakota Humanities Council, 1988); *Beach (ND) Advance*, October 1, 1909; Herbert E. Gaston, *The Nonpartisan League* (New York: Harcourt, Brace, and Howe, 1920), 45–50; *Wibaux (MT) Pioneer*, October 18, 1912.

44. *Iconoclast*, March 28, 1913; *Iconoclast*, February 20, 1914; *Iconoclast*, March 20, 1914.

45. *Bowman Citizen*, March 12, 1914; *Iconoclast*, March 20, 1914; *Bowman Citizen*, March 26, 1914.

46. *Bowman Citizen*, June 14, 1914.

47. Walter Eli Quigley, "Out Where the West Begins," typescript, ca. 1932, MNHS; *Iconoclast*, September 11, 1914; W. E. Quigley, "The Truth about the Non-Partisan League," *Lincoln (NE) Daily Star*, April 14, 1919; Harrison B. French, "Bowman Was Townley's First Testing Ground," in *Prairie Tales*, ed. Bowman Rural Area Development Committee, Bowman County, ND (Sioux Falls, SD: Midwest-Beach Printing, 1965), 287–288.

48. Socialist Party District Two (Williams and McKenzie Counties) Organization Department Accounting Book, 1914, State Historical Society of North Dakota, Bismarck, ND (hereafter SHSND).

49. *Iconoclast*, October 23, 1914; *Iconoclast*, December 11, 1914.

50. Despite insistent claims that Arthur Townley invented the Nonpartisan League, he did not. See *Ward County Independent* (Minot, ND), April 15, 1915; *Deering (ND) Enterprise*, May 13, 1915; "'I, Not Townley, Started League,' Non Party Candidate Declares," *Bismarck Tribune*, April 29, 1920; Leon Durocher to unknown, March 29, 1920, in Asher Howard, *The Leaders*

of the Nonpartisan League: Their Aims, Purposes and Records . . . (Minneapolis: published by author, 1920), 35; Larry Remele, "The Immaculate Conception at Deering," *North Dakota History* 47, no. 1 (Winter 1980): 28–31; Ellsworth, "Origins of the Nonpartisan League," 279–285; Homer Ayres, interview by Earl Hausle, June 28, 1971, Sturgis, South Dakota, interview #187, South Dakota Oral History Project, University of South Dakota, Vermillion, SD.

51. *Bowman Citizen*, November 19, 1914; *Iconoclast*, November 13, 1914.

52. Ellsworth, "Origins of the Nonpartisan League," 181–183; *Co-operator's Herald* (Fargo, ND), November 13, 1914; *Co-operator's Herald*, December 5, 1913; *Co-operator's Herald*, October 30, 1914.

53. *Iconoclast*, December 25, 1914; *Iconoclast*, January 8, 1915.

54. *Iconoclast*, December 25, 1914; *Iconoclast*, January 8, 1915; Ellsworth, "Origins of the Nonpartisan League," 280–281; S. Roy Weaver, *The Non-Partisan League in North Dakota* (Toronto: Canadian Reconstruction Association, 1921), 7–8.

55. *Journal of the House of Representatives* (Bismarck: State of North Dakota, 1915), 189; Usher L. Burdick, *History of the Farmers' Political Action in North Dakota* (Baltimore: Worth Brothers, 1944), 69.

56. *Iconoclast*, February 19, 1915.

57. *Iconoclast*, February 5, 1915; *Iconoclast*, February 12, 1915.

58. Ellsworth, "Origins of the Nonpartisan League," 234–238.

59. *Bismarck Tribune*, February 3, 1915; Bahmer, "Economic and Political Background," 439; *Co-operator's Herald*, February 5, 1915; "Fourth Man to Join the League," *Nonpartisan Leader*, January 27, 1916.

60. *Iconoclast*, February 26, 1915. The typesetter miscast one line of type in the resignation letter. The original reads: "I believe the time has come for those men in all parties who desire real progress (and no party has a monopoly on these men) to stand together in favor of a program that shall be so specific in character that difficulty connected with a useless an- it will appeal at once to the fair minded in all parties. And realizing the tagonism along the paths of partisan ship, I propose to teach FIRST a program and then follow this with investigation."

61. *Portland (ND) Republican*, February 18, 1915; *Douglas (ND) Herald*, February 25, 1915; Larry Remele, "The Nonpartisan League and the North Dakota Press: Organization Period, 1915–1916," *North Dakota Quarterly* 44, no. 4 (Autumn 1976): 35.

62. Durocher to Leonard Sackett, July 31, 1955, NDIRS; Otto M. Thomason, "The North Dakota Uprising," typescript, 1939, Otto M. Thomason Papers, MNHS; Arthur LeSueur, "The Nonpartisan League," ca. 1920, Correspondence and Miscellaneous Papers, ca. 1918 folder, box 2, Arthur LeSueur Papers, MNHS.

63. Ellsworth, "Origins of the Nonpartisan League," 243–246; Larry Remele, "The Immaculate Conception at Deering," *North Dakota History* 47, no. 1 (Winter 1980): 28–31.

64. Gaston, *The Nonpartisan League*, 60; contract, April 12, 1915, folder 1 (Scrapbook), box 2, John C. Hagan Papers, University of North Dakota, Grand Forks, ND (hereafter UND).

65. Ellsworth, "Origins of the Nonpartisan League," 254–255; Edward C. Blackorby, *Prairie Rebel: The Public Life of William Lemke* (Lincoln: University of Nebraska Press, 1963), 25.

66. Initial membership drives focused on Ward, McHenry, Wells, and Eddy Counties. See Larry Remele, "The Nonpartisan League and the North Dakota Press," 35–46; and Arthur C. Townley, interview by Richard C. Dobson, date and place unknown (ca. 1950s), folder 38, box 1, Nonpartisan League Collection, NDIRS. Lawrence Goodwyn famously posited that the People's Party systematically grew out of local cooperative movements. Critics suggest otherwise. See

Stanley B. Parsons, Karen Toombs Parsons, Walter Killilae, and Beverly Borgers, "The Role of Cooperatives in the Development of a Movement Culture of Populism," *Journal of American History* 69, no. 4 (March 1983): 866–885. The NPL may be a better example of cooperative movements developing into political forces than Goodwyn's Populists.

67. Remele, "The Nonpartisan League and the North Dakota Press," 37, 45.

68. Gerald Gamm and Robert D. Putnam, "The Growth of Voluntary Associations in America, 1840–1940," *Journal of Interdisciplinary History* 29 (Spring 1999): 511–557; Nelson, *Scrapbook Memoirs*, 61; Henke, *Prairie Politics*, 71–72, 78–81, 137–138, 141–143.

69. Martin J. Schiesl, *The Politics of Efficiency: Municipal Administration and Reform in America, 1880–1920* (Berkeley: University of California Press, 1980), 68–87; Carl H. Chrislock, *The Progressive Era in Minnesota, 1899–1918* (Saint Paul: Minnesota Historical Society Press, 1971), 60–61; *Farmers' Open Forum* (Washington, DC) 1, no. 1 (September 1, 1915): 12; *Farmers' Open Forum* 3, no. 8 (April 1918): 16; Hugh Lovin, "The Farmer Revolt in Idaho, 1914–1922," *Idaho Yesterdays* 20 no. 3 (Fall 1976): 5; W. G. Scholtz to Ray McKaig, December 27, 1916, reel 1, National Nonpartisan League Papers, MNHS.

70. The Nonpartisan League's origins and commitment to popular democracy stand out today, making it part of the tradition that Robert Johnston describes as the "radical middle class" and Robert Westbrook calls "petty bourgeois radicalism." See Johnston, *The Radical Middle Class*, x; and Robert Westbrook, *Democratic Hope: Pragmatism and the Politics of Truth* (Ithaca, NY: Cornell University Press, 2005), 135–136. Yet by the standards of their own time, these farmers were anything but radical. In 1939, Benton H. Wilcox noted that the bulk of the League's members were "ordinary business men . . . seeking to correct injustices in the marketing and credit systems . . . and endeavoring to build up the wealth of the community in which they were citizens." Benton H. Wilcox, "An Historical Definition of Northwestern Radicalism," *Mississippi Valley Historical Review* 26, no. 3 (December 1939): 394. See also John D. Hicks, "The Legacy of Populism in the Western Middle West," *Agricultural History* 23, no. 4 (October 1949): 225; Louis Geiger, "Conservative Reform and Rural Radicalism," *North Dakota Quarterly* 28, no. 1 (Winter 1960): 6; William C. Pratt, "Rural Radicalism on the Northern Plains, 1912–1950," *Montana: The Magazine of Western History* 42, no. 1 (Winter 1992): 44; Thomas P. Shilts, "Political Culture on the Northern Plains: North Dakota and the Nonpartisan Experience" (master's thesis, University of North Dakota, 1997), 6.

71. N. C. Abbott, "Social Center Development in North Dakota," *North Dakota Quarterly* 2, no. 3 (April 1912): 356.

72. Lynn Haines, *The Lindberghs* (New York: Vanguard, 1931), 275–276; Burdick, *History of the Farmers' Political Action*, 78.

73. *Nebraska State Journal*, October 31, 1919, Non-Partisan League Scrapbook, Nebraska State Historical Society, Lincoln, NE.

74. Glabb, "John Burke and the Progressive Revolt," 62–63; Robert D. Johnston, "The Possibilities of Politics: Democracy in America, 1877–1917," in *American History Now*, ed. Eric Foner and Lisa McGirr (Philadelphia: Temple University Press, 2011), 96–124.

75. Lawrence Goodwyn, *Democratic Promise: The Populist Moment in America* (New York: Oxford University Press, 1976). Theodore Saloutos and John Hicks imagined later farmer organizations as firmly derived from Populism. Their *Agricultural Discontent in the Middle West, 1900–1939* (Madison: University of Wisconsin Press, 1951) came out in paperback later that year with a new title: *Twentieth-Century Populism: Agricultural Discontent in the Middle West, 1900–1939* (Lincoln: University of Nebraska Press, 1951). Grant McConnell, *The Decline of Agrarian*

Democracy (Berkeley: University of California Press, 1953), reiterated this stance. David Danbom's *Born in the Country: A History of Rural America* (Baltimore: Johns Hopkins University Press, 1995) notes the shared orientation of the NPL and the Populists.

76. Postel, *The Populist Vision.*

77. "Farm Union Coming to Colorado to Emancipate Farmers of State," *Great Divide* (Denver, CO), January 24, 1917, reel 1, National Nonpartisan League Papers, MNHS.

78. Arthur LeSueur, untitled manuscript, 1918, Correspondence and Miscellaneous Papers ca. 1918 folder, box 2, Arthur LeSueur Papers, MNHS; Elisabeth S. Clemens, *The People's Lobby: Organizational Innovation and the Rise of Interest Group Politics in the United States, 1890–1925* (Chicago: University of Chicago Press, 1997), 145–183.

79. Frederick M. Davenport, "The Farmers' Revolution in North Dakota," *Outlook* 114 (October 11, 1916): 325–327. Philip Scranton's *Proprietary Capitalism: The Textile Manufacture at Philadelphia, 1800–1885* (New York: Cambridge University Press, 1983) points to the "historical coexistence of multiple successful paths to profit and accumulation" in the United States (4). For a careful exploration of "capitalists against capitalism," see Johnston, *The Radical Middle Class,* 74–89.

80. Michael Merrill, "The Anti-Capitalist Origins of the United States," *Review, a Journal of the Fernand Braudel Center* 13, no. 4 (Fall 1990): 465–497; Terry Bouton, *Taming Democracy: "The People," the Founders, and the Troubled Ending of the American Revolution* (New York: Oxford University Press, 2007); Ronald P. Formisano, *For the People: American Populist Movements from the Revolution to the 1850s* (Chapel Hill: University of North Carolina Press, 2008); Gretchen Ritter, *Goldbugs and Greenbacks: The Antimonopoly Tradition and the Politics of Finance in America, 1865–1896* (New York: Cambridge University Press, 1997); Bruce Palmer, *"Man over Money": The Southern Populist Critique of American Capitalism* (Chapel Hill: University of North Carolina Press, 1980); Christopher Lasch, *The True and Only Heaven: Progress and Its Critics* (New York: W. W. Norton, 1991); Johnston, *The Radical Middle Class.* Eric Foner suggests that the failure of socialism in America resulted from the ongoing presence of "an older republican tradition hostile to large accumulations of property, but viewing small property as the foundation of economic and civic autonomy." "Why Is There No Socialism in the United States?," *History Workshop Journal* 17, no. 1 (Spring 1984): 63.

81. During the early twentieth century, lines between leftists and liberals often blurred. See Doug Rossinow, *Visions of Progress: The Left-Liberal Tradition in America* (Philadelphia: University of Pennsylvania Press, 2008).

82. Patrick McGuire, "Death of a Myth: The Nonpartisan League and the Socialist Party of America, 1912–1920," in *13th Dakota History Conference, April 9, 10, 11, 1981, Papers* (Madison, SD: Dakota State College, 1982): 604–606.

83. *Nebraska State Journal,* October 31, 1919, Non-Partisan League Scrapbook, Nebraska State Historical Society, Lincoln, NE; "The Farmers' Crusade: Letters from George Cronyn, a Non-Partisan League Organizer," *Liberator* 1, no. 8 (October 1918): 7; Gaston, *The Nonpartisan League,* 62; Stock, *Main Street in Crisis,* 71–73; Quigley, "Out Where the West Begins," 22.

84. A partial list of Socialists employed by the Nonpartisan League includes Albert Bowen, Arthur Townley, Leon Durocher, Beecher Moore, O. M. Thomason, Henry Teigan, Joseph Gilbert, D. C. Coates, O. S. Evans, Henry Martinson, Arthur LeSueur, and Walter Thomas Mills. See Howard, *The Leaders of the Nonpartisan League*; Roy S. Weaver, *The Nonpartisan League in North Dakota* (Toronto: Canadian Reconstruction Association, 1921), 12; and Geiger, "Conservative Reform and Rural Radicalism," 7.

85. "Farmers Can Organize," *Nonpartisan Leader*, October 14, 1915; Charles Edward Russell, *The Story of the Nonpartisan League: A Chapter in American Evolution* (New York: Harper and Bros., 1920), 213–214.

86. H. J. Trelease, J. L. Morken, and I. P. Quam (Committee), *The Political Situation in Nelson County* (Fargo, ND: Cooperative Print Shop, 1916), NDIRS.

87. Ibid.

88. Many ascribe the emergence of the Nonpartisan League to a Scandinavian penchant for socialism and social democracy. See Saloutos, "The Rise of the Equity Cooperative Exchange," 37; Robinson, *History of North Dakota*, 329; Michael Rogin, *The Intellectuals and McCarthy: The Radical Specter* (Cambridge, MA: MIT Press, 1967), 106, 119; Jon Wefald, *A Voice of Protest: Norwegians in American Politics, 1890–1917* (Northfield, MN: Norwegian-American Historical Association, 1971); Kathleen Moum, "Harvest of Discontent: The Social Origins of the Nonpartisan League, 1880–1922" (PhD diss., University of California–Irvine, 1986), 109–110; R. Douglas Hurt, "Agricultural Politics in the Twentieth-Century American West," in *The Political Culture of the New West*, ed. Jeff Roche (Lawrence: University Press of Kansas, 2008), 55; Karen V. Hansen, *Encounter on the Great Plains: Scandinavian Settlers and the Dispossession of Dakota Indians, 1890–1930* (New York: Oxford University Press, 2014), 202–203. But Carl H. Chrislock and Daron William Olson both carefully refute this assumption. See Chrislock, *Ethnicity Challenged: The Upper Midwest Norwegian-American Experience in World War I* (Northfield, MN: Norwegian-American Historical Association, 1981), 95; and Olson, "Norwegians, Socialism and the Nonpartisan League in North Dakota, 1904–1920: How Red Was Their Protest?" (master's thesis, University of North Dakota, 1993).

89. Olson, "Norwegians, Socialism, and the Nonpartisan League," 17, 84–85; William C. Sherman and Playford V. Thorson, eds., *Plains Folk: North Dakota's Ethnic History* (Fargo: North Dakota Institute for Regional Studies, 1986), 196–200; Odd S. Lovoll, *Norwegian Newspapers in America: Connecting Norway and the New Land* (Saint Paul: Minnesota Historical Society Press, 2010), 269; Gaston, *The Nonpartisan League*, 77–78.

90. Era Bell Thompson, *American Daughter* (Chicago: University of Chicago Press, 1946), 42.

91. *Nonpartisan Leader*, November 18, 1915; Sherman and Thorson, *Plains Folk*, 400; boxes 15 and 16, Harry Lashkowitz Papers, NDIRS; Gregory Orfalea, "Mosque on the Prairie," in *Taking Root: Arab-American Community Studies*, vol. 2, ed. Eric Hooglund (Washington, DC: American Arab Anti-Discrimination Committee, 1985), 8; William C. Sherman, *Prairie Mosaic: An Ethnic Atlas of North Dakota* (Fargo: North Dakota Institute for Regional Studies, 1983), 48–50; *State of North Dakota 1919 Legislative Manual* (Bismarck: Bismarck Tribune, 1919), 256–261, 282–284.

92. Burdick, *History of the Farmers' Political Action in North Dakota*, 82.

93. *Nonpartisan Leader*, January 6, 1916; *Nonpartisan Leader*, September 30, 1915.

94. Ray McKaig to J. C. Kelley, February 4, 1918, folder K, "NPL Correspondence," box 1, Ray McKaig Papers, Idaho State Historical Society, Boise, ID.

95. Gaston, *The Nonpartisan League*, 61.

96. Trelease, Morken, and Quam (Committee), *The Political Situation in Nelson County*.

97. James A. McCullough, "Why the League Is," *Nonpartisan Leader*, April 5, 1917.

98. Quigley, "Out Where the West Begins," 17.

99. Burdick, *History of the Farmers' Political Action in North Dakota*, 83.

100. Thomason, "The North Dakota Uprising."

101. Charles Edward Russell, *In and Out of the Yoke: A Plain Story of the Farmer and the Nonpartisan League* (Saint Paul, MN: Allied Printing, 1917), 23; *Nonpartisan Leader*, September 22, 1915.

102. Edward C. Blackorby, *Prairie Populist: The Life and Times of Usher L. Burdick* (Fargo: North Dakota Institute for Regional Studies and State Historical Society of North Dakota, 2001), 142–144; William Lemke to William Langer, July 10, 1915, folder 17, box 1, and William Lemke to William Langer, November 1, 1915, folder 18, box 1, William Lemke Papers, UND; Blackorby, *Prairie Rebel,* 34.

103. "League Organizers Start Speaking Campaign," *Nonpartisan Leader*, December 16, 1915; Larry Remele, "Power to the People: The Nonpartisan League," in *The North Dakota Political Tradition*, ed. Thomas W. Howard (Ames: Iowa State University Press, 1981), 76; *Nonpartisan Leader*, January 27, 1916; Thomason, "The North Dakota Uprising."

104. *Nonpartisan Leader*, September 30, 1915.

105. *Fargo (ND) Forum*, January 24, 1916; Remele, "The Nonpartisan League and the North Dakota Press," 45; A. C. Townley to Nonpartisan League members, February 10, 1916, folder 8, box 1, Mayme Hokana Papers, SHSND.

106. Townley to Nonpartisan League members, February 10, 1916.

107. Gaston, *The Nonpartisan League*, 93–97; S. A. Olsness to P. O. Thorson, March 27, 1916, folder 3, box 1, S. A. Olsness Papers, NDIRS.

108. "The League's Caucuses Prove Political Sensation" and "Caucus Reports Show Spirit of United Farmers," *Nonpartisan Leader*, March 2, 1916; Gaston, *The Nonpartisan League*, 99–100.

109. Robert L. Morlan, *Political Prairie Fire: The Nonpartisan League, 1915–1922* (1955; repr., Saint Paul: Minnesota Historical Society Press, 1985), 49–50.

110. Nelson, *Scrapbook Memoirs*, 99.

111. Quigley, "Out Where the West Begins," 7; Townley, interview by Dobson; Gaston, *The Nonpartisan League*, 115.

112. Trelease, Morken, and Quam (Committee), *The Political Situation in Nelson County*.

113. Ellsworth, "The Origins of the Nonpartisan League," 270; *The Co-operator's Herald*, January 28, 1915; "L. J. Frazier," in *Walsh Heritage: A Story of Walsh County and Its Pioneers*, ed. Walsh County Bicentennial Historic Book Committee (Grafton, ND: Walsh County Historical Society, 1976), 366; Nels Erickson, *The Gentleman from North Dakota, Lynn J. Frazier* (Bismarck: State Historical Society of North Dakota, 1986), 7.

114. *Fargo Forum*, April 1, 1916; *Nonpartisan Leader*, April 6, 1916; "Official Program, Non-Partisan League, March 31, April 1, 1916, Fargo, ND," North Dakota NPL Collection, UND.

115. Victor Wardrope to William Lemke, April 1, 1916, and William Lemke to Victor Wardrope, April 11, 1916, folder 20, box 1, William Lemke Papers, UND; William Lemke to William Langer, February 17, 1916, and J. M. Anderson to William Lemke, March 2, 1916, folder 19, box 1, William Lemke Papers, UND.

116. *Nonpartisan Leader*, April 6, 1916; *Nonpartisan Leader*, May 18, 1916.

117. O. T. Rishoff to John Gillette, March 26, 1916, folder 19, box 1, John M. Gillette Papers, UND.

118. *Nonpartisan Leader*, April 13, 1916; Thomason, "The North Dakota Uprising."

119. Louis B. Hanna to O. J. Sorlie, May 12, 1916, Oscar J. Sorlie Papers, NDIRS.

120. E. Smith Peterson to O. J. Sorlie, May 10, 1916, Oscar J. Sorlie Papers, NDIRS.

121. *Nonpartisan Leader*, April 13, 1916; *Nonpartisan Leader*, April 27, 1916; Blackorby, *Prairie Populist*, 146.

122. Davenport, "The Farmers' Revolution in North Dakota," 325; H. C. Harty to J. M. Anderson, March 31, 1916, folder 19, box 1, William Lemke Papers, UND.

123. Trelease, Morken, and Quam (Committee), *The Political Situation in Nelson County*.

124. *Nonpartisan Leader*, June 1, 1916.

125. *Nonpartisan Leader*, June 15, 1916; *Nonpartisan Leader*, June 22, 1916.

126. *Nonpartisan Leader*, June 22, 1916.

127. A. E. Brine to Leonard Sackett, August 30, 1955, folder 2, box 1, Nonpartisan League Collection, 1917–1964, NDIRS; Davenport, "The Farmers' Revolution in North Dakota," 325.

128. Remele, "North Dakota's Forgotten Farmers Union"; *Bismarck Tribune*, June 23, 1916.

129. Charlie Barrer to *Nonpartisan Leader*, June 19, 1916, and June 30, 1916, reel 1, National Nonpartisan League Papers, MNHS.

130. Blackorby, *Prairie Rebel*, 44; Morlan, *Political Prairie Fire*, 80–81; Wilkins, "Alexander McKenzie and the Politics of Bossism," 35.

131. Robert P. Wilkins, "Referendum on War? The General Election of 1916 in North Dakota," *North Dakota History* 36, no. 4 (Fall 1969): 296–335.

132. Wilkins, "Referendum on War? The General Election of 1916 in North Dakota"; Remele, "Power to the People," 79; Russell, *The Story of the Nonpartisan League*, 221.

133. Remele, "Power to the People," 79–81; membership card, folder 1, Thorwald Mostad Papers, SHSND.

134. Thompson, *American Daughter*, 10, 56.

135. Morlan, *Political Prairie Fire*, 93; "Farmers Boss Government of North Dakota," *Chicago Tribune*, January 4, 1917; John Howard Todd, "Long, Hard Road Facing Dakota Farmers' League," *Minneapolis Tribune*, January 3, 1917.

136. "A Farmers' State," *New York Times*, January 4, 1917, 10; *Nonpartisan Leader*, January 4, 1917; Morlan, *Political Prairie Fire*, 92; "New Administration," *Bismarck Tribune*, December 9, 1916.

137. Morlan, *Political Prairie Fire*, 101–102; John Howard Todd, "Farmers Say People Demand Reform in N.D. Constitution," *Minneapolis Tribune*, January 5, 1917.

138. Arthur Guy Divet, "Biography," typescript, ca. November 1927, Arthur Guy Divet Papers, SHSND.

139. Remele, "Power to the People," 81; Morlan, *Political Prairie Fire*, 105.

140. "Equity and League Talk Peace Plans," *Bismarck Tribune*, December 11, 1916.

141. J. G. Grites to O. H. Olson, February 28, 1917, and J. M. Anderson to O. H. Olson, March 13, 1917, folder 16, box 9, Ole H. Olson Papers, NDIRS.

142. Remele, "Power to the People," 81; Marcella Andre, "They Won the Right to Vote . . . But Little Else," in *Women on the Move*, ed. Pearl Andre (Bismarck: North Dakota Democratic-NPL Women, 1975), 35–46.

143. "A Farmer's State," *New York Times*, January 4, 1917, 10; correspondence, December 1916, January 1917, and February 1917, reels 1 and 2, National Nonpartisan League Papers, MNHS.

144. Marion Butler to A. C. Townley, February 9, 1917, reel 2, National Nonpartisan League Papers, MNHS; James L. Hunt, *Marion Butler and American Populism* (Chapel Hill: University of North Carolina Press, 2003).

Chapter Two

1. "Non-Partisan Public Meeting," *Alberta Non-Partisan* (Calgary), February 8, 1918.

2. Louise C. McKinney, "A Message to Our Members: A Wider Democracy," *Alberta Non-Partisan*, February 8, 1918.

3. Hilda Ridley, "Pen Portraits of Progressive Women," *Christian Guardian*, February 16, 1921, clipping in file 7, box 1, Louise Crummy McKinney Fonds, Glenbow Archives, Calgary,

AB (hereafter GA); "Alberta Department of League Very Active," *Nonpartisan Leader of Western Canada* (Regina, SK), May 30, 1917.

4. A. B. Gilbert, "Out for a 'Solid West': The Coming Political Battle of the Nonpartisan League," *Forum* 60 (December 1918): 727.

5. Louis Levine, "Will Agrarian Movement Affect Our Politics?," *New York Times Magazine*, March 18, 1917, 6.

6. Ibid.

7. The crucial difference between mobilizing and organizing is outlined in Charles Payne, *I've Got the Light of Freedom: The Organizing Tradition in the Mississippi Freedom Struggle* (Berkeley: University of California Press, 1996); and Harry C. Boyte, *Everyday Politics: Reconnecting Citizens and Public Life* (Philadelphia: University of Pennsylvania Press, 2004), 34.

8. Louis Levine, "Farmers Causing Political Upheaval in the West," *New York Times Magazine*, February 27, 1917, 51.

9. "Silas E. Haight," Ancestry.com, *1900 United States Federal Census*, online database; "Silas E. Haight," Ancestry.com, *1906 Canada Census of Manitoba, Saskatchewan, and Alberta*, online database. Unfortunately, Haight cannot be found in the 1910 US census or the 1911 Canadian census. See also Paul F. Sharp, *The Agrarian Revolt in Western Canada: A Survey Showing American Parallels* (Minneapolis: University of Minnesota Press, 1948), 77.

10. "S. E. Haight Family," in Osnabrock Area Historical Committee, *The Spirit Lives On: Osnabrock, ND, 1887–1987* (Grafton–Grand Forks, ND: Associated Printers, 1987), 277; "Silas E. Haight," Ancestry.com and the Church of Jesus Christ of Latter-Day Saints, *1916 Canada Census of Manitoba, Saskatchewan, and Alberta* online database; Sharp, *The Agrarian Revolt in Western Canada*, 77.

11. Joseph E. Taylor III, "Boundary Terminology," *Environmental History* 13 (July 2008): 454–481; Sheila McManus, *The Line Which Separates: Race, Gender, and the Making of the Alberta-Montana Borderlands* (Lincoln: University of Nebraska Press, 2005), 83–141.

12. Paul F. Sharp, "When Our West Moved North," *American Historical Review* 55, no. 2 (January 1950): 286–300; John Herd Thompson, "Political Divergence between the U.S. and Canadian Northern Plains, c1900 to the 1960s," paper presented at Western History Association Conference, Lake Tahoe, NV, October 14, 2010; *Census of Prairie Provinces: Population and Agriculture, Manitoba, Saskatchewan, Alberta, 1916* (Ottawa: J. de Labroquerie Tache, 1918), xxx.

13. Paul F. Sharp, "The Northern Great Plains: A Study in Canadian-American Regionalism," *Mississippi Valley Historical Review* 39, no. 1 (June 1952): 72–73; Malcolm J. Morrison, "Sir Charles Edward Saunders, Dominion Cerealist," *Genome* 51, no. 6 (2008): 465–469.

14. By the late 1910s, wheat accounted for 55 percent of all crops grown in Saskatchewan and Alberta. *Census of Prairie Provinces*, 288, 292, 306; John C. Lehr, John Everitt, and Simon Evans, "The Making of the Prairie Landscape," *Prairie Forum* 33, no. 1 (Spring 2008): 1–38.

15. John Herd Thompson, *The Harvests of War: The Prairie West, 1914–1918* (Toronto: McClelland and Stewart, 1978), 59–60; Peter R. Sinclair, "Class Structure and Populist Protest: The Case of Western Canada," in *Society and Politics in Alberta: Research Papers*, ed. Carlo Caldarola (Toronto: Methuen, 1979), 78; Paul Voisey, *Vulcan: The Making of a Prairie Community* (Toronto: University of Toronto Press, 1988), 201–246; Donald G. Wetherell and Irene R. A. Kmet, *Town Life: Main Street and the Evolution of Small Town Alberta, 1880–1947* (Edmonton: University of Alberta Press, 1995).

16. Georgina Binnie-Clark, *Wheat and Woman* (1914; repr., Toronto: University of Toronto Press, 1979), 302; L. H. Neatby, *Chronicle of a Pioneer Prairie Family* (Saskatoon, SK: Western

Producer Prairie Books, 1979), 23; Bill Waiser, "'Land I Can Own': Settling in the Promised Land," in *The Prairie West as Promised Land*, ed. R. Douglas Francis and Chris Kitzan (Calgary: University of Calgary Press, 2007): 155–174; Ian McPherson and John Herd Thompson, "The Business of Agriculture: Prairie Farmers and the Adoption of 'Business Methods,' 1880–1950," in *Canadian Papers in Business History* (Victoria, BC: Public History Group, University of Victoria, 1989): 245–269.

17. James S. Woodsworth, *Strangers within Our Gates; or, Coming Canadians* (1909; repr., Toronto: University of Toronto Press, 1972); Thompson, *The Harvests of War*, 73–94; Sheila Mc-Manus, "Gender(ed) Tensions in the Work and Politics of Alberta Farm Women," in *Telling Tales: Essays in Western Women's History*, ed. Catherine A. Cavanaugh and Randi R. Warne (Vancouver: University of British Columbia Press, 2000), 123–146; Sarah Carter, *The Importance of Being Monogamous: Marriage and Nation Building in Western Canada to 1915* (Edmonton: University of Alberta Press, 2008); Sandra Rollings-Magnusson, "Canada's Most Wanted: Pioneer Women on the Western Prairies," *Canadian Journal of Sociology and Anthropology* 37, no. 2 (May 2000): 223–238; Jeremy Adelman, *Frontier Development: Land, Labour, and Capital on the Wheatlands of Canada and Argentina, 1890–1914* (New York: Oxford University Press, 1994).

18. John C. Everitt, "The Borderlands and the Early Canadian Grain Trade," in *Borderlands: Essays in Canadian-American Relations* (Toronto: ECW, 1991), 156–159, 171; John C. Everitt and Donna Shimanura Everitt, "American Influences in the Canadian Grain Trade: An Overview," in *Bulletin of the Association of North Dakota Geographers* 34 (1984): 1–9; "James Edward Gage" in *Barnes County (ND) History*, ed. Barnes County Historical Society (Dallas: Taylor, 1976), 76; John G. McHugh, *Thirty-Fifth Annual Report of the Minneapolis Chamber of Commerce of Minneapolis* (Minneapolis, 1918), 195–204.

19. Louis Aubrey Wood, *A History of Farmers' Movements in Canada* (1924; repr., Toronto: University of Toronto Press, 1975), 160–163, 170; John Everitt and Roberta Kempthorne, "The Flour Milling Industry in Manitoba since 1870," *Manitoba History* 26 (Autumn 1993), http://www.mhs.mb.ca/docs/mb_history/26/flourmilling.shtml, accessed October 21, 2011; Sharp, *The Agrarian Revolt in Western Canada*, 56, 64–69; Thompson, *The Harvests of War*, 65–66; *Grain Growers' Guide* (Winnipeg, MB), March 1, 1922, quoted in Sharp, "When Our West Moved North," 288–289; Joseph Santos, "Going against the Grain: Why Do Canada and the United States Market Wheat So Differently?," *American Review of Canadian Studies* 40, no. 1 (March 2010): 104–117.

20. Ian MacPherson, *Each for All: A History of the Co-operative Movement in English Canada, 1900–1945* (Toronto: Macmillan, 1979), 1–66.

21. Sharp, *The Agrarian Revolt in Western Canada*, 32–53; Wood, *A History of Farmers' Movements in Canada*, 173–174, 194, 197–198, 201.

22. Harald S. Patton, *Grain Growers' Cooperation in Western Canada* (Cambridge, MA: Harvard University Press, 1928), 143–146, 148, 151; Wood, *A History of Farmers' Movements in Canada*, 296; John F. Varty, "On Protein, Prairie Wheat, and Good Bread: Rationalizing Technologies and the Canadian State, 1912–1935," *Canadian Historical Review* 85, no. 4 (December 2004): 721–753.

23. Jeremy Adelman, "Prairie Farm Debt and the Financial Crisis of 1914," *Canadian Historical Review* 71, no. 4 (December 1990): 491–519.

24. *Official Synopsis of the Report of the Agricultural Credit Commission of the Province of Saskatchewan* (Regina: J. W. Reid, 1913), 15–16; Adelman, "Prairie Farm Debt," 511–519.

25. Cole Harris, *The Reluctant Land: Society, Space, and Environment in Canada before Confederation* (Seattle: University of Washington Press, 2008); Jonathan F. Vance, *Maple Leaf Empire:*

Canada, Britain, and Two World Wars (New York: Oxford University Press, 2011); Donald V. Smiley, "Canada and the Quest for a National Policy," *Canadian Journal of Political Science* 8, no. 1 (March 1975): 40–62.

26. John F. Conway, "Populism in the United States, Russia, and Canada: Explaining the Roots of Canada's Third Parties," *Canadian Journal of Political Science* 11, no. 1 (March 1978): 99–124; Doreen Barrie, *The Other Alberta: Decoding a Political Enigma* (Regina, SK: Canadian Plains Research Center, 2006).

27. Kenneth H. Norrie and Donald Harman Akenson, "The National Policy and Prairie Economic Discrimination, 1870–1930," *Canadian Papers in Rural History* 1 (1978): 13–32; Bradford James Rennie, *The Rise of Agrarian Democracy: The United Farmers and Farm Women of Alberta, 1909–1921* (Toronto: University of Toronto Press, 2000), 48–49; Sinclair, "Class Structure and Populist Protest," 82; C. B. Macpherson, *Democracy in Alberta: The Theory and Practice of a Quasi-Party System* (Toronto: University of Toronto Press, 1953), 20–25; D. S. Spafford, " 'Independent' Politics In Saskatchewan before the Nonpartisan League," *Saskatchewan History* 18, no. 1 (Winter 1965): 1–9.

28. "Saskatchewan Secretary's Report," *Grain Growers' Guide*, February 28, 1912, 10; John Conway, "The Prairie Populist Resistance to the National Policy: Some Reconsiderations," *Journal of Canadian Studies* 14, no. 3 (Autumn 1979): 78; Ian McKay, *Reasoning Otherwise: Leftists and the People's Enlightenment in Canada, 1890–1920* (Toronto: Between the Lines, 2008), 197–208; Sharp, *The Agrarian Revolt in Western Canada*, 59, 81.

29. Norman Penner, "The Western Canadian Left—In Retrospect," in *Western Canadian Politics: The Radical Tradition*, ed. Donald C. Kerr (Edmonton: NeWest Institute for Western Canadian Studies, 1981), 8–9; Rennie, *The Rise of Agrarian Democracy*, 16; Sharp, *The Agrarian Revolt in Western Canada*, 58–59; Bradford James Rennie, "The Utopianism of the Alberta Farm Movement, 1909–1923," in *The Prairie West as Promised Land*, ed. R. Douglas Francis and Chris Kitzan (Calgary: University of Calgary Press, 2007), 257; Patrick Joyce, *Industrial England and the Question of Class, 1848–1914* (New York: Cambridge University Press, 1991), 65–74.

30. Will Holmes, "The Church in Politics," *Alberta Non-Partisan*, April 12, 1918, 7; Richard Allen, "The Social Gospel as the Religion of the Agrarian Revolt," in *The West and the Nation: Essays in Honour of W. L. Morton*, ed., Carl Berger and Ramsay Cook (Toronto: McClelland and Stewart, 1976), 174–186; McKay, *Reasoning Otherwise*, 213–280.

31. R. Douglas Francis, "The Kingdom of God on the Prairies: J. S. Woodsworth's Vision of the Prairie West as Promised Land," in *The Prairie West as Promised Land*, ed. R. Douglas Francis and Chris Kitzan (Calgary: University of Calgary Press, 2007), 226, 237; Allen Mills, *Fool for Christ: The Political Thought of J. S. Woodsworth* (Toronto: University of Toronto Press, 1991) 38–95.

32. Ramsay Cook, "Ambiguous Heritage: Wesley College and the Social Gospel Reconsidered," *Manitoba History* 19 (Spring 1990), http://www.mhs.mb.ca/docs/mb_history/19/wesleycollege.shtml, accessed October 24, 2011; William Irvine, *The Farmers in Politics* (Toronto: McClelland and Stewart, 1920), 5, 52, 56.

33. Linda Rasmussen, Lorna Rasmussen, Candace Savage, and Anne Wheeler, ed., *A Harvest Yet to Reap: A History of Prairie Women* (Toronto: Canadian Women's Educational Press, 1976), 124; Alvin Finkel, "Populism and Gender: The UFA and Social Credit Experiences," *Journal of Canadian Studies* 27, no. 4 (Winter 1992): 76–97; Veronica Strong-Boag, "Pulling in Double Harness or Hauling a Double Load: Women, Work and Feminism on the Canadian Prairie," *Journal of Canadian Studies* 21, no. 3 (Fall 1986): 32–52.

34. R. G. Marchildon, "Improving the Quality of Rural Life in Saskatchewan: Some Activities of the Women's Section of the Saskatchewan Grain Growers, 1913–1920," in *Building beyond the Homestead: Rural History on the Prairies*, ed., David C. Jones and Ian MacPherson (Calgary: Calgary University Press), 88–109; Rasmussen et al., *A Harvest Yet to Reap*, 175; Sharp, *The Agrarian Revolt in Western Canada*, 78.

35. Edward Alexander Partridge, *Manifesto of the No-Party League of Western Canada* (Winnipeg: De Monfort, 1913), 6; Spafford, "'Independent' Politics in Saskatchewan."

36. Sidney Godwin, "Farmer's Political League," *Grain Growers' Guide*, October 4, 1916, 19; "The Nonpartisan Political League," *Grain Growers' Guide*, September 5, 1917, 8, 24; Frederick W. Laidley, *The Why of the Farmers' Non-Partisan League of Canada* (Swift Current, SK: Farmers' Non-Partisan League of Canada, 1916), 8, file 28, series 4, John Ford Fonds, GA.

37. Laidley, *The Why of the Farmers' Non-Partisan League of Canada*, 7, 9.

38. Sidney Godwin, "Farmer's Political League," *Grain Growers' Guide*, October 4, 1916, 19; Laidley, *The Why of the Farmers' Non-Partisan League of Canada*, 5, 10; "The Nonpartisan Political League," 8, 24; W. L. Morton, *The Progressive Party in Canada* (Toronto: University of Toronto Press, 1950), 47; Jason McCollom, "Political Harvests: Transnational Farmers' Movements in North Dakota and Saskatchewan" (PhD diss., University of Arkansas, 2014).

39. "North Dakota Farmers Win All Along the Line," *Nonpartisan Leader of Western Canada* (Swift Current, SK), November 8, 1916; "Burning Up the Trail," *Nonpartisan Leader of Western Canada*, May 30, 1917.

40. "A Tribute: In Memoriam," *Alberta Non-Partisan*, November 6, 1918; "A New War Baby," *Nutcracker* (Calgary, AB), December 15, 1916.

41. Anthony Mardiros, *William Irvine: The Life of a Prairie Radical* (Toronto: James Lorimer, 1979), 1–22.

42. Ibid., 59; "'Mrs. McKinney, M.L.A., Assists," *Nonpartisan Leader of Western Canada*, June 20, 1917.

43. "The Non-Partisan Political League of Alberta Balance Sheet," *Alberta Non-Partisan*, March 29, 1918, 9; Lorne Proudfoot to William Irvine, April 10, 1917, file 1a, series 11, Lorne Proudfoot Fonds, GA; Sharp, *The Agrarian Revolt in Western Canada*, 78; Mardiros, *William Irvine*, 53–57.

44. *The Alberta Non-Partisan*, October 12, 1917; "Life and Work: Non-Partisan Politics," *Nutcracker*, January 15, 1917; "First-Born Child of New Canadian Political World," undated newspaper clipping, ca. 1917, file 4, series 11, Lorne Proudfoot Fonds, GA; Louise McKinney, *The Farmers' Opportunity* (Calgary: Non Partisan League, 1919), 23. David Laycock suggests the Alberta Nonpartisan League introduced "a non-British parliamentary, and hence 'radical,' democratic representation" to Canadian political life. *Populism and Democratic Thought in the Canadian Prairies* (Toronto: University of Toronto Press, 1990), 70, 72. McKay credits Irvine with developing something new in Canadian politics: "the concept of group government—in essence, an attempt to rethink democracy under conditions of corporate, post-proprietorial capitalism. . . . Each producing occupational group would organize itself . . . [and] ultimately the groups would come together to form a co-operative commonwealth." *Reasoning Otherwise*, 206.

45. Roderick McKenzie, "How Farmers Captured a State," *Grain Growers' Guide*, September 6, 1916, 18; "Farmers Capture Government," *Grain Growers' Guide*, September 6, 1916, 5.

46. Sharp, *The Agrarian Revolt in Western Canada*; Seymour Martin Lipset, *Agrarian Socialism: The Cooperative Commonwealth Federation in Saskatchewan* (Berkeley: University of California Press, 1959); Thompson, "Political Divergence between the Canadian and U.S. Northern

Plains," 8–11; Seymour Martin Lipset, *Continental Divide: The Values and Institutions of the United States and Canada* (New York: Routledge, 1991); Jason Kaufman, *The Origin of Canadian and American Political Differences* (Cambridge, MA: Harvard University Press, 2009). In *Code Politics: Campaigns and Cultures on the Canadian Prairies* (Vancouver: University of British Columbia Press, 2011), Jared J. Wesley examines the persistence of different kinds of distinct politics emanating from the region. He identifies specific, historically derived "codes" that structured political discourse and culture in Alberta, Saskatchewan, and Manitoba. Unfortunately, he limits himself to politics from the 1930s on.

47. W. E. Quigley, "The Truth about the Non-Partisan League: Chapter IV, I Enter the League Employ," *Lincoln Daily Star*, April 16, 1919, reel 4, National Nonpartisan League Papers, Minnesota Historical Society, Saint Paul, MN (hereafter MNHS).

48. Charles Lamb, "The Nonpartisan League and Its Expansion into Minnesota," *North Dakota Quarterly* 49, no. 3 (Summer 1981): 118.

49. Theodore Saloutos, "The Rise of the Equity Cooperative Exchange," *Mississippi Valley Historical Review* 32, no. 1 (June 1945): 31–62; Quigley, "The Truth about the Non-Partisan League: Chapter IV"; Frank H. Jonas, "The Art of Political Dynamiting," *Western Political Quarterly* 10, no. 2 (June 1957): 385–386.

50. Walter Eli Quigley, "Out Where the West Begins," typescript, ca. 1932, MNHS, 25; Walter Day, interviewed by Russell Fridley, November 27, 1967, Bagley, MN, MNHS.

51. Quigley, "Out Where the West Begins," 26; Lamb, "The Nonpartisan League and Its Expansion into Minnesota," 132–133; "Minnesotan Sees League Victory," *Nonpartisan Leader* (Fargo, ND), April 5, 1917; Ralph Lee Kloske, "Nonpartisan Leaguers in Minnesota: A Consideration of Organizers, Members, and Voters" (master's thesis, University of Wisconsin–Madison, 1976), 205–218.

52. "Report of Ray McKaig, and His Itinerary," January 1917, 1, reel 1, National Nonpartisan League Papers, MNHS.

53. Richard White, *Railroaded: The Transcontinentals and the Making of America* (New York: W. W. Norton, 2011), 482–486; Howard Lamar, *Dakota Territory, 1861–1889: A Study of Frontier Politics* (New Haven, CT: Yale University Press, 1956); Jon Lauck, *Prairie Republic: The Political Culture of Dakota Territory, 1879–1889* (Norman: University of Oklahoma Press, 2010), 24–56; R. Alton Lee, *Principle over Party: The Farmers' Alliance and Populism in South Dakota, 1880–1900* (Pierre: South Dakota Historical Society Press, 2011); Paula G. Nelson, *After the West Was Won: Homesteaders and Town-Builders in Western South Dakota, 1900–1917* (Iowa City: University of Iowa Press, 1986).

54. William C. Pratt, "Socialism on the Northern Plains, 1900–1924," *South Dakota History* 18, no. 1–2 (Spring–Summer 1988): 1–35; A. E. Brine to Leonard Sackett, August 30, 1955, folder 2, box 1, Nonpartisan League Collection, North Dakota Institute for Regional Studies, North Dakota State University, Fargo, ND (hereafter NDIRS); "Union Indorses League," *Nonpartisan Leader*, March 29, 1916; Carl J. Hofland, "The Nonpartisan League in South Dakota" (master's thesis, University of South Dakota, 1940), 15. NPL leaders drew on their relationship with Emil Sudan, a prominent member of South Dakota's Socialist Party. See Clarence Sharp, interview by Carl Ross, Minneapolis, MN, December 9 and December 16, 1987, 20th Century Radicalism in Minnesota Oral History Project, MNHS; Homer Ayres, interview by Earl Hausle, Sturgis, SD, June 28, 1971, South Dakota Oral History Project, tape 187, South Dakota Oral History Center, University of South Dakota, Vermillion, SD (hereafter USD).

55. "Why Farmers Failed," *Nonpartisan Leader*, March 15, 1917; and "Recruiting in Several

States," *Nonpartisan Leader*, April 5, 1917; Ayres, interview by Hausle; Hofland, "The Nonpartisan League in South Dakota," 83; Hubert H. Humphrey, *The Education of a Public Man: My Life and Politics* (New York: Doubleday, 1976), 6–12.

56. A. E. Brine to Leonard Sackett, August 30, 1955, folder 2, box 1, Nonpartisan League Collection, NDIRS.

57. Alan M. Clem, *Prairie State Politics: Popular Democracy in South Dakota* (Washington, DC: Public Affairs, 1967), 27–31; John E. Miller, "Politics since Statehood," in *A New South Dakota History*, ed. Harry F. Thompson (Sioux Falls, SD: Center for Western Studies, 2009), 199–201.

58. George W. Dixon to Ray McKaig, December 19, 1916, reel 1, National Nonpartisan League Papers, MNHS; Ray McKaig to J. D. Ream, August 18, 1917, folder R, box 1, Ray McKaig Papers, Idaho State Historical Society, Boise, ID (hereafter ISHS); Lee, *Principle over Party*, 175.

59. Gilbert C. Fite, "South Dakota's Rural Credit System: A Venture in State Socialism, 1917–1946," *Agricultural History* 21, no. 4 (October 1947): 239–249; Gilbert C. Fite, "Peter Norbeck and the Defeat of the Nonpartisan League in South Dakota," *Mississippi Valley Historical Review* 33, no. 2 (September 1946): 220–221.

60. B. B. Haugan to Peter Norbeck, January 29, 1917, "Dalton-Holsope, 1917" folder, box 59, Peter Norbeck Papers, USD; "James O. Berdahl," interview by O. A. Rothlisberger, June 17, 1974, Sioux Falls, SD, tape 1161, South Dakota Oral History Project, South Dakota Oral History Center, USD; L. W. Robinson to Harry King, January 3, 1917, "1918 Political Campaign" folder, box 84, Peter Norbeck Papers, USD; "What's Up in South Dakota," *Nonpartisan Leader*, March 15, 1917.

61. Rex C. Myers, "Homestead on the Range: The Emergence of Community in Eastern Montana, 1900–1925," *Great Plains Quarterly* 10 (Fall 1990): 218–227.

62. Theodore Saloutos, "The Montana Society of Equity," *Pacific Historical Review* 14, no. 4 (December 1945): 394.

63. Ibid., 395; "The Struggle in Montana," *Nonpartisan Leader*, May 3, 1917, 13. See also Michael P. Malone, *The Battle for Butte: Mining and Politics on the Northern Frontier, 1864–1906* (Seattle: University of Washington Press, 1981); Donald MacMillan, *Smoke Wars: Anaconda Copper, Montana Air Pollution, and the Courts, 1890–1920* (Helena: Montana Historical Society Press, 2000); and Thomas A. Clinch, *Urban Populism and Free Silver in Montana: A Narrative of Ideology in Political Action* (Missoula: University of Montana Press, 1970).

64. Levine, "Farmers Causing Political Upheaval in West," 51; "Report of Ray McKaig, and His Itinerary," January 1917, and "Non-Partizan League: Montana Farmers Federation Endorses New Political Organization for State," newspaper clippings, December 1916, both reel 1, National Nonpartisan League Papers, MNHS; "The Corporations' Nonpartisan League," *Nonpartisan Leader*, February 22, 1917.

65. John M. Gillette to George E. Vincent, October 16, 1916, folder 21, box 1, John M. Gillette Papers, University of North Dakota, Grand Forks, ND (hereafter UND); "League Not to Lose Townley's Services," *National Leader* (Minneapolis) 14, no. 10 (June 1922): 7.

66. "The Farmers' Crusade: Letters from George Cronyn, a Non-Partisan League Organizer," *Liberator* 1, no. 8 (October 1918): 10.

67. Ibid., 9, 8.

68. Ibid., 9.

69. Quoted in Kloske, "Nonpartisan Leaguers in Minnesota," 14.

70. "N.P. League in Kansas," *Kansas City Star*, April 7, 1918, in IWW and NPL Clippings, vol. 1, Kansas State Historical Society, Topeka, KS; Joseph Gilbert, interview by Kathryn Johnson,

January 27, 1954, Minneapolis, MN, transcript, Joseph Gilbert Papers, MNHS; Herbert E. Gaston, *The Nonpartisan League* (New York: Harcourt, Brace, and Howe, 1920), 7.

71. William Cronon, *Nature's Metropolis: Chicago and the Great West* (New York: W. W. Norton, 1991), 335–340; Timothy B. Spears, *100 Years on the Road: The Travelling Salesman in American Culture* (New Haven, CT: Yale University Press, 1995), 1–191.

72. Spears, *100 Years on the Road*, 192–220; Gaston, *The Nonpartisan League*, 62; "Minnesota, the Nonpartisan League, and the Future," *Nation* 117, no. 3030 (August 1, 1923): 102.

73. Gaston, *The Nonpartisan League*, 63; W. E. Quigley, "The Truth about the Non-Partisan League: Chapter V, I Meet Townley," April 17, 1919, *Lincoln Daily Star*, reel 4, National Nonpartisan League Papers, MNHS.

74. Alfred Knutson, "The Nonpartisan League," undated, Alfred Knutson Papers and Interview Transcripts (Lowell Dyson), box 21, Larry S. Remele Papers, State Historical Society of North Dakota, Bismarck, ND (hereafter SHSND).

75. Charles Edward Russell, "The Non-Partisan League" in *Papers and Proceedings, Eleventh Annual Meeting of the American Sociological Society* (Chicago: University of Chicago Press, 1917), 35–36; Knutson, "The Nonpartisan League"; W. A. Salchenberger to Nonpartisan League of Canada, February 9, 1918, file 58, series 4, John Ford Fonds, GA; Cronyn, "The Farmers' Crusade," 9.

76. Knutson, "The Nonpartisan League"; Henry R. Martinson, "Some Memoirs of a Nonpartisan League Organizer," *North Dakota History* 42, no. 2 (Spring 1975): 20; "This Minnesota Farmer Decided It Was Time for Him to Get Busy," *Nonpartisan Leader*, March 22, 1917.

77. Quigley, "The Truth about the Non-Partisan League: Chapter IV"; "Letters from Our Readers," *Nonpartisan Leader*, November 4, 1918.

78. Gaston, *The Nonpartisan League*, 62; "Organize," *South Dakota Leader* (Mitchell, SD), August 16, 1919; "The Nonpartisan League of Canada, Alberta Branch, Program," 1917, file 4, series 11, Lorne Proudfoot Fonds, GA.

79. Otto M. Thomason, "The North Dakota Uprising," ca. 1939, Otto M. Thomason Papers, MNHS; Gaston, *The Nonpartisan League*, 63.

80. Gaston, *The Nonpartisan League*, 63; James Manahan, *Trials of a Lawyer* (Minneapolis: Farnham, 1933), 219–220.

81. Martinson, "Some Memoirs of a Nonpartisan League Organizer," 20; Knutson, "The Nonpartisan League"; Quigley, "The Truth about the Non-Partisan League: Chapter V."

82. Mr. and Mrs. Herman Haven, interview by O. A. Rothlisberger, Brentford, SD, August 1, 1972, tape 411, South Dakota Oral History Project, South Dakota Oral History Center, USD; and Martinson, "Some Memoirs of a Nonpartisan League Organizer," 20, 19.

83. A. B. Gilbert to J. E. Cutler, January 7, 1917, National Nonpartisan League: 1917–1931 folder, box 4, Arthur Benson Gilbert Papers, MNHS; W. E. Quigley, "The Truth about the Non-Partisan League: Chapter XVII, League Educational Department," *Lincoln Daily Star*, April 30, 1919, reel 4, National Nonpartisan League Papers, MNHS; Eugene V. Debs, "The School for the Masses: The People's College of Fort Scott, KS," *American Socialist* 2, no. 10 (September 18, 1915), 2; Julia M. Allen, "'Dear Comrade': Marian Wharton of the People's College, Fort Scott, Kansas, 1914–1917," *Women's Studies Quarterly* 22 (Spring–Summer 1994): 119–133.

84. "The League Correspondence Course: Farm Economics, Lesson XI and XII," folder 2, North Dakota Nonpartisan League Collection, UND.

85. "The League Correspondence Course: Farm Economics, Lesson V and VI," 7; "The League Correspondence Course: Farm Economics, Lesson XI and XII," 1; "The League Correspondence Course: Organization Work, Lesson VII and VIII," 8, MNHS.

86. Alfred Knutson to Paul J. Dovre, September 19, 1961, Alfred Knutson Papers and Interview Transcripts (Lowell Dyson), box 21, Larry Remele Papers, SHSND. For employee resistance to correspondence training, see Paul John Dovre, "A Study of Nonpartisan League Persuasion, 1915–1920" (PhD diss., Northwestern University, 1963), 65.

87. Cronyn, "The Farmers' Crusade," 6; A. E. Bowen to Henry Teigan, December 6, 1919, reel 5, National Nonpartisan League Papers, MNHS; Albert Dakan to Ray McKaig, February 27, 1918, folder D, box 1, Ray McKaig Papers, ISHS.

88. Gaston, *The Nonpartisan League*, 65; Martinson, "Some Memoirs of a Nonpartisan League Organizer," 20.

89. W. E. Quigley, "The Truth about the Non-Partisan League: Chapter XVI, House Bill 44," *Lincoln Daily Star*, April 21, 1919, reel 4, National Nonpartisan League Papers, MNHS.

90. Leon Durocher to Leonard Sackett, July 31, 1955, and A. E. Brine to Leonard Sackett, August 30, 1955, folder 2, box 1, Nonpartisan League Collection, NDIRS; Kloske, "Nonpartisan Leaguers in Minnesota," 38; Quigley, "The Truth about the Non-Partisan League: Chapter IV."

91. Quigley, "The Truth about the Non-Partisan League: Chapter IV"; John Ford to W. H. Tompkins, March 9, 1918, file 59, series 4, John Hooper Ford Fonds, GA. For an example of an NPL organizer's contract, see "Organizer's Contract," ca. 1920, Minnesota Nonpartisan League, reel 5, National Nonpartisan League Papers, MNHS.

92. Leon Durocher to Arthur LeSueur, April 11, 1915, folder 9, box 1, Arthur LeSueur Papers, MNHS; Quigley, "The Truth about the Non-Partisan League: Chapter IV."

93. Quigley, "The Truth about the Non-Partisan League: Chapter IV."

94. Ibid.; Ray McKaig to Walter Caddell, February 22, 1918, folder C, box 1, Ray McKaig Papers, ISHS.

95. Sharp, *The Agrarian Revolt in Western Canada*, 83; John Ford to Frank C. Simpson, January 10, 1918, file 58, series 4, John Ford Fonds, GA.

96. A. C. Townley to Ray McKaig, September 16, 1918, folder T, box 1, Ray McKaig Papers, ISHS; "Organizer's Daily Report," Correspondence September 1919 folder, box 2, Christian Abraham Sorensen Papers, Nebraska State Historical Society, Lincoln, NE (hereafter NSHS).

97. "Minnesota, the Nonpartisan League, and the Future."

98. W. E. Quigley, "The Truth about the Non-Partisan League: Chapter VII, Reorganizing North Dakota," *Lincoln Daily Star*, April 19, 1919, reel 4, National Nonpartisan League Papers, MNHS; Quigley, "The Truth about the Non-Partisan League: Chapter XVI."

99. W. E. Quigley, "The Truth about the Non-Partisan League: Chapter XV, Miscellaneous," *Lincoln Daily Star*, April 29, 1919, reel 4, National Nonpartisan League Papers, MNHS.

100. Henry Teigan to L. N. Sheldon, June 4, 1917, reel 2, National Nonpartisan League Papers, MNHS; Gaston, *The Nonpartisan League*, 71.

101. Quigley, "The Truth about the Non-Partisan League: Chapter XV."

102. *Nebraska Leader* (Lincoln), October 4, 1919.

103. Charles Edward Russell, *The Story of the Nonpartisan League: A Chapter in American Evolution* (New York: Harper and Bros., 1920), 203; "C. S. Townley Killed in Accident," *National Leader* 16, no. 2 (February 1923): 2.

104. Christopher Wells, "The Changing Nature of Country Roads: Farmers, Reformers, and the Shifting Uses of Rural Space, 1880–1905," *Agricultural History* 80, no. 2 (Spring 2006): 143–166.

105. Wells, "The Changing Nature of Country Roads," 157–158; Wayne E. Fuller, *RFD: The Changing Face of Rural America* (Bloomington: Indiana University Press, 1964), 177–198.

106. Carl F. W. Larson, "A History of the Automobile in North Dakota to 1911," *North Dakota History* 54, no. 4 (Fall 1980): 3–24; *Bowman (ND) Citizen*, August 26, 1915; Roy S. Weaver,

The Nonpartisan League in North Dakota (Toronto: Canadian Reconstruction Association, 1921), 3.

107. Joseph Interrante, "You Can't Go to Town in a Bathtub: Automobile Movement and the Reorganization of Rural American Space, 1900–1930," *Radical History Review* 21 (1979): 151–168; Michael L. Berger, *The Devil Wagon in God's Country: The Automobile and Social Change in Rural America, 1893–1929* (Hamden, CT: Archon Books, 1979), 77–102; Ronald R. Kline, *Consumers in the Country: Technology and Social Change in Rural America* (Baltimore: Johns Hopkins University Press, 1994), 55–86.

108. Reynold M. Wik, *Henry Ford and Grass-Roots America* (Ann Arbor: University of Michigan Press, 1972), 36; Lindsay Brooke, *Ford Model T: The Car That Put the World on Wheels* (Minneapolis: Motorbooks, 2008), 90; Christopher W. Wells, "The Road to the Model T: Culture, Road Conditions, and Innovation at the Dawn of the American Motor Age," *Technology and Culture* 48, no. 3 (July 2007): 497–523; Berger, *The Devil Wagon in God's Country*, 47–51; David Roberts, *In The Shadow of Detroit: Gordon M. McGregor, Ford of Canada, and Motoropolis* (Detroit: Wayne State University Press, 2006).

109. W. G. Scholtz to Ray McKaig, April 24, 1918, folder S, box 1, Ray McKaig Papers, ISHS; Alfred Knutson, interview notes by Lowell Dyson, August 5, 1965, Minot, ND, in Alfred Knutson Papers and Interview Transcripts folder, box 21, Larry Remele Papers, SHSND; "Christian Lee (Lia)," *Griggs County History, 1879–1976* (Cooperstown, ND: The Committee, 1976), 202.

110. Wik, *Henry Ford and Grass-Roots America*, 78; Cronyn, "The Farmers' Crusade," 10; John A. Jakle and Keith A. Sculle, *The Gas Station in America* (Baltimore: Johns Hopkins University Press, 1994), 48–50.

111. Stephen Meyer, *The Five Dollar Day: Labor Management and Social Control in the Ford Motor Company, 1908–1921* (Albany: State University of New York Press, 1981); Clarence Hooker, *Life in the Shadows of the Crystal Palace, 1910–1927: Ford Workers in the Model T Era* (Bowling Green, OH: Bowling Green State University Press, 1997).

112. "History of the Postal System of Canada," in *The Encyclopedia of Canada*, ed. W. Stewart Wallace (Toronto: University Associates of Canada, 1948), 5:148–150.

113. "Support Your Own Organization," *Nonpartisan Leader*, November 23, 1915.

114. *The Farmer-Owned County Newspaper* (Saint Paul: Northwestern Service Bureau, 1918), 6.

115. Northwestern Service Bureau file, reel 12, National Nonpartisan League Papers, MNHS; Ferdinand A. Teigen, *The Nonpartisan League: Its Origin, Development, and Secret Purposes* (Saint Paul: Economic Research and Publishing, 1918); W. C. Coates, "A Correction, Note the Date of the Newspaper Meeting to Be Held at Verndale, Sunday, Feb. 9, at 2 P.M.," broadside, February 3, 1919, MNHS.

116. Frank J. Prochaska, *Autobiography and Family History* (North Dakota: published by author, 1932), 2:67–68.

117. The *Producers News* survived the NPL and became famous as an agrarian voice for the Communist Party in the 1930s. Verlaine Stoner McDonald, *The Red Corner: The Rise and Fall of Communism in Northeastern Montana* (Helena: Montana Historical Society Press, 2010), 43–70.

118. Robert Morlan, *Political Prairie Fire: The Nonpartisan League, 1915–1922* (1955; repr., Saint Paul: Minnesota Historical Society Press, 1985), 119.

119. Otto M. Thomason, "The North Dakota Uprising," Otto M. Thomason Papers, MNHS.

120. Robert P. Wilkins, "Referendum on War? The General Election of 1916 in North Dakota," *North Dakota History* 36, no. 4 (Fall 1969): 313; Bill Reid, "John Miller Baer: Nonpartisan League Cartoonist and Congressman," *North Dakota History* 44, no. 1 (Winter 1977): 4–13.

121. Gaston, *The Nonpartisan League*, 172; Worth Robert Miller, *Populist Cartoons: An*

Illustrated History of the Third-Party Movement in the 1890s (Kirksville, MO: Truman State University Press, 2011); Glen Jeansonne, "Goldbugs, Silverites, and Satirists: Caricature and Humor in the Presidential Election of 1896," *Journal of American Culture* 11, no. 2 (Summer 1988): 1–8; Michael Cohen, "'Cartooning Capitalism': Radical Cartooning and the Making of American Popular Radicalism in the Early Twentieth Century," *International Review of Social History* 52, supplement 15 (December 2007): 35–58.

122. See, for example, John Miller Baer, "What Are You Going to Do About It?," *Nonpartisan Leader*, February 8, 1917.

123. The character first appeared in the *Nonpartisan Leader* on November 11, 1915, and recurred frequently in the years that followed.

124. Norris C. Hagen, *Vikings of the Prairie: Three North Dakota Settlers Reminisce* (New York: Exposition, 1958), 168–169; Thomas Bivins, "The Body Politic: The Changing Shape of Uncle Sam," *Journalism Quarterly* 64, no. 1 (Spring 1987): 13–20; Alton Ketchum, "The Search for Uncle Sam," *History Today* 40, no. 4 (April 1990): 20–25.

125. John Miller Baer, "I'll Get Some of the Bills Cleaned Up, But I Can't Get Much Through This Wringer," *Nonpartisan Leader*, March 1, 1917; Baer, "By The Dawn's Early Light," *Nonpartisan Leader*, May 20, 1918.

126. Newspaper clipping, *Grand Forks Herald*, October 8, 1969, folder 1, box 1, John Baer Papers, UND; Eric Moen, "The Farmers, They Say, 'Can't Stick Together,'" *Nonpartisan Leader*, October 7, 1915; "They Can't Beat You at the Primaries," *Nonpartisan Leader*, June 22, 1916; "Frazier Speaking from Car in a Rainstorm," *Nonpartisan Leader*, June 29, 1916.

127. Quoted in Reid, "John Miller Baer," 10; Gaston, *The Nonpartisan League*, 171–172; Morlan, *Political Prairie Fire*, 129–134.

128. See, for example, Arch Dale, "Our National Shame," *Grain Growers' Guide*, July 19, 1916; Dale, "One By One He is Breaking His Bonds," *Grain Growers' Guide*, October 25, 1916; Dale, "A Great Team for Killing Thistles," *Grain Growers' Guide*, October 10, 1917. See also Sharp, *Agrarian Revolt in Western Canada*, 37; John Miller Baer, "Everybody Works the Farmer," *Alberta Non-Partisan*, October 26, 1917; and Baer, "No More 'Dummy' Legislators," *Alberta Non-Partisan*, April 26, 1918.

129. Mrs. Alberta Cundal to Jack Ford, March 18, 1918, file 32, series 4, John Ford Fonds, GA; Kelly Hannan, "The Nonpartisan League in Alberta and North Dakota: A Comparison," *Alberta History* 52, no. 1 (Winter 2004): 21.

130. William Irvine, interview by Una MacLean, Edmonton, AB, June 6 and 7, September 12 and 13, November 30, 1960, 5, M4077a, United Farmers of Alberta Oral History Project, GA; "Three Prominent Men in the Non-partisan Movement Now Sweeping the Province," *Nutcracker*, June 7, 1917; Lorne Proudfoot to J. H. Ford, June 19, 1917, file 1a, series 11, Lorne Proudfoot Fonds, GA.

131. *Mosquito Creek Roundup: Nanton-Parkland* (Nanton, AB: Nanton and District Historical Society, 1976), 188, 219, 585; Rennie, *The Rise of Agrarian Democracy*, 132–133.

132. "Women in the Legislature," *Nutcracker*, June 21, 1917.

133. Jean Stevenson to Jack Ford, June 12, 1918, and Jean Stevenson to Jack Ford, June 3, 1918, file 57, series 4, John Ford Fonds, GA.

134. Irvine, interview by Una MacLean, 5; "The Labor Representation League," 1917, file 28, series 4, John Ford Fonds, GA; Sharp, *Agrarian Revolt in Western Canada*, 93–94.

135. The Mail Bag, *Grain Growers' Guide*, March 21, 1917; The Mail Bag, *Grain Growers' Guide*, October 10, 1917; L. D. Courville, "The Conservatism of the Saskatchewan Progressives," *Historical Papers / Communications Historiques* 9, no. 1 (1974): 159–160; Evelyn Eager, *Saskatchewan*

Government: Politics and Pragmatism (Saskatoon, SK: Western Producer Prairie Books, 1980), 47; Morton, *The Progressive Party in Canada*, 47.

136. D. A. Mumby to Samuel V. Haight, June 12, 1917, file 6, Zoa Haight Papers, Saskatchewan Archives Board, University of Saskatchewan, Saskatoon, SK (hereafter SABS); "Meeting of Members of Nonpartisan Political League," June 16, 1917, file 6, Zoa Haight Papers, SABS; *Regina (SK) Daily Post*, June 18, 1917; "Horatius at the Land Bridge" *Nonpartisan Leader of Western Canada*, August 15, 1917; Sharp, *Agrarian Revolt in Western Canada*, 91–93.

137. Louise McKinney to Zoa Haight, June 21, 1917, "Six Reasons Why You Should Vote for Mrs. S. V. Haight, Nonpartisan Candidate," June 1917, and "Six Reasons Why You Should Vote for Mrs. S. V. Haight," postcard, no date (ca. 1917), all file 6, Zoa Haight Papers, SABS.

138. "Non-Partisan League Annual Convention," *Alberta Non-Partisan*, March 29, 1918; Sharp, *Agrarian Revolt in Western Canada*, 95–98.

139. A. B. Gilbert to J. E. Cutler, January 7, 1917, National Nonpartisan League, 1917–1931 folder, box 4, Arthur Benson Gilbert Papers, MNHS.

140. Hugh T. Lovin, "Ray McKaig: Nonpartisan League Intellectual and Raconteur," *North Dakota History* 47, no. 3 (Summer 1980): 12–20.

141. "Report of Ray McKaig, and His Itinerary," January 1917, 2, and J. D. Ream to Ray McKaig, January 26, 1917, reel 1, National Nonpartisan League Papers, MNHS.

142. "Report of Ray McKaig, and His Itinerary," 2–3.

143. Ibid.; "Colorado Farmers Indorse North Dakota Political Plan," February 14, 1917, newspaper clipping, reel 2, National Nonpartisan League Papers, MNHS.

144. Hugh T. Lovin, "The Farmer Revolt in Idaho, 1914–1922," *Idaho Yesterdays* 20 no. 3 (Fall 1976): 2–15; and Matthew C. Godfrey, *Religion, Politics, and Sugar: The Mormon Church, the Federal Government, and the Utah-Idaho Sugar Company, 1907–1921* (Logan: Utah State University Press, 2007), 93–126; "Report of Ray McKaig, and His Itinerary," 4; A. E. Lycan to Lynn J. Frazier, January 31, 1917, reel 1, National Nonpartisan League Papers, MNHS.

145. "Report of Ray McKaig, and His Itinerary," 4–6; "Constitution of the Nonpartisan League," *Herald* (Seattle, WA), March 10, 1916; Henry McCormick to Ray McKaig, February 12, 1917, reel 2, National Nonpartisan League Papers, MNHS; Mary M. Cronin, "Fighting for the Farmers: The Pacific Northwest's Nonpartisan League Newspapers," *Journalism History* 23, no. 3 (Autumn 1997): 126–136.

146. "Report of Ray McKaig, and His Itinerary," 4–6; Steven L. Piott, *Giving Voters a Voice: The Origins of the Initiative and Direct Referendum in America* (Columbia: University of Missouri Press, 2003), 188–189; Marilyn P. Watkins, *Rural Democracy: Family Farmers and Politics in Western Washington, 1890–1925* (Ithaca, NY: Cornell University Press, 1995), 85–123.

147. "Report of Ray McKaig, and His Itinerary," 5–6.

148. Alfred Knutson to Paul J. Dovre, September 19, 1961, Alfred Knutson Papers and Interview Transcripts folder, box 21, Larry Remele Papers, SHSND; Asher Howard, *The Leaders of the Nonpartisan League: Their Aims, Purposes and Records . . .* (Minneapolis: published by author, 1920), 41; R. W. Morser to Ray McKaig, June 5, 1917, folder M, box 1, Ray McKaig Papers, ISHS; Albert Dakan to Ray McKaig, January 31, 1917, reel 1, National Nonpartisan League Papers, MNHS.

149. Knutson to Dovre, September 19, 1961; Lovin, "The Farmer Revolt in Idaho, 1914–1922," 7–9.

150. Ray McKaig to Joseph Gilbert, July 31, 1917, folder G, box 1, and Ray McKaig to A. F. Moore, August 7, 1917, folder M, box 1, Ray McKaig Papers, ISHS.

151. Knutson to Dovre, September 19, 1961; Gaston, *The Nonpartisan League*, 167; Alfred Knutson to Ray McKaig, December 11, 1917, folder K, box 1, Ray McKaig Papers, ISHS.

152. Elmo Bryant Phillips, "The Non-Partisan League in Nebraska" (master's thesis, University of Nebraska, 1931), 6, 9; Howard, *The Leaders of the Nonpartisan League*, 41; Douglas Bakken, "NPL in Nebraska, 1917–1920," *North Dakota History* 39, no. 2 (Spring 1972): 26–31; C. A. Sorenson, untitled profiles of Nebraska Nonpartisan League leaders, undated, Correspondence August 1922 folder, box 3, Charles Abraham Sorensen Papers, NSHS; Robert W. Cherny, *Populism, Progressivism, and the Transformation of Nebraska Politics, 1885–1915* (Lincoln: University of Nebraska Press, 1981), 149–166.

153. C. E. Spence to Ray McKaig, January 20, 1917, folder S, box 1, Ray McKaig Papers, ISHS.

154. Ray McKaig to C. E. Spence, December 10, 1917, C. E. Spence to Ray McKaig, January 13, 1918, C. E. Spence to Ray McKaig, April 4, 1918, and Ray McKaig to C. E. Spence, April 27, 1918, all folder S, box 1, Ray McKaig Papers, ISHS; Ray McKaig to J. D. Brown, February 26, 1918, and Ray McKaig to J. D. Brown, March 20, 1918, folder B, box 1, Ray McKaig Papers, ISHS; "Spence Idol of State Grange," *Bend (OR) Bulletin*, June 13, 1918, 4; "Not the Grange," *National Grange Monthly*, April 1918, clipping, reel 14, National Nonpartisan League Papers, MNHS.

155. M. L. Amos to Lynn J. Frazier, January 26, 1917, reel 1, National Nonpartisan League Papers, MNHS; "The League Answers Kansas' Call," *Nonpartisan Leader*, April 5, 1917, 9; James Malin, *Winter Wheat in the Golden Belt of Kansas* (Lawrence: University of Kansas Press, 1944); George E. Ham and Robin Higham, eds., *The Rise of the Wheat State: A History of Kansas Agriculture, 1861–1986* (Manhattan, KS: Sunflower University Press, 1987).

156. Ray McKaig, list of state granges and membership numbers, no date (ca. 1917), folder G, box 1, Ray McKaig Papers, ISHS; B. Needham to Ray McKaig, July 30, 1917, folder N, box 1, Ray McKaig Papers, ISHS; Ray McKaig to J. F. Buehler, July 31, 1917, folder B, box 1, Ray McKaig Papers, ISHS; Ray McKaig to J. D. Ream, August 18, 1917, folder R, box 1, Ray McKaig Papers, ISHS.

157. "The League Answers Kansas' Call"; "Kansas Goes to Work," *Nonpartisan Leader*, May 3, 1917, 15; "Starts League in Kansas," *Kansas City Star*, September 7, 1917, in IWW and NPL Clippings, vol. 1, Kansas State Historical Society, Topeka, KS; Howard, *The Leaders of the Nonpartisan League*, 41.

158. "Badger State Also Enlists," *Nonpartisan Leader*, April 5, 1917; "Wisconsin Farmers Desire Organization," *Nonpartisan Leader*, February 22, 1917; Theodore Saloutos, "The Decline of the Wisconsin Society of Equity," *Agricultural History* 15, no. 3 (July 1941): 137–150; Theodore Saloutos, "The Wisconsin Society of Equity," *Agricultural History* 14, no. 2 (April 1940): 78–98; Elizabeth S. Clemens, *The People's Lobby: Organizational Innovation and the Rise of Interest Group Politics in the United States, 1890–1925* (Chicago: University of Chicago Press, 1997), 173–174; Frederick I. Olson, "The Socialist Party and the Union in Milwaukee, 1900–1912," *Wisconsin Magazine of History* 44, no. 2 (Winter 1960–1961): 110–116.

159. Davis Douthit, *Nobody Owns Us: Story of Joe Gilbert, Midwestern Rebel* (Chicago: Cooperative League of the USA, 1948), 100–101; "County Equity Society Has Meeting in Plymouth and H. Herbst Succeeds Peterson," *Sheboygan (WI) Press*, November 8, 1917; Saloutos, "The Decline of the Wisconsin Society of Equity," 140–141.

160. John W. Gunn, *The Emballotted Farmers: A Story of the Nonpartisan League* (Girard, KS: Appeal to Reason, 1918), 108–109.

Chapter Three

1. Carlos A. Schwantes, "Making the World Unsafe for Democracy: Vigilantes, Grangers and the Walla Walla 'Outrage' of June 1918," *Montana: The Magazine of Western History* 31, no. 1

(Winter 1981): 25; "Memorial to the President of the United States by the Executive Committee of the Washington State Grange," June 10, 1918, newspaper clipping, folder 1, Correspondence and Miscellaneous Papers: January–December 1918, box 2, Arthur LeSueur Papers, Minnesota Historical Society, Saint Paul, MN (hereafter MNHS); "State Grange Backs Wilson," *Spokane (WA) Spokesman-Review*, June 7, 1918.

2. Schwantes, "Making the World Unsafe for Democracy," 25.

3. Ibid.; "Grangers Expelled from Hall," *Tacoma Times*, June 7, 1918, 1; Lena S. Bouck to Ray McKaig, June 15, 1918, folder B, box 1, Ray McKaig Papers, Idaho State Historical Society, Boise, ID (hereafter ISHS).

4. "To Take Stand on Nonpartizans," *Spokane Spokesman-Review*, June 6, 1918; Schwantes, "Making the World Unsafe for Democracy," 20–26.

5. "Memorial to the President of the United States," Arthur LeSueur Papers, MNHS; "State Grange Backs Wilson"; Schwantes, "Making the World Unsafe for Democracy," 25; Marilyn P. Watkins, *Rural Democracy: Family Farmers and Politics in Western Washington, 1890–1925* (Ithaca, NY: Cornell University Press, 1995), 174; "Squelch Speakers for Nonpartizans," *Spokane Spokesman-Review*, June 8, 1918.

6. Schwantes, "Making the World Unsafe for Democracy," 22–28; *Memorial to the Congress of the Untied States Concerning Conditions in Minnesota 1918* (Saint Paul: Nonpartisan League, 1918), 5–7; D. D. Lemond, report, February 25, 1918, National Archives and Records Administration, M1085, Investigative Case Files of the Bureau of Investigation, 1908–1922 (hereafter ICFBOI), Bureau Section files, 1909–1921, case no. 46889, roll 876, 113, http://www.fold3.com/image /3141863/, accessed December 29, 2013.

7. *Report of the State Council of Defense to the Governor of Washington, June 16, 1917, to January 19, 1919* (Olympia, WA: Frank Lamborn, 1919), 61; Schwantes, "Making the World Unsafe for Democracy," 22–28; ICFBOI, Old German Files, 1909–1921, case no. 8000-217184, General Matters, Washington State Grange, Non-Partisan League, Walla Walla Citizens, 1–293, http://www .fold3.com/image/2402633/, accessed December 29, 2013.

8. "Where Does the Grange Stand?," *Spokane Spokesman-Review*, June 9, 1918; Carlos A. Schwantes, "The Ordeal of William Morley Bouck, 1918–1919: Limits to the Federal Suppression of Agrarian Dissidents," *Agricultural History* 59, no. 3 (July 1985): 417–428; National Civil Liberties Bureau, *War-Time Prosecutions and Mob-Violence* (New York: National Civil Liberties Bureau, 1919), 17; "Will Bouck in Trouble," *Princeton (MN) Union*, August 22, 1918.

9. William Bouck to A. C. Townley, December 21, 1918, folder T, box 1, Ray McKaig Papers, ISHS; Martin County (MN) Attorney Albert Allen quoted in Robert Morlan, *Political Prairie Fire: The Nonpartisan League, 1915–1922* (1955; repr., Saint Paul: Minnesota Historical Society Press, 1986), 168.

10. Katherine A. Benton-Cohen, *Borderline Americans: Racial Division and Labor War in the Arizona Borderlands* (Cambridge, MA: Harvard University Press, 2009); Colleen O'Neill, "Domesticity Deployed: Gender, Race, and the Construction of Class Struggle in the Bisbee Deportation," *Labor History* 34, nos. 2–3 (Spring–Summer 1993): 256–273; Ernest Freeberg, *Democracy's Prisoner: Eugene V. Debs, the Great War, and the Right to Dissent* (Cambridge, MA: Harvard University Press, 2008).

11. "93 Years Old But a Staunch Leaguer," *Nebraska Leader* (Lincoln), September 20, 1919; Frank G. Moorhead, "The Non-Partisan League in Politics," *Nation* 107, no. 2779 (October 5, 1918): 364; John W. Gunn, *The Emballotted Farmers: A Story of the Nonpartisan League* (Girard, KS: Appeal to Reason, 1918), 110; Catherine McNichol Stock, *Main Street in Crisis: The Great*

Depression and the Old Middle Class on the Northern Plains (Chapel Hill: University of North Carolina Press, 1992), 72–73.

12. "Local Business Men and the Farmers," *Nonpartisan Leader* (Fargo, ND), December 2, 1915.

13. John L. Miklethun, undated speech (ca. late 1920s), Political Organization Papers folder, box 3, John L. Miklethun Papers, MNHS; Carl H. Chrislock, *Watchdog of Loyalty: The Minnesota Commission of Public Safety during World War I* (Saint Paul: Minnesota Historical Society Press, 1991), 181.

14. "Insurance Men to Fight," *Nonpartisan Leader*, March 23, 1916; *Spectator: An American Weekly Review of Insurance* 97, no. 3 (July 20, 1916): 23; "Obnoxious Bills in Many States Beaten," *National Underwriter* (Chicago), March 8, 1917, 1.

15. "Some Bankers Still at Dirty Work," *Nonpartisan Leader*, February 3, 1916; "Banks Still Misleading," *Nonpartisan Leader*, December 9, 1915; North Dakota Bankers Association to members, June 8, 1916, reprinted in *Nonpartisan Leader*, June 22, 1916.

16. Larry Remele, "The Nonpartisan League and the North Dakota Press: Organization Period, 1915–1916," *North Dakota Quarterly* 44, no. 4 (Autumn 1976): 30–46; Odd S. Lovoll, *Norwegian Newspapers in America: Connecting Norway and the New Land* (Saint Paul: Minnesota Historical Society Press, 2010), 164–165, 269; "League Rank and File Answer Unfair Attack," *Nonpartisan Leader*, March 16, 1916.

17. Dorothy St. Arnold, memoir, folder 3, Dorothy St. Arnold Family Reminiscences, MNHS.

18. "The Farmer and the Business Man," *Nonpartisan Leader*, February 24, 1916.

19. "Business Men for Frazier," *Nonpartisan Leader*, May 18, 1916; "What Business Men Should Know," *Nonpartisan Leader*, June 15, 1916; John L. Miklethun, undated speech (ca. late 1920s), Political Organization Papers folder, box 3, John L. Miklethun Papers, MNHS.

20. Clement A. Lounsberry, *North Dakota History and People: Outlines of American History* (Chicago: S. J. Clarke, 1917), 3:154; "A Letter from the Grand Forks Herald," *Nonpartisan Leader*, May 4, 1916.

21. "More of the Big Business Campaign of Attack on the League," *Nonpartisan Leader*, June 22, 1916; "Before and after the 'G. G. League' Reformed," *Nonpartisan Leader*, July 20, 1916; Morlan, *Political Prairie Fire*, 68–70.

22. "Fake League in Minnesota," *Nonpartisan Leader*, March 29, 1917; "They Don't Want Fake League," *Nonpartisan Leader*, May 10, 1917.

23. Phil Zimmerman to Charles Sessions, October 12, 1919, W. A. White to P. E. Zimmerman, June 14, 1916, Philip E. Zimmerman to Dr. L. T. Guild, July 1, 1916, and Philip E. Zimmerman to Henry J. Allen, July 17, 1916, all reel 1, Philip E. Zimmerman Papers, Kansas State Historical Society, Topeka, KS (hereafter KSHS); "Zimmerman Heads Kansas Trade Body," *Automobile Topics: The Trade Authority* 65, no. 8 (July 8, 1922): 685.

24. Woodrow Wilson, *War Messages*, 65th Cong., 1st Sess., Senate Doc. No. 5, Serial No. 7264, Washington, DC, 1917, 3–8; Robert P. Wilkins, "Referendum on War? The General Election of 1916 in North Dakota," *North Dakota History* 36, no. 4 (Fall 1969): 296–335; Ernest R. May, *The World War and American Isolation, 1914–1917* (Cambridge, MA: Harvard University Press, 1959); Frederick A. Clements, "Woodrow Wilson and World War I," *Presidential Studies Quarterly* 34, no. 1 (March 2004): 62–82.

25. Quoted in Robert P. Wilkins, "The Non-Ethnic Roots of North Dakota Isolationism," *Nebraska History* 44, no. 3 (September 1963): 208; Herbert E. Gaston, *The Nonpartisan League* (New York: Harcourt, Brace and Howe, 1920), 175–176; R. M. Black, ed., *A History of Dickey County, North Dakota* (Ellendale, ND: Dickey County Historical Society, 1930), 321–322.

26. Robert P. Wilkins, "The Nonpartisan League and Upper Midwest Isolationism," *Agricultural History* 39, no. 2 (April 1965): 103; "Won't Back a War for Private Profit," *New York Times*, March 4, 1917, 19.

27. "Resolutions on War by National Nonpartisan League," in *National Nonpartisan League: Origins, Purpose, and Method of Operation, War Program and Statement of Principles* (Saint Paul: National Nonpartisan League, 1917), 24–26; Larry Remele, "The Tragedy of Idealism: The National Nonpartisan League and American Foreign Policy, 1917–1919," *North Dakota Quarterly* 42, no. 4 (Autumn 1974): 78–95.

28. Quoted in Gaston, *The Nonpartisan League*, 182–183.

29. "A Vile Assault in the Name of Patriotism," *Nonpartisan Leader*, June 14, 1917; "Resisting the Government" and "Townley and the War Crisis," *Minneapolis Tribune*, June 13, 1917; "A New National Party," *Literary Digest* 55, no. 6 (August 11, 1917): 13.

30. "Northwest Doing Its Full Share in Liberty Loans," *Minneapolis Tribune*, June 10, 1917; Oliver S. Morris, "North Dakota's Part," *New York Times*, July 9, 1917, 7; "Non-Partisan Men Deny Opposition to Liberty Loan," *Minneapolis Tribune*, June 12, 1917; Gaston, *The Nonpartisan League*, 191. By the end of the first Liberty Loan drive, North Dakotans raised more than twice their $1,500,000 quota. See "The Nonpartisan League: Loyal or Disloyal," broadside, 1918, MNHS.

31. Carol Jensen, "Loyalty as a Political Weapon: The 1918 Campaign in Minnesota," *Minnesota History* 43, no. 2 (Summer 1972): 43–57.

32. Gaston, *The Nonpartisan League*, 211–212; Melvyn Dubofsky, *We Shall Be All: A History of the Industrial Workers of the World* (Chicago: Quadrangle Books, 1969), 314–317; Philip Taft, "The I.W.W. in the Grain Belt," *Labor History* 1, no. 1 (Winter 1960): 53–67; Clemens P. Work, *Darkest before Dawn: Sedition and Free Speech in the American West* (Albuquerque: University of New Mexico Press, 2005), 200–204.

33. "League Plans Supply of Farm Labor," *Nonpartisan Leader*, June 7, 1917.

34. T. E. Campbell, report, January 22, 1920, ICFBOI, Bureau Section files, 1909–1921, case no. 46889, roll 877, 92–93, http://www.fold3.com/image/5220382/, accessed December 29, 2013; Charles James Haug, "The Industrial Workers of the World in North Dakota, 1913–1917," *North Dakota Quarterly* 31, no.1 (Winter 1971): 101–102.

35. Stock, *Main Street in Crisis*, 41–85; Patrick McGuire, "Death of a Myth: The Nonpartisan League and the Socialist Party of America, 1912–1920," in *13th Dakota History Conference, April 9, 10, 11, 1981, Papers* (Madison, SD: Dakota State College, 1982): 604–606; Henry Martinson to Henry Teigan, March 22, 1919, reel 4 National Nonpartisan League Papers, MNHS.

36. J. Edmund Kerr, *The Psychology of Suspicion, as Demonstrated in North Dakota* (Saint Paul: J. Edmund Kerr, 1918), 14, Political Literature: 1917–1920 folder, box 6, John L. Miklethun Papers, MNHS.

37. "Purpose of Meeting Stated by League," *Nonpartisan Leader*, September 7, 1917.

38. *Address of A. C. Townley, President of the National Non-Partisan League at the Farmers and Workers Conference Held In St. Paul, September 18, 19, and 20* (Saint Paul: Nonpartisan League, 1917), 24–25.

39. "Mr. Townley's Loyalty Promise" and "Patriotism Plus Protest Is Urged on Nonpartisans," *Minneapolis Tribune*, September 19, 1917; *Resolutions Adopted by the Nonpartisan League Conference, Held at St. Paul, Sept. 18–19–20, 1917* (Saint Paul: Riverside, 1917), 1.

40. "Two Socialist Editors Arrested in St. Paul" and "Blaze of Disloyalty Ends Conference of Nonpartisan League," *Minneapolis Tribune*, September 21, 1917; Chrislock, *Watchdog of Loyalty*, 170–171.

41. Chrislock, *Watchdog of Loyalty*, 171.

42. T. E. Campbell, report, September 25, 1917, ICFBOI, Bureau Section files, 1909–1921, case no. 46889, roll 875, 389, http://www.fold3.com/image/5199592/, accessed December 28, 2013; "Blaze of Disloyalty Ends Conference of Nonpartisan League," *Minneapolis Tribune*, September 21, 1917; Jensen, "Loyalty as a Political Weapon," 47.

43. "Blaze of Disloyalty Ends Conference of Nonpartisan League," *Minneapolis Tribune*, September 21, 1917; "Lundeen and Gronna Receive Full Shares of Rooseveltian Ire," *Minneapolis Tribune*, September 29, 1917; A. C. Townley, "The 'Nonpartisan League,'" *New York Times*, October 16, 1917, 12.

44. "LaFollette Faces Expulsion Inquiry," *New York Times*, October 6, 1917, 1; Carol E. Jensen, *Agrarian Pioneer in Civil Liberties: The Nonpartisan League in Minnesota during World War I* (New York: Garland, 1986), 34.

45. William J. Breen, *Uncle Sam at Home: Civilian Mobilization, Wartime Federalism, and the Council of National Defense, 1917–1919* (Westport, CT: Greenwood, 1984), 3–20.

46. Kevin Mattson, *Creating a Democratic Public: The Struggle for Urban Participatory Democracy during the Progressive Era* (University Park: Pennsylvania State University Press, 1998), 108; Breen, *Uncle Sam at Home*, 17–50; Seward W. Livermore, *Politics Is Adjourned: Woodrow Wilson and the War Congress, 1916–1918* (Middletown, CT: Wesleyan University Press, 1966), 42–43.

47. George Creel, *How We Advertised America* (New York: Harper and Bros., 1920); James R. Mock and Cedric Larson, *Words That Won the War: The Story of the Committee on Public Information, 1917–1919* (New York: Russell and Russell, 1968); Alan Axelrod, *Selling the Great War: The Making of American Propaganda* (New York: Palgrave Macmillan, 2009); Breen, *Uncle Sam at Home*, 71–96.

48. Chrislock, *Watchdog of Loyalty*, 55 (see also 40–64); "Safety Commission to Handle Minnesota Defense Is Proposed," *Minneapolis Tribune*, April 1, 1917; "Public Safety Bill Backed for Passage Today in Senate," *Minneapolis Tribune*, April 5, 1917.

49. Julius Rosen, report, March 15, 1918, ICFBOI, Bureau Section files, 1909–1921, case no. 46889, roll 876, 24, http://www.fold3.com/image/3141151/, accessed December 30, 2013; *Report of Minnesota Commission of Public Safety*, (Saint Paul: Lewis F. Dow, 1919); William Millikan, *A Union against Unions: The Minneapolis Citizens Alliance and Its Fight against Organized Labor, 1903–1947* (Saint Paul: Minnesota Historical Society Press, 2001), 103–104, 117; "Lawlessness in Minnesota," *The Public: A Journal of Democracy* 21, no. 1058 (July 13, 1918): 876; Jensen, *Agrarian Pioneer in Civil Liberties*, 53, 63.

50. *Report of Minnesota Commission of Public Safety*, 164; Charles Ames to Otto H. Kahn, November 19, 1917, quoted in Jensen, *Agrarian Pioneer in Civil Liberties*, 35.

51. Morlan, *Political Prairie Fire*, 146–147; Julius Rosen, report, March 15, 1918, ICFBOI, Bureau Section files, 1909–1921, case no. 46889, roll 876, 35, http://www.fold3.com/image/3141334/, accessed December 28, 2013; Millikan, *A Union against Unions*, 118.

52. State Executive Committee, Nonpartisan League, to Governor Joseph A. A. Burnquist, February 26, 1918, reel 4, National Nonpartisan League Papers, MNHS; Joseph A. A. Burnquist to Arthur LeSueur, March 11, 1918, folder 1, box 2, Arthur LeSueur Papers, MNHS; Chrislock, *Watchdog of Loyalty*, 182, 299.

53. Breen, *Uncle Sam at Home*, 83; Robert N. Manley, "The Nebraska State Council of Defense: Loyalty Programs and Policies during World War I" (master's thesis, University of Nebraska, 1959); J. L. Albert to Charles Wooster, 1918 folder, box 11, Charles Wooster Papers, Nebraska State Historical Society, Lincoln, NE (hereafter NSHS).

54. O. S. Evans to the Nebraska State Council of Defense, January 18, 1918, Correspondence: January 1918 folder, box 2, Christian Abraham Sorensen Papers, NSHS; C. A. Sorensen to Arthur L. Weatherly, April 11, 1918, and C. A. Sorensen to Elizabeth Freeman, April 18, 1918, both Correspondence: April 1918 folder, box 2, Christian Abraham Sorensen Papers, NSHS.

55. Raymond J. Beach, "An Open Letter to the Nebraska State Council of Defense," Correspondence: June 1918 folder, box 2, Christian Abraham Sorensen Papers, NSHS.

56. C. A. Sorensen to Arthur LeSueur, June 28 and June 29, 1918, Correspondence: June 1918 folder, box 2, Christian Abraham Sorensen Papers, NSHS; Phillips, "The Non-Partisan League in Nebraska," 33–34; *Merriam (NE) Reporter*, July 10, 1918, in Non-Partisan League Scrapbook, NSHS.

57. F. D. Becker to Gov. S. V. Stewart, June 27, 1917, folder 15, box 2, Montana Council of Defense Records, Montana Historical Society, Helena, MT (hereafter MHS); An Act Providing for the Creation and Appointment of the Montana Council of Defense and County Councils, 15th Leg., Ex. Sess. (February 20, 1918).

58. Work, *Darkest before Dawn*, 90–94, 135, 212–217; Nancy Rice Fritz, "The Montana Council of Defense" (master's thesis, University of Montana, 1966); Anna Zelick, "Mob Action in Lewistown, 1917–1918: Patriots on the Rampage," *Montana: The Magazine of Western History* 31, no. 1 (Winter 1981): 30–43; Will A. Campbell to Joseph M. Dixon, September 15, 1920, folder 5, box 24, Joseph M. Dixon Papers, K. Ross Toole Archives / Special Collections, University of Montana, Missoula; Charles Sackett Johnson, "An Editor and a War: Will A. Campbell and the Helena Independent, 1914–1921" (master's thesis, University of Montana, 1977); Chairman, Prairie County Council of Defense, to Charles Greenfield, March 30, 1918, folder 22, box 2, Montana Council of Defense Records, MHS.

59. Stuart Baratta, "Promoting Propaganda: The Washington State Council of Defense, the University of Washington, Its Journalism Department, and Washington Newspapers during World War I" (master's thesis, University of Washington, 1995); Breen, *Uncle Sam at Home*, 72, 81, 93; Carl F. Reuss, "The Farm Labor Problem in Washington, 1917–18," *Pacific Northwest Quarterly* 34, no. 4 (October 1943): 339–352.

60. *Report of the State Council of Defense to the Governor of Washington*, 58–60.

61. Ibid., 5, 61.

62. Gilbert C. Fite, *Peter Norbeck, Prairie Statesman* (Columbia: University of Missouri Press, 1948), 62–65; Peter Norbeck to Fred Bletvold, January 21, 1918, folder 3, box 84, and Peter Norbeck to W. Harry King, May 2, 1918, folder 2, box 84, in Peter Norbeck Papers, Archives and Special Collections, University of South Dakota, Vermillion, SD (hereafter USD).

63. W. Harry King to Olaf Eidem, May 8, 1918, folder 9, box 52, W. Harry King to Thomas Sterling, May 20, 1918, folder 4, box 82, W. Harry King to Lewis Benson, February 11, 1918, folder 1, box 84, R. B. Warne to W. Harry King, March 4, 1918, folder 2, box 84, all in Peter Norbeck Papers, Archives and Special Collections, USD; Gilbert C. Fite, "Peter Norbeck and the Defeat of the Nonpartisan League in South Dakota," *Mississippi Valley Historical Review* 33, no. 2 (September 1946): 227–228.

64. Carl J. Hofland, "The Nonpartisan League in South Dakota" (master's thesis, University of South Dakota, 1940), 49–50; Fite, *Peter Norbeck*, 65–66; Fite, "Peter Norbeck and the Defeat of the Nonpartisan League," 230.

65. *Governor Norbeck Accepts Suggestion of Mr. Bates and Will Accompany Him* (Redfield, SD: Redfield Press, 1918); Fite, "Peter Norbeck and the Defeat of the Nonpartisan League," 231–232.

66. Frank W. Blackmar, ed., *History of the Kansas State Council of Defense* (Topeka: Kansas

State Printing Plant, 1921), 20; Philip Zimmerman to Chairman, State Council of Defense, February 5, 1918, reel 1, Philip Zimmerman Papers, KSHS; "It's a Vile Oath," *Topeka Journal*, March 16, 1918, IWW and NPL Clippings, KSHS; *Kansas Farmer's Union* (Salina), March 21, 1918, and May 9, 1918, National Farmers' Union Papers, series 16, vol. 41, Western Americana Collection, University of Colorado at Boulder, Boulder, CO.

67. J. L. Cross to Arthur Capper, February 27, 1918, and Arthur Capper to J. L. Cross, March 4, 1918, folder 222, box 14, Governor Arthur Capper Papers, KSHS; Homer E. Socolofsky, *Arthur Capper: Publisher, Politician, and Philanthropist* (Lawrence: University of Kansas Press, 1962), 102; M. L. Amos to Arthur Capper, March 23, 1918, folder 222, box 14, Governor Arthur Capper Papers, KSHS.

68. Arthur Capper to B. A. Belt, April 1, 1918, Arthur Capper to P. E. Zimmerman, March 30, 1918, and P. E. Zimmerman to Arthur Capper, April 6, 1918, all folder 222, box 14, Governor Arthur Capper Papers, KSHS; P. E. Zimmerman to Chairman, State Council of Defense, February 5, 1918, Secretary, State Council of Defense, to P. E. Zimmerman, February 6, 1918, and P. E. Zimmerman to C. P. Eklund, August 8, 1918, reel 1, Philip E. Zimmerman Papers, KSHS.

69. Arthur Capper to L. M. Rogers, April 16, 1918, folder 222, box 14, Governor Arthur Capper Papers, KSHS; Socolofsky, *Arthur Capper*, 103; Blackmar, *History of the Kansas State Council of Defense*, 100.

70. Breen, *Uncle Sam at Home*, 93; Hugh T. Lovin, "Moses Alexander and the Idaho Lumber Strike of 1917: The Wartime Ordeal of a Progressive," *Pacific Northwest Quarterly* 66, no. 3 (July 1975): 115–122; A. F. Parker to Moses Alexander, March 5, 1918, State Council of Defense 1917/1918 folder, box 17, Moses Alexander Papers, ISHS; "Beware Rumors," *Colorado Council of Defense Weekly Newsletter* no. 9 (December 1917): 1, Digital World War I Collection, University of Colorado at Boulder Library, Boulder, CO, http://libcudl.colorado.edu/wwi/index.asp, accessed February 6, 2012.

71. Robert M. Caulkins Jr., "The Non Partisan League: Minnesota, North Dakota, Civil Liberties and the Struggle for Survival During World War I" (master's thesis, University of North Dakota, 2008); *North Dakota State Council of Defense Plan and Purpose of Organization: Organized May 28, 1917* (Bismarck: State Council of Defense, 1917); "The Nonpartisan League: Loyal or Disloyal," broadside, 1918, MNHS; Breen, *Uncle Sam at Home*, 75; H. G. Garben to A. Bruce Bielaski, February 19, 1918, ICFBOI, Bureau Section files, 1909–1921, case no. 46889, roll 876, 56–64, http://www.fold3.com/image/3141469/, accessed December 28, 2013; Kenneth Smemo, *Against the Tide: The Life and Times of Federal Judge Charles F. Amidon, North Dakota Progressive* (New York: Garland, 1986), 98–154.

72. Caulkins, "The Non Partisan League," 175–197.

73. Charles Edward Russell, *The Story of the Nonpartisan League: A Chapter in American Evolution* (New York: Harper and Bros., 1920), 246; Creel, *How We Advertised America*, 180; "Our Cover This Week," *Nonpartisan Leader*, June 23, 1919.

74. *Aiding the Enemies of Our Nation! A Timely Warning!* (Grand Forks: Jeremiah Bacon, 1918); R. B. Warne to W. Harry King, March 4, 1918, folder 2, box 84, Peter Norbeck Papers, USD; P. E. Zimmerman to Arthur Capper, April 6, 1918, folder 222, box 14, Arthur Capper Papers, KSHS; J. D. Bacon to Peter Norbeck, November 11, 1918, folder 5, box 84, Peter Norbeck Papers, USD.

75. W. Harry King to J. D. Bacon, February 2, 1918, folder 2, box 84, W. Harry King to J. D. Bacon, May 15, 1918, folder 1, box 84, and J. D. Bacon to Peter Norbeck, November 11, 1918, folder 5, box 84, all in Peter Norbeck Papers, USD. Bacon also offered Walter Quigley—a longtime NPL organizer—$3,000 for a tell-all book smearing the League. Quigley refused the offer. Finally,

Bacon supported the prominent insurance man Harry Curran Wilbur, who kept close ties with anti-League businesses in Minneapolis. See Ralph A. Moore, deposition, February 18, 1919, Campaign Materials 1920 folder, box 2, Christian Abraham Sorensen Papers, NSHS; and Millikan, *A Union against Unions*, 152.

76. Peter Norbeck to Nicolay Grevstad, August 27, 1918, folder 4, box 2, Nicolay A. Grevstad Papers, Norwegian-American Historical Association Archives, Saint Olaf College, Northfield, MN (hereafter NAHA); Rudolph Lee to Norbeck Committee, February 22, 1918, folder 3, box 82, Peter Norbeck Papers, USD; W. Harry King to Nebraska State Council of Defense, October 24, 1918, and Vice Chairman of Nebraska State Council of Defense to W. Harry King, October 26, 1918, folder 44, box 6, Nebraska State Council of Defense, NSHS; J. L. Dobell to W. Harry King, March 4, 1918, folder 5, box 82, Peter Norbeck Papers, USD.

77. Ralph A. Moore, deposition, February 18, 1919, Campaign Materials 1920 folder, box 2, Christian Abraham Sorensen Papers, NSHS; Phillips, "The Non-Partisan League in Nebraska," 12–17; "Plot to Break Up the League," *Nebraska Leader*, July 12, 1919, 1. See also J. P. McGrath to W. H. Ware, March 21, March 30, April 3, June 3, July 28, 1918, Correspondence 1918 folders, box 2, Christian Abraham Sorensen Papers, NSHS.

78. Millikan, *A Union against Unions*, 107, 117, 152; Chrislock, *Watchdog of Loyalty*, 323; Luke W. Boyce to A. L. Porter, September 15, 1920, Background Materials folder, box 1, Northern Information Bureau Records, MNHS.

79. I. B. Bruce, report, August 12, 1918, ICFBOI, Bureau Section files, 1909–1921, case no. 46889, roll 876, 248–252, http://www.fold3.com/image/3142763/, accessed December 29, 2013.

80. Charles Ames to A. Bruce Bielaski, December 8, 1917, ICFBOI, Bureau Section files, 1909–1921, case no. 46889, roll 876, 120, http://www.fold3.com/image/3141905/, accessed December 30, 2013. See the hundreds of pages related to the NPL in ICFBOI, Bureau Section files, 1909–1921, cases 46889, 9-5-409, and 202600-27; and ICFBOI, Old German Files, 1909–1921.

81. W. E. Quigley, "The Truth about the Non-Partisan League: Chapter XV, Miscellaneous," *Lincoln Daily Star*, April 29, 1919, reel 4, National Nonpartisan League Papers, MNHS; Carl H. Chrislock, *The Progressive Era in Minnesota, 1899–1918* (Saint Paul: Minnesota Historical Society Press, 1971), 154; Carl Vrooman to Ray McKaig, April 18, 1918, folder U–V, box 1, Ray McKaig Papers, ISHS.

82. George Creel to Woodrow Wilson, February 19, 1918, in *The Papers of Woodrow Wilson*, vol. 46, *January 16–March 12, 1918*, ed. Arthur S. Link (Princeton, NJ: Princeton University Press, 1984), 386–387; John Miller Baer to Woodrow Wilson, February 7, 1918, in *The Papers of Woodrow Wilson*, 46:270.

83. Woodrow Wilson to George Creel, February 18, 1918, in *The Papers of Woodrow Wilson*, 46:369; Woodrow Wilson to John Miller Baer, February 18, 1918, in *The Papers of Woodrow Wilson*, 46:370; Woodrow Wilson to Joseph Patrick Tumulty, April 5, 1918, in *The Papers of Woodrow Wilson*, vol. 47, *March 13–May 12, 1918*, ed. Arthur S. Link (Princeton, NJ: Princeton University Press, 1984), 260; Livermore, *Politics Is Adjourned*, 154.

84. Ray McKaig to A. J. Tabor, January 8, 1918, folder T, and Ray McKaig to James A. Geer, April 27, 1918, folder G, both box 1, Ray McKaig Papers, ISHS; "Help with the Campaign," advertisement, *Idaho Leader*, August 17, 1918; Walter Eli Quigley, "Out Where the West Begins," typescript, ca. 1932, 69, MNHS; R. J. Gaston to Woodrow Wilson, April 5, 1918, in *The Papers of Woodrow Wilson*, 47:260.

85. Woodrow Wilson to George Creel, April 1, 1918, in *The Papers of Woodrow Wilson*, 47:215–216; George Creel to Woodrow Wilson, April 2, 1918, ibid., 47:226.

86. David Franklin Houston to Woodrow Wilson, March 28, 1918, in *The Papers of Woodrow Wilson*, , 47:177–179; Woodrow Wilson to Joseph Patrick Tumulty, March 20, 1918, ibid., 47:88.

87. "Court Martial for Spies," *New York Times*, April 17, 1918, 12; "Wilson Opposes New Spy Bill," *New York Times*, April 23, 1918, 6; "Townley Denies Charges," *New York Times*, May 2, 1918, 5.

88. "Lundeen and Gronna Receive Full Shares of Rooseveltian Ire," *Minneapolis Tribune*, September 29, 1917; Theodore Roosevelt, "The Ghost Dance of the Shadow Huns," October 1, 1917, in *Roosevelt in the Kansas City Star: War-Time Editorials* (New York: Houghton Mifflin, 1921), 5–6.

89. Oliver Burdon to Theodore Roosevelt, April 30, 1918 (citing a telegram from Charles Ames, member of the Minnesota Commission of Public Safety), April 1918 folder, and Theodore Roosevelt to Donald Cotton, May 14, 1918, May 1918 folder, both box 2, Donald Cotton Papers, MNHS; "Roosevelt and the Kaiser," *Nonpartisan Leader*, May 6, 1918.

90. Theodore Roosevelt, "Murder, Treason, and Parlor Anarchy," July 18, 1918, in *Roosevelt in the Kansas City Star*, 182–183; James F. Vivian, "The Last Round-Up: Theodore Roosevelt Confronts the Nonpartisan League, October 1918," *Montana: The Magazine of Western History* 36, no. 1 (Winter 1986): 38–39; Theodore Roosevelt, "Good Luck to the Anti-Bolshevists of Kansas," September 12, 1918, in *Roosevelt in the Kansas City Star*, 215–216.

91. Theodore Roosevelt, "Spies and Slackers," September 24, 1918, in *Roosevelt in the Kansas City Star*, 222–223; Vivian, "The Last Round-Up," 40–42; " 'Trust No Party Except One of America First,' " *Chicago Tribune*, October 6, 1918, 7.

92. Vivian, "The Last Round-Up," 44, 46; Theodore Roosevelt, *The Great Adventure: Present Day Studies in American Nationalism* (New York: Scribners, 1918), 41.

93. Vivian, "The Last Round-Up," 46–47; Theodore Roosevelt, "An American Congress," November 18, 1918, in *Roosevelt in the Kansas City Star*, 266.

94. Woodrow Wilson to William Kent, May 1, 1918, in *The Papers of Woodrow Wilson*, 47:475; and Livermore, *Politics Is Adjourned*, 279n19.

95. *War-Time Prosecutions and Mob Violence, Involving the Rights of Free Speech, Free Press and Peaceful Assemblage, from April 1, 1917, to March 1, 1919* (New York: National Civil Liberties Bureau, 1919), 12.

96. Ray McKaig to C. W. Booth, March 20, 1918, folder B, box 1, Ray McKaig Papers, ISHS; Ray McKaig, "The Farmers Mob the Mobbers," *The Public: A Journal of Democracy* 21, no. 1069 (September 18, 1918): 1241–1242; Moorhead, "The Non-Partisan League in Politics," 365.

97. Work, *Darkest before Dawn*, 144; C. A. Sorensen to Arthur L. Weatherly, April 11, 1918, Correspondence: April 1918 folder, box 2, Christian Abraham Sorensen Papers, NSHS; Phillips, "The Nonpartisan League in Nebraska," 25; *Wahoo (NE) Wasp*, August 1, 1918, in Non-Partisan League Scrapbook, NSHS.

98. *War-Time Prosecutions and Mob Violence*, 13; Homer Ayres, interview by Earl Hausle, June 28, 1971, Sturgis, SD, interview 187, South Dakota Oral History Project, USD.

99. Gaston, *The Nonpartisan League*, 230; Ayres, interview by Hausle; Work, *Darkest before Dawn*, 144; Moorhead, "The Non-Partisan League in Politics," 364.

100. Gaston, *The Nonpartisan League*, 227; *War-Time Prosecutions and Mob Violence*, 13; H. C. Peterson and Gilbert C. Fite, *Opponents of War, 1917–1918* (Madison: University of Wisconsin Press, 1957), 199; Clarence Sharp, interview by Carl Ross, Minneapolis, MN, December 9 and December 16, 1987, 1, 20th Century Radicalism in Minnesota Oral History Project, MNHS.

101. Lena L. Borchardt, "I Lived a Full Life: Reminiscences," folder 1, Lena Borchardt Collection, MNHS; Gaston, *The Nonpartisan League*, 227; Steven J. Keillor, *Hjalmar Peterson of*

Minnesota: The Politics of Provincial Independence (Saint Paul: Minnesota Historical Society Press, 1987), 48.

102. "Non-Partisan League Organizers Not Wanted," *Evening Herald* (Klamath, OR), April 27, 1918; "Second Non-Partisan Leagues Is Tarred," *Evening Record* (Ellensburg, WA), May 1, 1918; *War-Time Prosecutions and Mob Violence*, 13; Gaston, *The Nonpartisan League*, 225.

103. "Presbyterian Pastor and League Organizer Assaulted by Banker," *Montana Nonpartisan* (Great Falls), August 14, 1920; Daniel McCorkle to R. A. Haste, December 13, 1922, folder H, box 1, Daniel S. McCorkle Papers, MHS.

104. *War-Time Prosecutions and Mob Violence*, 13; Kim E. Nielsen, "'We All Leaguers by Our House': Women, Suffrage, and Red-Baiting in the National Nonpartisan League," *Journal of Women's History* 6, no. 1 (Spring 1994): 33; Henry R. Martinson, "Some Memoirs of a Nonpartisan League Organizer," *North Dakota History* 42, no. 2 (Spring 1975): 20.

105. *War-Time Prosecutions and Mob Violence*, 37–38; L. J. Duncan to Peter Norbeck, February 6, 1918, telegram, folder 1, box 84, Peter Norbeck Papers, USD; *Lincoln (NE) Trade Review*, July 13, 1918, newspaper clipping, Non-Partisan League Scrapbook, NSHS.

106. *War-Time Prosecutions and Mob Violence*, 12, 37; Quigley, "Out Where the West Begins"; Moorhead, "The Non-Partisan League in Politics," 365; C. A. Sorenson to Arthur Weatherby, June 15, 1918, Correspondence: June 1918 folder, box 2, Christian Abraham Sorensen Papers, NSHS.

107. Fred Hoppe, "History of the League in Colfax County," *Nebraska Leader*, November 1, 1919, 1; "The Heart Cry of a Farm Wife," *Nonpartisan Leader*, June 3, 1918; Profile of Nebraska NPL Leaders, John O. Schmidt, no date, Correspondence: August 1922 folder, box 3, Christian Abraham Sorensen Papers, NSHS.

108. "Lawlessness in Minnesota," 877–878; *War-Time Prosecutions and Mob Violence*, 13; Chrislock, *Watchdog of Loyalty*, 290; "What the Organized Farmer Is Doing," *Nonpartisan Leader*, November 24, 1919; Memorandum for Mr. Allen, March 26, 1919, ICFBOI, Bureau Section files, 1909–1921, case no. 46889, roll 877, 115, http://www.fold3.com/image/5220441/, accessed December 30, 2013; William H. Thomas, *Unsafe for Democracy: World War I and the U.S. Justice Department's Covert Campaign to Suppress Dissent* (Madison: University of Wisconsin Press, 2008), 167–168; "All Luverne Greets 32 Citizens Freed in Tar-Feather Case," *Minneapolis Tribune*, November 16, 1919.

109. Christian Sorensen to James M. Pierce, July 11, 1918, Correspondence: July 1918 folder, box 2, Christian Abraham Sorensen Papers, NSHS; Republican Ransom County Central Committee, resolution, July 17, 1918, folder 21, box 6, Martin O. Thompson Papers, North Dakota Institute for Regional Studies, North Dakota State University, Fargo, ND (hereafter NDIRS); *Memorial to the Congress of the Untied States Concerning Conditions in Minnesota 1918* (Saint Paul: Nonpartisan League, 1918).

110. Millard L. Gieske, "The Politics of Knute Nelson, 1912–1920" (PhD diss., University of Minnesota, 1965), 93–98; Ray McKaig to Albert Dakin, March 2, 1918, folder D, Ray McKaig to William Bouck, March 2, 1918, folder B, H. P. Coder to Ray McKaig, March 8, 1918, folder C, Ray McKaig to Ralph Drake, March 20, 1918, folder D, all box 1, Ray McKaig Papers, ISHS; Chrislock argues that "to a considerable extent, bitter memories of the excesses of the loyalty crusade enhanced the strength" of the Nonpartisan League in Minnesota. *Watchdog of Loyalty*, 181.

111. Peterson and Fite, *Opponents of War*, 195–196; Frederick C. Luebke, *Bonds of Loyalty: German Americans and World War I* (DeKalb: Northern Illinois University Press, 1974).

112. "Many Meetings Are Planned by the League," *Nonpartisan Leader*, January 25, 1917; "League Meetings and Announcements," *Nonpartisan Leader*, March 8, 1917. Meitzen first joined

the League in March or April 1917. See E. O. Meitzen to W. H. Howard, December 20, 1919, pp. 26–36 folder, box 1, Materials Related to the Nonpartisan League, 1905–1920 Collection, comp. Asher Howard, MNHS; "Texas Investigates the League," *Nonpartisan Leader*, April 26, 1917.

113. Kathleen Neils Conzen, *Germans in Minnesota* (Saint Paul: Minnesota Historical Society Press, 2003); Chrislock, *Watchdog of Loyalty*, 133–156; La Vern J. Rippley, "Conflict in the Classroom: Anti-Germanism in Minnesota Schools, 1917–19," *Minnesota History* 47, no. 5 (Spring 1981): 170–183; Charles Lamb, "The Nonpartisan League and Its Expansion into Minnesota," *North Dakota Quarterly* 49, no. 2 (Summer 1981): 130–135; Phillips, "The Non-Partisan League in Nebraska," 39–43; Robert N. Manley, "Language, Loyalty, and Liberty: The Nebraska State Council of Defense and the Lutheran Churches," *Concordia Historical Institute Quarterly* 37, no. 1 (April 1964): 1–16; Niel M. Johnson, "The Missouri Synod Lutherans and the War against the German Language, 1917–1923," *Nebraska History* 56, no. 1 (Spring 1975): 136–144.

114. Richard Lofthus, "'Over Here, Over There': The World War I Correspondence of the Pvt. John Warns Family," *South Dakota History* 36, no. 1 (Spring 2006): 9; Peter Norbeck to Henry J. Allen, May 7, 1919, folder 8, box 2, Henry J. Allen Papers, KSHS.

115. Chrislock, *The Progressive Era in Minnesota*, 169–170; Bruce Larson, *Lindbergh of Minnesota: A Political Biography* (New York: Houghton Mifflin, 1973).

116. Burton W. Folsom Jr., "Immigrant Voters and the Nonpartisan League in Nebraska, 1917–1920," *Great Plains Quarterly* 1 (Summer 1981): 159–168.

117. "Lindsborg Has Record Crowd," *Ellsworth County (KS) Leader*, October 9, 1919; "Lindsborg for the Farmers," *Ellsworth County Leader*, June 10, 1920; Philip Zimmerman to Lacey Simpson, November 3, 1919, reel 1, Philip E. Zimmerman Papers, KSHS; *Ellsworth County Leader*, January 27, 1921; Wayne Wheeler, *An Analysis of Social Change in a Swedish-Immigrant Community: The Case of Lindsborg, Kansas* (New York: AMS, 1986), 152.

118. Nicolay Grevstad to J. A. O. Preus, July 25, 1917, folder 5, box 1, Nicolay A. Grevstad Papers, NAHA; Lowell J. Soilke, *Norwegian Americans and the Politics of Dissent, 1880–1924* (Northfield, MN: Norwegian-American Historical Association, 1991), 156–164.

119. Carl H. Chrislock, *Ethnicity Challenged: The Upper Midwest Norwegian-American Experience in World War I* (Northfield, MN: Norwegian-American Historical Association, 1981), 68–74, 95–96; Nicolay Grevstad to A. C. Weiss, March 4, 1918, folder 6, box 1, Nicolay A. Grevstad Papers, NAHA; Lovoll, *Norwegian Newspapers in America*, 268–271; Knute Nelson to Nicolay Grevstad, December 16, 1918, and Knute Nelson to Nicolay Grevstad, October 21, 1920, folder 3, box 2, Nicolay A. Grevstad Papers, NAHA.

120. *Denver Catholic Register*, March 4, 1920, newspaper clipping, and *Catholic Echo*, April 8, 1920, newspaper clipping, Bishop Vincent Wehrle Papers, State Historical Society of North Dakota, Bismarck, ND (hereafter SHSND).

121. G. M. Bruce to Peter Norbeck, November 7, 1918, folder 5, box 84, Peter Norbeck Papers, USD; *Clergymen Endorse Farmer Government of North Dakota* (Fargo, ND, 1920), 29; *Ellsworth County Leader*, March 17, 1921.

122. Chrislock, *Ethnicity Challenged*, 85, 97; Carl H. Chrislock, "Name Change and the Church," *Norwegian American Studies* 27 (1977): 194–223; "The State as a Business Organization," *Lutheran Companion* 27, no. 5 (February 1, 1919): 55; R. F. Nelson, "Some Comments by a Reader," *Lutheran Companion* 27, no. 12 (March 22, 1919): 151.

123. C. A. Sorensen to William Oeschger, October 7, 1920, Correspondence: October 1920 folder, box 2, Christian Abraham Sorensen Papers, NSHS; R. N. McKaig, "League Is Real Christianity," newspaper clipping, no date, folder S, box 1, Ray McKaig Papers, ISHS.

124. Ray McKaig to Joseph Gilbert, December 21, 1917, and January 4, 1918, folder G, Ray McKaig to Albert Dakin, March 2, 1918, folder D, all box 1, Ray McKaig Papers, ISHS.

125. *Clergymen Endorse Farmer Government of North Dakota* (Fargo, 1920); Mercer Green Johnston, *An Account of My Stewardship* (Newark, NJ: Trinity Church, 1916); F. Halsey Ambrose, *A Sermon on Socialism, Preached by Rev. F. Halsey Ambrose . . .* (Grand Forks, ND: Grand Forks Herald, 1919); Stock, *Main Street in Crisis,* 67.

126. Carol E. Jenson, *Agrarian Pioneer in Civil Liberties: The Nonpartisan League in Minnesota during World War II* (New York: Garland, 1986), 83.

127. Hugh G. Lovin, "Disloyalty, Libel, and Litigation: Ray McKaig's Ordeal, 1917–1920," *Idaho Yesterdays* 27, no. 2 (Summer 1983): 13–24.

128. William Bouck to Ray McKaig, April 1, 1918, folder B, box 1, Ray McKaig Papers, ISHS.

129. Morlan, *Political Prairie Fire,* 169–171.

130. James Manahan, *Trials of a Lawyer* (Minneapolis: Farnham, 1933), 232–238; Morlan, *Political Prairie Fire,* 256–261; "Townley Tried by Prejudiced Judge Says George Creel," *Nebraska Leader,* August 23, 1919.

131. Morlan, *Political Prairie Fire,* 171–172; Judson King, "The Prosecution of Arthur Townley," *Nation* 109, no. 2822 (August 2, 1919): 143.

132. Jenson, *Agrarian Pioneer in Civil Liberties,* 156–157.

133. *Gilbert v. State of Minnesota,* 254 U.S. 325, 41 S.Ct. 125, no. 79, http://www.soc.umn .edu/~samaha/cases/gilbert_v_minnesota.html, accessed March 11, 2012; Jenson, *Agrarian Pioneer in Civil Liberties,* 160–164. For the 1925 reversal, see *Gitlow v. New York,* 268 U.S. 652 (1925), http://openjurist.org/268/us/652/gitlow-v-people-of-the-state-of-new-york, accessed March 11, 2012; Davis Douthit, *Nobody Owns Us: The Story of Joe Gilbert, Midwestern Rebel* (Chicago: Cooperative League of the USA, 1948), 196.

134. Charles M. Gardner to Ray McKaig, February 7, 1918, folder G, box 1, Ray McKaig Papers, ISHS.

135. Ray McKaig, list of state granges and membership numbers, no date (ca. 1917), folder G, box 1, Ray McKaig Papers, ISHS; Quigley, "Out Where the West Begins." Winnebago County was a hotbed of NPL activity. See A. A. Sandin to A. C. Townley, January 10, 1917, reel 1, National Nonpartisan League Papers, MNHS; and Soilke, *Norwegian Americans and the Politics of Dissent,* 239n46.

136. Robert P. Howard, memoir, 1979, 37, 42, Archives / Special Collections, University of Illinois at Springfield, Springfield, IL, http://library.uis.edu/archives/collections/oral/pdf /HOWARDRvI.pdf, accessed March 2, 2012; "Editorial," *Iowa Magazine* 2, no. 6 (December 1918): 6; "Editorial," *Iowa Magazine* 2, no. 6 (December 1918): 6; Tom McMahon, interview by Greg Zieren, February 22, 1980, Des Moines, IA, 2–3, Iowa Labor Oral History Collection, State Historical Society of Iowa, Iowa City, IA; Henry C. Wallace, editorial, manuscript, undated, 240–242, reel 4, Henry A. Wallace Collection, Iowa Digital Library, http://wallace.lib.uiowa.edu/, accessed March 2, 2012; A. B. Gilbert, "Out for a 'Solid West': The Coming Political Battle of the Nonpartisan League," *Forum* 60 (December 1918): 734.

137. "Editorial," *Iowa Magazine* 2, no. 6 (December 1918): 6–7; M. Eberstein, report, January 29, 1918, ICFBOI, Bureau Section files, 1909–1921, case no. 46889, roll 877, 24, http://www.fold3.com /image/5220196/, accessed December 30, 2013; "Feeding Iowa on Anti-Farmer Prejudice," *Nonpartisan Leader,* April 1, 1918; Ray McKaig to L. J. Duncan, March 2, 1918, folder D, box 1, Ray McKaig Papers, ISHS; James M. Pierce to C. A. Sorensen, July 8, 1918, Correspondence: July 1918 folder, box 2, Christian Abraham Sorensen Papers, NSHS; *Oelwein (IA) Daily Register,* July 24, 1918.

138. Moorhead, "The Non-Partisan League in Politics," 365; James M. Pierce, "Some Farmers' Views of the Greater Iowa Association," *Iowa Homestead* (Des Moines), March 21, 1918, 1; "Iowa," *Chicago Tribune*, March 10, 1918, 8; Gaston, *The Nonpartisan League*, 229–230; "Arrested in the Dead of Night," *Iowa Homestead*, August 22, 1918.

139. For Sheldon, see Asher Howard, *The Leaders of the Nonpartisan League: Their Aims, Purposes and Records . . .* (Minneapolis: published by author, 1920), 40. See also James R. Green, *Grass-Roots Socialism: Radical Movements in the Southwest, 1895–1943* (Baton Rouge: Louisiana State University Press, 1978); Gavin Burbank, *When Farmers Voted Red: The Gospel of Socialism in the Oklahoma Countryside, 1910–1924* (Westport, CT: Greenwood, 1977); Jim Bissett, *Agrarian Socialism in America: Marx, Jefferson, and Jesus in the Oklahoma Countryside, 1904–1920* (Norman: University of Oklahoma Press, 2002). For the earliest days of the League in Oklahoma, see Gilbert C. Fite, "The Nonpartisan League in Oklahoma," *Chronicles of Oklahoma* 24, no. 2 (Summer 1946): 146–148. While the presence of socialism in Oklahoma attracted the NPL, there was also concern about the Socialist Party's "extreme ORTHODOXY," which "kept out of the ranks hundreds of thousands of men." See Farmers Non-Partisan League to Hugh E. Peters, November 27, 1916, reel 1, National Nonpartisan League Papers, MNHS.

140. James Fowler, "Creating an Atmosphere of Suppression, 1914–1917," *Chronicles of Oklahoma* 59, no. 2 (June 1981): 202–223; Nigel Sellars, " 'With Folded Arms? Or With Squirrel Guns?': The Green Corn Rebellion," *Chronicles of Oklahoma* 77, no. 2 (Summer 1999): 150–169; James Fowler, "Tar and Feather Patriotism: The Suppression of Dissent in Oklahoma during World War I," *Chronicles of Oklahoma* 56, no. 4 (December 1978): 409–430; O. A. Hilton, "The Oklahoma Council of Defense and the First World War," *Chronicles of Oklahoma* 20, no. 1 (March 1942): 36.

141. Gaston, *The Nonpartisan League*, 238; Fite, "The Nonpartisan League in Oklahoma," 149–151, 154; L. N. Sheldon to Henry Teigan, May 30, 1917, reel 2, National Nonpartisan League Papers, MNHS.

142. Kyle Grant Wilkison, *Yeoman, Sharecroppers, and Socialists: Plain Folk Protest in Texas, 1870–1914* (College Station: Texas A&M University Press, 2008), 201; E. O. Meitzen to W. H. Howard, December 20, 1919, pp. 26–36 folder, box 1, Materials Related to the Nonpartisan League, 1905–1920 Collection, comp. Asher Howard, MNHS; "Texas Investigates the League," *Nonpartisan Leader*, April 26, 1917; Thomas Hickey to Clara Hickey, June 16, 1917, folder 22, and M. S. Goodyear to Clara Hickey, November 26, 1917, folder 15, both Box 1, Papers of Thomas A. Hickey, Special Collections, Texas Tech University, Lubbock, TX (hereafter TTU).

143. "The National Nonpartisan League—Texas Branch," no date (ca. 1918), 3K423, Thomas A. Hickey Papers, Center for American History, University of Texas at Austin, Austin, TX; Green, *Grass-Roots Socialism*, 376.

144. Thomas Hickey to Clara Hickey, November 24, 1918, folder 14, box 1, Papers of Thomas A. Hickey, TTU; E. O. Meitzen to W. H. Howard, December 20, 1919, pp. 26–36 folder, box 1, Materials Related to the Nonpartisan League, 1905–1920 Collection, comp. Asher Howard, MNHS; Green, *Grass-Roots Socialism*, 376; Gaston, *The Nonpartisan League*, 222–224; "A Roman Holiday in Texas," *Nonpartisan Leader*, April 29, 1918, 3; C. E. Breniman, report, April 15, 1918, ICF-BOI, Bureau Section files, 1909–1921, case no. 46889, roll 876, 180, http://www.fold3.com/image /3142342/, accessed December 30, 2013.

145. Jack Ford to Charles England, January 15, 1918, file 35, and J. W. Leedy to Jack Ford, March 18, 1918, file 43, both series 4, John Ford Fonds, Glenbow Archives, Calgary, AB (hereafter GA).

146. Jack Ford to *Nonpartisan Leader*, April 13, 1918, file 52, Jack Ford to Henry Ross, May 28,

1918, file 63, Jack Ford to C. W. McDonnell, April 13, 1918, file 50, all series 4, John Ford Fonds, GA; C. W. McDonnell, "The Non-Partisan Movement: Methods Used to Oppose League in Minnesota," *Alberta Non-Partisan* (Calgary), May 24, 1918.

147. "What the League Will Do," *Nonpartisan Leader*, November 4, 1915; Morlan, *Political Prairie Fire*, 98–99.

148. Theodore G. Nelson, *Scrapbook Memoirs* (Salem, OR: Your Town, 1957), 71; D. Jerome Tweton, "The Anti-League Movement: The IVA," in *The North Dakota Political Tradition*, ed. Thomas W. Howard (Ames: Iowa State University Press, 1981), 93–94.

149. "Nelson's Little Game Foiled," *Nonpartisan Leader*, April 12, 1917; Theodore G. Nelson to *Nonpartisan Leader*, April 17, 1917, folder 4, box 1, S. A. Olsness Papers, NDIRS.

150. "Gang Was Nearly All There," *Nonpartisan Leader*, April 19, 1917; Theodore G. Nelson, "Independent Voters' Association History," in *A Volume of Truth* (Bismarck: Independent Voters Association, 1918), Printed Materials, 1913–1919 folder, box 14, Arthur LeSueur Papers, MNHS.

151. Nelson, *Scrapbook Memoirs*, 99.

152. Theodore G. Nelson to W. J. Church, February 1, 1918, and Theodore G. Nelson to William Lemke, February 1, 1918, both folder 4, box 1, S. A. Olsness Papers, NDIRS.

153. Nelson, "Independent Voters' Association History"; undated petition, reproduced in Nelson, *Scrapbook Memoirs*, 71. Internal evidence suggests the petition was created in late 1917 or early 1918.

154. "Proceedings," Lincoln Republican League, folder 4, box 1, Theodore Nelson Papers, SHSND.

155. Nelson, "Independent Voters' Association History"; Richard Whaley, "The Other Side of the Mountain: Stephen Joseph Doyle and Opposition to the NPL in 1918," *North Dakota Quarterly* 56, no. 4 (Fall 1988): 192–210; Martin O. Thompson to the Republican County Committee, October 10, 1918, folder 3, box 1, Martin O. Thompson Papers, NDIRS.

156. "Minutes of the Meeting of the Plain Citizens' Political Association, Cooperstown, ND, December 5, 1918," folder 1, box 1, Theodore Nelson Papers, SHSND.

157. "Minutes of the Statewide Mass Meeting Held at Bismarck, North Dakota on January 28th and 29th, 1919," folder 1, box 1, Theodore Nelson Papers, SHSND.

158. Thomas Contois, "The Fight against the Nonpartisan League: The Independent Voters Association" (senior thesis, Duke University, 1986).

Chapter Four

1. "To the Voters of Griggs and Steele Countys, Dover Township," Griggs County, ND, June 10, 1918, Correspondence and Related Papers, 1909–1954 folder, box 1, John L. Miklethun Papers, Minnesota Historical Society, Saint Paul, MN (hereafter MNHS).

2. Ibid.

3. Jacob L. Miklethun, "Nonpartisan League Meeting, February 22, 1918, Dover Precinct, Griggs County N.D.," March 11, 1918, Political Organization Papers, undated and 1918–1921 folder, box 3, John L. Miklethun Papers, MNHS; "To The Voters of Griggs and Steele Countys."

4. John L. Miklethun to H. E. More, June 26, 1918, Correspondence and Related Papers, 1909–1954 folder, box 1, John L. Miklethun Papers, MNHS; Primary Election Returns, June 26, 1918, and General Election Returns, November 5, 1918, "North Dakota Election Results Portal," Secretary of State's Office, Bismarck, ND, https://vip.sos.nd.gov/ElectionResultsPortal.aspx, accessed March 26, 2012.

5. A. B. Gilbert, "The Farmers' Experiment in Democracy," *World Tomorrow* 2, no. 8 (August 1918): 201–205.

6. Henry Teigen to O. J. Arness, October 3, 1918, National Nonpartisan League Papers, reel 4, MNHS; Theodore Saloutos, "The Expansion and Decline of the Nonpartisan League in the Western Middle West, 1917–1921," *Agricultural History* 20, no. 4 (October 1946): 236. For the numbers of NPL members (188,365 total at the end of 1918, with 50,162 in Minnesota and 35,062 in North Dakota), see Henry Teigan, "The National Nonpartisan League," *The American Labor Year Book*, ed. Alexander Trachtenberg (New York: Rand School of Social Science, 1920), 285. One contemporary journalist claimed that the League numbered 300,000 total members in late 1918, with 130,000 in Minnesota, North Dakota, and Idaho. See Frank G. Moorhead, "The Non-Partisan League in Politics," *Nation* 107, no. 2779 (October 5, 1918): 365. NPL secretary Teigan's numbers are more trustworthy.

7. Louis Levine, "Politics in Montana," *Nation* 107, no. 2783 (November 2, 1918): 507–508; Teigan, "The National Nonpartisan League," 284–285; W. Clavier to C. A. Sorensen, May 20, 1919, Correspondence May 1919 folder, box 2, C. A. Sorensen Papers, Nebraska State Historical Society, Lincoln, NE (hereafter NSHS); Jerry W. Calvert, *The Gibraltar: Socialism and Labor in Butte, Montana, 1895–1920* (Helena: Montana Historical Society Press, 1988), 128–129.

8. Elmo Bryant Philips, "The Non-Partisan League in Nebraska" (master's thesis, University of Nebraska–Lincoln, 1931), 47; *Merriam (NE) Reporter*, July 10, 1918, *Petersburg (NE) Index*, July 5, 1918, *Wahoo (NE) Wasp*, July 11, 1918, *Lexington (NE) Citizen*, July 19, 1918, all Non-Partisan League Scrapbook, NSHS; "Farmers Endorse Norris," October 1918, Correspondence October 1918 folder, box 2, Christian Abraham Sorensen Papers, NSHS; Robert Morlan, *Political Prairie Fire: The Nonpartisan League, 1915–1922* (1955; repr., Saint Paul: Minnesota Historical Society Press, 1986), 214.

9. Bruce L. Larson, *Lindbergh of Minnesota: A Political Biography* (New York: Harcourt Brace Jovanovich, 1971), 217.

10. Lynn Haines and Dora B. Haines, *The Lindberghs* (New York: Vanguard, 1931), 282.

11. Robert L. Morlan, "The Nonpartisan League and the Minnesota Campaign of 1918," *Minnesota History* 34, no. 6 (Summer 1955): 224, 229–231; Mary Lethert Wingerd, *Claiming the City: Politics, Faith, and the Power of Place in St. Paul* (Ithaca, NY: Cornell University Press, 2001), 215–218; Knud Wefald to Charles Barnes, June 22, 1918, Correspondence and Miscellaneous Papers, 1917–1918 folder, box 9, Knud Wefald Papers, MNHS.

12. Morlan, "The Nonpartisan League and the Minnesota Campaign of 1918," 232; Teigan, "The National Nonpartisan League," 284.

13. Walter Day, interview by Russell Fridley, November 27, 1967, Bagley, MN, 14, MNHS; H. B. Danielson to William I. Nolan, December 6, 1918, and William I. Nolan to H. B. Danielson, December 13, 1918, Correspondence and Other Papers: Dec. 6–11, 1918 folder, box 2, William I. Nolan Papers, MNHS.

14. "The Idaho State Platform" and "Advisory Ballot for League Members and Organized Labor, Democratic Primaries," *Idaho Leader* (Boise), August 17, 1918; Hugh T. Lovin, "The Farmer Revolt in Idaho, 1914–1922," *Idaho Yesterdays* 20, no. 3 (Fall 1976): 10.

15. "League Rallies," *Idaho Leader*, August 17, 1918; Moorhead, "The Non-Partisan League in Politics," 365.

16. Merle W. Wells, "Fred T. DuBois and the Nonpartisan League in the Idaho Election of 1918," *Pacific Northwest Quarterly* 56, no. 1 (January 1965): 23–25; Lovin, "The Farmer Revolt in Idaho," 10–11.

17. Lovin, "The Farmer Revolt in Idaho," 11; Morlan, *Political Prairie Fire*, 214.

18. Oliver S. Morris, "The Electoral Gains of the Nonpartisan League," *Liberator* 1, no. 10 (December 1918): 38–40; Steven L. Piott, *Giving Voters a Voice: The Origins of the Initiative and Referendum in America* (Columbia: University of Missouri Press, 2003), 224–225.

19. *The Hundred and Sixteen Nonpartisan League Members of the Sixteenth Legislative Assembly of North Dakota to the Farmers and Other Workers of America* (Bismarck, ND, 1919).

20. William MacDonald, "North Dakota's Experiment," *Nation* 108, no. 2803 (March 22, 1919): 420.

21. Ibid., 420–421; Alvin S. Tostlebee, "The Bank of North Dakota: An Experiment in Agrarian Banking," *Studies in History, Economic, and Public Law* 64, no. 1 (1924): 65.

22. Herbert E. Gaston, *The Nonpartisan League* (New York: Harcourt, Brace and Howe, 1920), 291.

23. Charles Edward Russell, *The Story of the Nonpartisan League: A Chapter in American Evolution* (New York: Harper and Bros., 1920), 251–252.

24. Industrial Commission of North Dakota, *The North Dakota Industrial Program: A Report on the Organization and Progress of the North Dakota State Industries, and the Administration of Related Laws, Protecting and Promoting Agriculture and Other Industries in the State, Enacted and Established by the Sixteenth Session of the North Dakota Legislative Assembly* (Bismarck: Commissioner of Immigration, 1920), 6–7, 9, 52.

25. Ibid., 19; Tostlebee, "The Bank of North Dakota," 76; Russell, *The Story of the Nonpartisan League*, 272.

26. Morlan, *Political Prairie Fire*, 235–236; Joseph H. Mader, "The Political Influence of the Nonpartisan League on the Press of North Dakota" (master's thesis, University of Minnesota, 1937), 126–127.

27. Elwyn B. Robinson, *History of North Dakota* (Lincoln: University of Nebraska Press, 1966), 343.

28. Quoted in Tostlebee, "The Bank of North Dakota," 20. See also Walter W. Liggett, "N.D. State Bank Has Many Precedents," *Nonpartisan Leader* (Fargo, ND), February 10, 1919. The Federal Reserve banks loomed larger as a model for the NPL than the many failed state banks of nineteenth-century America. But the Bank of North Dakota moved well beyond the Federal Reserve model, which initially depended on a private-public partnership that drew together the resources of regional corporate interests. Only later did the Fed become an instrument of centralized state governance. See James Livingston, *The Origins of the Federal Reserve System: Money, Class, and Corporate Capitalism, 1890–1913* (Ithaca, NY: Cornell University Press, 1986); "The Federal Reserve System—Its Purpose and Work," special issue, *Annals of the American Academy of Political and Social Science* 99, no. 188 (January 1922); and Allan H. Metzler, *A History of the Federal Reserve*, vol. 1, *1913–1951* (Chicago: University of Chicago Press, 2003). See also Rozanne Enerson Junker, *The Bank of North Dakota: An Experiment in State Ownership* (Santa Barbara, CA: Fithian, 1989), 11–12.

29. Junker, *The Bank of North Dakota*, 16–17, 26; Early Settlers Organization Committee of Seven, *Dauntless Dunn: Stories of Some of Its Early Settlers* (Dunn County, ND: The Committee, 1970), 5; Robinson, *History of North Dakota*, 342.

30. Tostlebee, "The Bank of North Dakota," 82–83.

31. Junker, *The Bank of North Dakota*, 19–21; and William Millikan, *A Union against Unions: The Minneapolis Citizen's Alliance and Its Fight against Organized Labor, 1903–1947* (Saint Paul: Minnesota Historical Society Press, 2001), 240, 439n49.

32. Kenneth Smemo, *Against the Tide: The Life and Times of Federal Judge Charles F. Amidon,*

North Dakota Progressive (New York: Garland, 1986), 162–166; Junker, *The Bank of North Dakota,* 36–37.

33. *Green v. Frazier,* 253 U.S. 233 (1920), 240, http://caselaw.lp.findlaw.com/cgi-bin/getcase .pl?court=us&vol=253&invol=233, accessed April 5, 2012; Tostlebee, "The Bank of North Dakota," 111; "North Dakota Laws Upheld," *Nonpartisan Leader,* June 14, 1920.

34. Industrial Commission of North Dakota, *The North Dakota Industrial Program,* 36, 46.

35. Junker, *The Bank of North Dakota,* 55; Industrial Commission of North Dakota, *The North Dakota Industrial Program,* 21; Steven R. Hoffbeck, National Register of Historic Places Registration Form, "North Dakota Mill and Elevator," January 8, 1981, 2, Special Collections, University of North Dakota, Grand Forks, ND (hereafter UND); Gilbert Cooke, "The North Dakota State Mill and Elevator," *Journal of Political Economy* 46, no. 1 (February 1938): 23–51.

36. Herbert E. Gaston, "Where Merchants and Farmers Agree," *Nonpartisan Leader,* February 2, 1920, 5; Hoffbeck, National Register of Historic Places Registration Form, "North Dakota Mill and Elevator," 2–5.

37. Hoffbeck, National Register of Historic Places Registration Form, "North Dakota Mill and Elevator," 5–7.

38. Paul R. Fossum, "The Agrarian Movement in North Dakota," *Johns Hopkins University Studies in Historical and Political Science* 43 (1925): 111–112.

39. Industrial Commission of North Dakota, *The North Dakota Industrial Program,* 37–39; "North Dakota to Boom Home Owning," *Nonpartisan Leader,* February 3, 1919.

40. Michelle L. Dennis, National Register of Historic Places Registration Form, "Nonpartisan League's Home Building Association Resources in North Dakota," February 2006, 15–16. The association likely built bungalows because they best fit middle-class aspirations and financial reach. See Janet Ore, *The Seattle Bungalow: People and Houses, 1900–1940* (Seattle: University of Washington Press, 2006), 1–17.

41. Dennis, "Nonpartisan League's Home Building Association Resources in North Dakota," 17–29.

42. Walter W. Liggett, "Fair Tax Principles for North Dakota," *Nonpartisan Leader,* February 17, 1919; Industrial Commission of North Dakota, *The North Dakota Industrial Program,* 52; Russell, *The Story of the Nonpartisan League,* 264–266.

43. Gaston, *The Nonpartisan League,* 296–297; Industrial Commission of North Dakota, *The North Dakota Industrial Program,* 51–55.

44. "North Dakotan Scores Editor," *Minnesota Leader* (Saint Paul), April 10, 1920.

45. W. E. H. Porter, Henry A. Wibert, and Chris Orton to William Langer, March 23, 1919, folder 1, box 15, William Langer Papers, UND; Carl Olson to editor, *Independent* (Fargo), May 26, 1919, folder 11, box 1, Theodore Nelson Papers, State Historical Society of North Dakota, Bismarck, ND (hereafter SHSND).

46. A. H. Aaserude to *Independent,* June 17, 1919, and George Schonberger to Independent Voters Association, June 21, 1919, both folder 11, box 1, Theodore Nelson Papers, SHSND; "The Big North Dakota Victory," *Nonpartisan Leader,* July 14, 1919.

47. Louis Levine, "Will Agrarian Movement Affect Our Politics?" *New York Times Magazine,* March 18, 1917, 10.

48. Harry C. Boyte, *Everyday Politics: Reconnecting Citizens and Public Life* (Philadelphia: University of Pennsylvania Press, 2004), 24–27; Dana R. Fisher, *Activism, Inc.: How the Outsourcing of Grassroots Campaigns Is Strangling Progressive Politics in America* (Stanford, CA: Stanford University Press, 2006).

49. Richard Christen, "Personal Reflections on the Non-Partisan League" (senior thesis, Minot State College, 1975), 5; Lawrence Goodwyn, *The Populist Moment: A Short History of the Agrarian Revolt in America* (New York: Oxford University Press, 1978), 20–54.

50. Nellie Stone Johnson, *Nellie Stone Johnson: The Life of an Activist* (Saint Paul: Ruminator Books, 2000), 39, 107.

51. Arthur C. Townley, interview by Richard C. Dobson, date and place unknown (ca. 1950s), folder 38, box 1, Nonpartisan League Collection, North Dakota Institute for Regional Studies, North Dakota State University, Fargo, ND (hereafter NDIRS); Terrence J. Tully to William Langer, May 1, 1919, folder 2, box 15, William Langer Papers, UND; Christen, "Personal Reflections on the Non-Partisan League," 11; Johnson, *Nellie Stone Johnson*, 39.

52. Ed Braun to editor, *Independent*, June 15, 1919, folder 11, box 1, Theodore Nelson Papers, SHSND; Neal Dilley, "Win We Must," *Colorado Leader* (Denver), July 2, 1920.

53. "Why They Joined the League," *Idaho Leader*, July 19, 1919; "News from State Office," *Nebraska Leader* (Lincoln), August 13, 1921.

54. In deciphering this shift in the League's membership, I have been aided by Francesca Polletta, *Freedom Is an Endless Meeting: Democracy in American Social Movements* (Chicago: University of Chicago Press, 2004); and Michael K. Briand, *Practical Politics: Five Principles for a Community That Works* (Urbana: University of Illinois Press, 1999).

55. "Returned Soldier Organizer for League Writes Letter," *Idaho Leader*, July 12, 1919; "Walks 110 Miles to League Gathering," *Colorado Leader*, August 20, 1920.

56. S. A. Olsness to L. B. Garnaas, September 27, 1920, folder 6, box 1, S. A. Olsness Papers, SHSND.

57. "Notice to League Members," *South Dakota Leader* (Mitchell, SD), August 30, 1919; "Sincere Purpose of League Farmers Is Like That of the Nation's Farmers," *Nonpartisan Leader*, March 30, 1916.

58. Kathleen Moum, "Harvest of Discontent: The Social Origins of the Nonpartisan League, 1880–1922" (PhD diss., University of California–Irvine, 1986), 140–143.

59. Goodwyn, *The Populist Moment*, 20–54; "Everybody's Coming to the League Picnics," *Nonpartisan Leader*, June 16, 1916.

60. "Attention League Members," *Montana Nonpartisan* (Great Falls), June 12, 1920.

61. Announcements of upcoming picnics filled NPL newspapers. For a sampling, see "Series of League Picnics in South Dakota Sept. 3 to 20," *South Dakota Leader*, August 30, 1919; "Big League Picnics," *Idaho Leader*, September 13, 1919; "League Meetings," *Nebraska Leader*, July 5, 1919; and "One of the Greatest League Speakers to Visit Colorado Will Talk at Twelve Meetings," *Colorado Leader*, August 6, 1920. See also Norris C. Hagen, *Vikings of the Prairie: Three North Dakota Settlers Reminisce* (New York: Exposition, 1958), 168.

62. "Three Thousand Nonpartisans Spend 4th at Manzanola, Colo.," *Colorado Leader*, July 9, 1920; "Farmer's Picnic Monday Was a Huge Success," *Colorado Leader*, July 16, 1920; "Big Picnic N.P.L. Lincoln County," *Colorado Leader*, July 16, 1920; "Hugo Welcomed Nonpartisan League Picnic Last Tuesday," *Colorado Leader*, July 23, 1920; "Leaguers, Attention!," *Colorado Leader*, July 30, 1920.

63. Robert H. Johnson, "Where Is the Money?" *Nonpartisan Leader*, May 4, 1916; "To the Hyphenates," *Nonpartisan Leader*, August 31, 1916; "Farming North Dakota," *Nonpartisan Leader*, October 5, 1916.

64. "On to Bismarck," *Nonpartisan Leader*, January 6, 1916; "Marching to Bismarck," *Nonpartisan Leader*, February 10, 1916.

65. "A Cry Out from the Wilderness," in Frederick W. Laidley, *The Why of the Farmers' Non-Partisan League of Canada* (Swift Current, SK: Farmers' Non-Partisan League of Canada, 1916), inside front cover, file 28, series 4, John Ford Fonds, Glenbow Archives, Calgary, AB; "Members and Friends of the Non-Partisan League," *Colorado Leader*, July 16, 1920.

66. "Helen Township League Chorus," *Nonpartisan Leader*, April 7, 1919; "Big League Picnics," *Idaho Leader*, September 13, 1919; Mrs. Ray McKaig, "Idaho's League Song," ca. 1918, folder M, box 1, Ray McKaig Papers, Idaho State Historical Society, Boise, ID.

67. "Hanna's Goat Cavorts at Farmers' Convention," *Nonpartisan Leader*, February 24, 1916. *Life in Sing Sing* (Indianapolis, Bobbs-Merrill, 1904) offered the first printed use of "goat" as a reference to someone else's anger or exasperation.

68. "Hanna's Goat at Deering Picnic," *Nonpartisan Leader*, August 17, 1916; "Training for the Summer Scrap," *Montana Nonpartisan*, February 8, 1919.

69. "Billican's 'Goat That Can't Be Got' Made Official League Emblem," *Idaho Leader*, November 1, 1919; Arthur Warner, "The Farmer Butts Back," *Nation* 111, no. 2878 (August 28, 1920): 240. League papers advertised a "gold and white enamel" goat "watch fob" for a dollar. See, for example, the advertisement in the *Montana Nonpartisan*, April 7, 1920.

70. *The Sentinel*, pamphlet, undated, no author, Citizens Alliance of Minneapolis Records, reel 12, MNHS.

71. "The Donkey, the Elephant, and the Goat," *Montana Nonpartisan*, September 11–18, 1920; J. P. Buschlen, *The Donkey, the Elephant, and the Goat at a Public Meeting* (Great Falls: Montana Printing, 1920).

72. *The Goat* (Fargo, ND) 1, no. 7 (October 1920): 1.

73. Here I depend on the attention given to "micropolitical enactments" in the "fabric of everyday political speech and action" in Jason Frank, *Constituent Moments: Enacting the People in Postrevolutionary America* (Durham, NC: Duke University Press, 2010), 33.

74. Karen Starr, "Fighting for a Future: Farm Women of the Nonpartisan League," *Minnesota History* 48, no. 6 (Summer 1983): 255–262; Kim E. Nielsen, "'We All Leaguers by Our House': Women, Suffrage, and Red-Baiting in the National Nonpartisan League," *Journal of Women's History* 6, no. 1 (Spring 1994): 31–50; Catherine E. Rymph, *Republican Women: Feminism and Conservatism from Suffrage through the Rise of the New Right* (Chapel Hill: University of North Carolina Press, 2006), 14–39.

75. Annie Pike Greenwood, *We Sagebrush Folks* (New York: D. Appleton–Century, 1934), 315. See also Barbara Handy Marchello, *Women of the Northern Plains: Gender and Settlement on the Homestead Frontier, 1870–1930* (Saint Paul: Minnesota Historical Society Press, 2005), 85–115; and Mary Neth, *Preserving the Family Farm: Women, Community, and the Foundations of Agribusiness in the Midwest, 1900–1940* (Baltimore: Johns Hopkins University Press, 1998), 40–94.

76. "Some Ideas of Country Women," *Nonpartisan Leader*, December 28, 1916.

77. "Keep Your Eyes on Washington State," *Nonpartisan Leader*, August 5, 1918; Greenwood, *We Sagebrush Folks*, 473, 401; Donald B. Marti, *Women of the Grange: Mutuality and Sisterhood in Rural America, 1866–1920* (New York: Greenwood, 1991).

78. "New Department for Women Readers," *Nonpartisan Leader*, May 4, 1916.

79. "Should Farm Women Unite?," *Nonpartisan Leader*, June 15, 1916.

80. "Women and Votes," *Nonpartisan Leader*, August 17, 1916; and "Is the Vote Enough?," *Nonpartisan Leader*, September 7, 1916.

81. "Canadian Women's Work," *Nonpartisan Leader*, September 28, 1916.

82. Here the NPL deviated from the example of the Farmers' Alliance and the People's Party. In the 1880s and 1890s, both attracted rural women and promised to include them in the

movement, especially in the Midwest and Rocky Mountains. See Rebecca Edwards, *Angels in the Machinery: Gender in American Party Politics from the Civil War to the Progressive Era* (New York: Oxford University Press, 1997), 91–110.

83. Minnie Boyer Davis to Farmers' Nonpartisan League, May 2, 1917, and Nonpartisan League to Minnie Boyer Davis, May 5, 1917, both reel 2, National Nonpartisan League Papers, MNHS.

84. "Nonpartisan League Infection," *Nonpartisan Leader*, April 5, 1917; Christen, "Personal Reflections on the Non-Partisan League," 16.

85. George E. Akerson, "Class Hatred Thrives in Stronghold Claimed by Nonpartisan League," *Minneapolis Tribune*, March 10, 1918; Mrs. Edwin A. Schacht, "Says Leader Worth All League Dues," *Minnesota Leader*, March 30, 1918.

86. Nielsen, "'We All Leaguers by Our House,'" 34; A. C. Townley to "Sir," no date (ca. 1918), folder 2, box 1, Nonpartisan League Collection, NDIRS.

87. Sara M. Evans, *Born for Liberty: A History of Women in America* (1989; New York: Free Press, 1997), 145–174; "Women's Part in the Farmers' Alliance," *Nonpartisan Leader*, April 1, 1918; "From Woman That Values the League," *Nonpartisan Leader*, April 8, 1918.

88. "From Woman That Values the League."

89. "Pay for the Wageless Years," *Nonpartisan Leader*, April 1, 1918.

90. Jim McMilan, "The Macdonald-Nielson Imbroglio: The Politics of Education in North Dakota, 1918–1921," *North Dakota History* 52, no. 4 (1985): 2–11.

91. Martin Thompson to Municipal Voter's League, May 8, 1919, folder 3, box 1, Martin O. Thompson Papers, NDIRS; "Mrs. Minnie D. Craig—Geneological Data (Mrs. E. O. Craig)," autobiography, no date, folder 3, box 1, Minnie D. Craig Papers, both NDIRS.

92. Amy G. Edmunds, "Farm Women and Nonpartisan League," *Nonpartisan Leader*, September 8, 1919, 8; McMilan, "The Macdonald-Nielson Imbroglio," 5; "Suffragists Organize," *Nonpartisan Leader*, July 13, 1916, 19; "Bismarck Welcomes Delegates to Annual Federation of Women's under Most Ideal Conditions," *Bismarck Tribune*, October 6, 1920; "Statements by the President and the Secretary of the North Dakota WCTU," broadside, November 1920, folder 2, box 1, Theodore Nelson Papers, SHSND; Eagle Glassheim, "To Fuel a Fire: Gender, Class, and Ethnicity in the North Dakota Nonpartisan League, 1915–1921" (senior thesis, Dartmouth College, 1992), 85–86.

93. Minnie J. Nielson, *A Message to Minnesota Womanhood* (Saint Paul: Minnesota Sound Government Association, 1920).

94. Ibid.; Warner, "The Farmer Butts Back," 240.

95. Larry Remele, "The North Dakota State Library Scandal of 1919," *North Dakota History* 44, no. 1 (Winter 1977): 21–29; Cheri Register, "Motherhood at Center: Ellen Key's Social Vision," *Women's Studies International Forum* 5, no. 6 (1982): 599–610.

96. A. C. Townley, speech at Forman, ND, May 28, 1919, transcribed by W. L. Divet, court reporter, Lisbon, ND, folder 2, box 2, Theodore Nelson Papers, SHSND.

97. Amy G. Edmunds, "Farm Women and Nonpartisan League," *Nonpartisan Leader*, September 8, 1919; "Viola Liessman," in *Women on the Move*, ed. Pearl Andre (Bismarck: North Dakota Democratic-NPL Women, 1975), 60; "What Woman's Auxiliary Is and Its Purpose," *Nebraska Leader*, November 29, 1919.

98. Edmunds, "Farm Women and Nonpartisan League."

99. A. C. Townley to "Madam," no date (ca. 1919), folder 21, box 6, Martin O. Thompson Papers, NDIRS; "National Federation of League Women," *Nonpartisan Leader*, January 5, 1920.

100. "Woman League Member Writes Letter about Organization," *Idaho Leader*, January 11, 1919; "'Big Business' Is Afraid Says Farmers [*sic*] Wife," *Nebraska Leader*, September 20, 1919.

101. "2,000 Woman in N.D. Auxiliary," *Nebraska Leader*, November 29, 1919; "Urges the Formation of Women's Auxiliary," *Nebraska Leader*, October 11, 1919; "Ellen Dahlsten Second Member to Join Auxiliary," *Nebraska Leader*, December 6, 1919; "How to Organize a Nonpartisan League Women's Club," *Colorado Leader*, September 24, 1920; "Women in Four Counties Start Organization Work," *Montana Nonpartisan*, May 15, 1920.

102. Mrs. B. T. White to F. J. Sullivan, January 28, 1919, folder 1, box 15, William Langer Papers, UND; Ray McKaig to A. R. Thomas, August 17, 1917, folder T, box 1, and "Twin Falls County List of Post-Date Checks," Non-Partisan League—Financial I: 1917–1920 folder, box 10, both Ray McKaig Papers, Idaho State Historical Society, Boise, ID.

103. "What Woman's Auxiliary Is and It's Purpose," *Nebraska Leader*, November 29, 1919; "A Page of, by, and for the Farm Women," *Nonpartisan Leader*, January 19, 1920.

104. Ruby Kraft to Mrs. Cart, no date (ca. 1945), speech to McLean County Convention of NPL Women's Clubs, no date (ca. 1922), Ruby Kraft to Josephine Leben, December 9, 1921, all folder 2, Ruby Kraft Papers, UND.

105. Ruby Kraft to Josephine Leben, August 23, 1922, folder 2, Ruby Kraft Papers, UND; Susie W. Stageberg, "The Kitchen Column," *Minnesota Herald* (Mora), March 15, 1922.

106. Margaret A. Hannah, "Women's N.P. Clubs," *Montana Nonpartisan*, June 12, 1920; "Subjects for Study," *Nonpartisan Leader*, October 18, 1920; "A Colorado Club," *Nonpartisan Leader*, June 27, 1921; "D.W.N.P.C. No. 16," *Minnesota Herald*, March 15, 1922; Beatrice E. Butler, "Introductory," *Minnesota Herald*, March 15, 1922.

107. *Minnesota Herald*, March 15, 1922; "What One Club Did," *Nonpartisan Leader*, April 18, 1921.

108. Lynn Wolfe Gentzler, "Kate Leila Gregg (1883–1954)," in *Dictionary of Missouri Biography*, ed. Lawrence O. Christensen, William E. Foley, Gary R. Kremer, and Kenneth H. Winn (Columbia: University of Missouri Press, 1999), 354–355; *Minnesota Herald*, April 15, 1922; "Third Annual Convention of Women's Nonpartisan Clubs," program, box 1, Susie Stageberg Papers, MNHS.

109. *Minnesota Herald*, May 15, 1922; and Lena L. Borchardt, "I Lived a Full Life: Reminiscences," folder 1, Lena Borchardt Collection, MNHS.

110. Aldyth Ward, "What One State Has Done for Women," *Nonpartisan Leader*, December 29, 1919; "Nonpartisan Women Busy," *Montana Nonpartisan*, February 1, 1919.

111. "Club 147 Getting Ready for Fall Campaign," *Minnesota Herald*, May 15, 1922.

112. Mary Lethert Wingerd, *Claiming the City: Politics, Faith, and the Power of Place in St. Paul* (Ithaca, NY: Cornell University Press, 2001) 146, 175–211, 219–220.

113. Richard M. Valelly, *Radicalism in the States: The Minnesota Farmer-Labor Party and the American Political Economy* (Chicago: University of Chicago Press, 1989), 26–29; David Paul Nord, "Minneapolis and the Pragmatic Socialism of Thomas Van Lear," *Minnesota History* 45, no. 1 (Spring 1976): 2–10; "Mayor Van Lear of Minneapolis at Open Air Meeting, Glencoe, Minn.," August 7, 1917, transcript, reel 14, National Nonpartisan League Papers, MNHS.

114. Gene Stanchfield and Jean Spielman to "The Local Unions, Minneapolis, Minn," no date (ca. February 1918), Political Undated: 1918–1943 folder, box 36, Central Labor Union of Minneapolis and Hennepin County Papers, MNHS; "Radical Socialists Ally Themselves with Nonpartisan League," *Minneapolis Tribune*, April 23, 1918, 13; George E. Akerson, "Townley–Van Lear Alliance Realized and Resented by Loyal Minnesota Farmers," *Minneapolis Tribune*, June 1, 1918; "A Non-Partisan Leaguer," *New York Times*, May 29, 1918, 12.

115. Wingerd, *Claiming the City*, 220; Luke W. Boyce to J. F. Gould, October 28, 1919, Americas

Committee of Minneapolis: J. F. Gould, Aug.–Oct. 1919 folder, Northern Information Bureau Records, MNHS.

116. Millard L. Gieske, *Minnesota Farmer-Laborism: The Third-Party Alternative* (Minneapolis: University of Minnesota Press, 1979), 52–53; Valelly, *Radicalism in the States*, 34–35.

117. "Miss Rankin," ca. 1917, vol. 1, transcript, reel 14, National Nonpartisan League Papers, MNHS; Calvert, *The Gibraltar*, 103–114.

118. Calvert, *The Gibraltar*, 116–117, 124–125.

119. Ibid., 127–134; Kurt Wetzel, "The Defeat of Bill Dunne: An Episode in the Montana Red Scare," *Pacific Northwest Quarterly* 64, no. 1 (January 1973): 14–15.

120. Hamilton Cravens, "The Emergence of the Farmer-Labor Party in Washington Politics, 1919–20," *Pacific Northwest Quarterly* 57, no. 4 (October 1966): 148–151; Carlos Schwantes, *Radical Heritage: Labor, Socialism, and Reform in Washington and British Columbia, 1885–1917* (Seattle: University of Washington Press, 1979), 159–162.

121. "State Mine Operation a Big Success," *Nonpartisan Leader*, February 23, 1920; Lynn Frazier, "Gov. Frazier's Own Story of the Non Partisan League," *New York Times Magazine*, May 16, 1920, 3; Thomas Shilts, "'To Prevent a Calamity Which Is Imminent': Governor Frazier and the Fuel Crisis of 1919," *North Dakota History* 63, no. 1 (Winter 1996): 6–20.

122. Jeffrey Sklansky, *The Soul's Economy: Market Society and Selfhood in American Thought, 1820–1920* (Chapel Hill: University of North Carolina Press, 2002), 180; Thorstein Veblen, "Using the I.W.W. to Harvest Grain," *Journal of Political Economy* 40, no. 6 (December 1932): 796–807; Christopher Cappozola, "Thorstein Veblen and the Politics of War, 1914–1920," *International Journal of Politics, Culture and Society* 13, no. 2 (1999): 255–271; Sylvia E. Bartley, "Intellect Surveilled: Thorstein Veblen and the Organs of State Security," paper presented at Second Conference of the International Thorstein Veblen Association, Northfield, MN, 1996, at http://www.elegant-technology.com/TVbarSI.html, accessed May 8, 2012.

123. Thorstein Veblen, *The Vested Interests and the Common Man* (1919; repr., New Brunswick, NJ: Transaction Publishers, 2002), 128–129.

124. Quoted in Leon Fink, *Progressive Intellectuals and the Dilemmas of Democratic Commitment* (Cambridge, MA: Harvard University Press, 1997), 192.

125. "Upton Sinclair," *Nonpartisan Leader*, May 13, 1918; Upton Sinclair, *The Brass Check: A Study of American Journalism* (Pasadena, CA: published by author, 1920), 269; Upton Sinclair, *The Goslings: A Study of America's Schools* (Pasadena, CA: published by author, 1924), 357–358; H. G. Teigan to Upton Sinclair, October 13, 1919, reel 5, National Nonpartisan League Papers, MNHS.

126. Casey Nelson Blake, *Beloved Community: The Cultural Criticism of Randolph Bourne, Van Wyck Brooks, Waldo Frank, and Lewis Mumford* (Chapel Hill: University of North Carolina Press, 1990), 176; Walter Bates Rideout, *Sherwood Anderson: A Writer in America*, (Madison: University of Wisconsin Press, 2006), 1:342.

127. Quoted in Blake, *Beloved Community*, 176–177.

128. Leo Hurst to H. E. Teigan, October 11, 1919, reel 5, National Nonpartisan League Papers, MNHS; Ronald Briley, "The Artist as Patron: Gutzon Borglum and North Dakota Politics, 1922," *South Dakota History* 20, no. 2 (Summer 1990): 120–145.

129. Briley, "The Artist as Patron," 139; list of election campaign speakers in North Dakota, no date, January–July 1920 folder, box 2, Arthur LeSueur Papers, MNHS.

130. Richard Lingeman, *Sinclair Lewis: Rebel from Main Street* (Saint Paul: Borealis Books, 2002), 122, 127, 155; Grace Hegger Lewis, *With Love from Gracie: Sinclair Lewis, 1912–1925* (New

York: Harcourt, Brace, 1951), 116; John J. Koblas, *Sinclair Lewis: Home at Last* (Bloomington, MN: Voyageur, 1981).

131. Sinclair Lewis, *Main Street* (New York: Harcourt, Brace and World, 1920), 244, 413; James Marshall, "Pioneers of *Main Street*," *Modern Fiction Studies* 31, no. 3 (Autumn 1985): 529–545.

132. Lewis, *Main Street*, 229, 244, 269, 330, 413, 418–420.

133. Davis Douthit, *Nobody Owns Us: The Story of Joe Gilbert, Midwestern Rebel* (Washington, DC: Cooperative League of the USA, 1948), 188.

134. Sinclair Lewis to Alfred Harcourt, June 12, 1919, and Sinclair Lewis to Alfred Harcourt, April 1920, in *From Main Street to Stockholm: Letters of Sinclair Lewis, 1919–1930*, ed. Harrison Smith (New York: Harcourt, Brace, 1952) 3, 26; Marshall, "Pioneers of *Main Street*."

135. Henry Teigan to Herbert Iverson, December 8, 1919, and Henry Teigan to Henry Raknerud, December 22, 1919, reel 5, National Nonpartisan League Papers, MNHS. Teigan told Raknerud that based on unexpired subscriptions, he counted nearly 250,000 members.

Chapter Five

1. Charles Merz, "The Nonpartisan League: A Survey," *New Republic* 22, no. 284 (May 12, 1920): 336–337.

2. Ibid., 337.

3. Walter Lippmann, *Public Opinion* (New York: Macmillan, 1922); Walter Lippmann, *The Phantom Public* (New York: Macmillan, 1927), 145; Robert B. Westbrook, *John Dewey and American Democracy* (Ithaca, NY: Cornell University Press, 1991), 280–286; Christopher Lasch, *The True and Only Heaven: Progress and Its Critics* (New York: W. W. Norton, 1991), 360–368; Leon Fink, *Progressive Intellectuals and the Dilemmas of Democratic Commitment* (Cambridge, MA: Harvard University Press, 1997), 13–51.

4. Walter Thomas Mills, *The Articles of Association of the National Nonpartisan League* (Saint Paul: National Nonpartisan League, ca. 1918), 11, 24.

5. Merle Curti, *The Growth of American Thought* (New York: Harper and Bros., 1943). Curti notes that "these criticisms of democracy did not go unchallenged. . . . The plain people themselves were probably little affected by the criticisms of democracy" (695–696).

6. Herbert E. Gaston, *The Nonpartisan League* (New York: Harcourt, Brace, and Howe, 1920), 304; Job Wells Brinton, *Wheat and Politics* (Minneapolis: Publication Office, 1931), 45.

7. *Farmers Independent* (Bagley, MN), May 20, 1920; Brinton, *Wheat and Politics*, 47–49; "Townley Sees Plot to Shut Fargo Bank," *New York Times*, October 4, 1919, 5; J. Edmond Buttree, *The Despoilers: Stories of the North Dakota Grain Fields* (Boston: Christopher, 1920), 267–271; "The Fargo Bank Failure," *Financier* (New York) 15 (October 15, 1919): 648.

8. Alvin S. Tostlebee, "The Bank of North Dakota: An Experiment in Agrarian Banking," *Studies in History, Economic, and Public Law* 64, no. 1 (1924): 105.

9. "Loans to Townley Close Fargo Bank," *New York Times*, October 3, 1919, 1; Brinton, *Wheat and Politics*, 45–46; Gaston, *The Nonpartisan League*, 307–309.

10. Brinton, *Wheat and Politics*, 68–69.

11. Sterling Evans, *Bound in Twine: The History and Ecology of the Henequen-Wheat Complex for Mexico and the American and Canadian Plains, 1880–1950* (College Station: Texas A&M University Press, 2007).

12. W. W. Liggett to J. W. Brinton, August 25, 1919, folder 1: Correspondence, 1918–1920, Job Wells Brinton Papers, State Historical Society of North Dakota, Bismarck, ND (hereafter SHSND).

13. "N.D. State Bank Ready to Open Soon," *Nonpartisan Leader* (Fargo, ND), April 21, 1919; *Sisal the Coming Industry of Florida*, brochure, ca. 1919, James F. Jaudon, Papers Related to Economic Endeavors, 1917–1919, Reclaiming the Everglades: South Florida's Natural History, 1884–1934, Florida International University Libraries, http://purl.fcla.edu/fcla/tc/rte/RTJJ00150007, accessed July 26, 2013.

14. J. W. Brinton to T. P. Harvey, January 17, 1920, folder 1: Correspondence, 1918–1920, Job Wells Brinton Papers, SHSND; Brinton, *Wheat and Politics*, 69; "Receiver for Sisal Trust," *New York Times*, June 21, 1921, 24.

15. Brinton, *Wheat and Politics*, 56–58; *Consumers United Stores Company of North Dakota*, brochure, ca. 1918, reel 4, National Nonpartisan League Papers, Minnesota Historical Society, Saint Paul, MN (hereafter MNHS).

16. Jerry Dempster Bacon, *A Warning to the Farmer against Townleyism as Exploited in North Dakota* (Grand Forks, ND: Bacon, 1918), 29; "Consumers United Stores Company of North Dakota."

17. J. W. Brinton to "Sir," June 10, 1918, folder 2, Thorwald Mostad Papers, SHSND; Leon Durocher to Henry Teigan, September 29, 1918, and October 6, 1918, and *Consumers United Stores Company of North Dakota*, brochure, ca. 1918, all reel 4, National Nonpartisan League Papers, MNHS.

18. *Consumers United Stores Company of North Dakota*, brochure; Brinton, *Wheat and Politics*, 57. NPL organizers who sold memberships to the Consumers United Stores Company included Leon Durocher, Walter Quigley, and H. P. Richardson.

19. *Consumers United Stores Company of North Dakota.*

20. Ibid.; J. G. Ingle, *"You Farmers Are a Set of G—D— Hogs!" Said Townley*, pamphlet, Dawson, ND, March 9, 1918, 8, both reel 4, National Nonpartisan League Papers, MNHS.

21. Buttree, *The Despoilers*, 151; Brinton, *Wheat and Politics*, 59–65; Larry Remele, "The North Dakota Farmers Union and the Nonpartisan League: Breakdown of a Coalition," *North Dakota Quarterly* 46, no. 4 (Autumn 1978): 40–50.

22. Remele, "The North Dakota Farmers Union and the Nonpartisan League," 46–47; "Grange and Union Refuse to Embark in Blue Sky Scheme," *Bismarck (ND) Tribune*, January 29, 1918; "That Chain Store Scheme," *Bismarck Tribune*, February 1, 1918.

23. Brinton, *Wheat and Politics*, 63; Gaston, *The Nonpartisan League*, 242; "A Correction," *Nonpartisan Leader*, January 12, 1920.

24. Alfred Knutson, "The Nonpartisan League," undated, Alfred Knutson Papers and Interview Transcripts (Lowell Dyson), box 21, Larry S. Remele Papers, SHSND.

25. "Farmers, Beware!!!" *Idaho Leader* (Boise), October 18, 1919.

26. Ingle, *"You Farmers Are a Set of G—D— Hogs!" Said Townley.*

27. Gaston, *The Nonpartisan League*, 63; and A. C. Townley, speech at Forman, ND, May 28, 1919, folder 2, box 2, Theodore Nelson Papers, SHSND.

28. Gaston, *The Nonpartisan League*, 86; "Townley's League Aims to Russianize America," *Commercial West* 33, no. 2 (June 1, 1918): 33; "The Non-Partisan League," *Review* 1, no. 10 (July 19, 1919): 207; William Henry Talmage, *Two Non-Partisan League Lectures: "The Mad Captain" and "We'll Stick"* (Redfield, SD, 1918), 4.

29. Gaston, *The Nonpartisan League*, 88.

30. S. R. Maxwell, *The Nonpartisan League from the Inside* (Saint Paul: Dispatch, 1918), 20–22, 25.

31. Davis Douthit, *Nobody Owns Us: The Story of Joe Gilbert, Midwestern Rebel* (Chicago: Cooperative League of the USA, 1948), 145; Maxwell, *The Nonpartisan League from the Inside*, 24–26; "Farmer Repudiates Nonpartisan League," *Eugene (OR) Register*, August 14, 1918.

32. W. E. Quigley, "An Introduction and a Challenge," *Lincoln (NE) Daily Star*, April 4, 1919, reel 4, National Nonpartisan League Papers, MNHS.

33. Ibid.

34. Will Wasson to S. A. Olsness, May 30, 1918, and S. A. Olsness to Will Wasson, June 4, 1918, folder 4, box 1, S. A. Olsness Papers, North Dakota Institute for Regional Studies, North Dakota State University, Fargo, ND (hereafter NDIRS).

35. William Langer, no date (ca. 1920), memo, folder 3, box 16, William Langer Papers, University of North Dakota, Grand Forks, ND (hereafter UND); petition, no date (ca. 1918), reproduced in Theodore G. Nelson, *Scrapbook Memoirs* (Salem, OR: Your Town, 1957), 71.

36. Maxwell, *The Nonpartisan League from the Inside*, 31–33, 35.

37. Ibid., 34–35.

38. Douthit, *Nobody Owns Us*, 148; O. M. Thomason to Henry Teigan, November 11, 1918, and Henry Teigan to O. M. Thomason, November 12, 1918, reel 4, National Nonpartisan League Papers, MNHS.

39. A. B. Gilbert, "Who Killed the Nonpartisan League?," *Nation* 123, no. 3189 (August 1926): 151.

40. W. E. Quigley, "The Truth about the Non-Partisan League: Chapter XX, Articles of Association," *Lincoln Daily Star*, May 5, 1919, reel 4, National Nonpartisan League Papers, MNHS; "Members to Pass on Choice of Leader," *Nonpartisan Leader*, December 16, 1918; "Delegates Approve the League's Books" and "League Members—Important," *Nonpartisan Leader*, December 23, 1918; "Membership Sustains A. C. Townley," *Nonpartisan Leader*, February 10, 1919.

41. Arthur LeSueur, "The Governor and Members of the Legislature," November 30, 1920, Correspondence Aug.–Dec. 1920 folder, box 2, Arthur LeSueur Papers, MNHS; Arthur LeSueur, "The Nonpartisan League: A Criticism," *Socialist Review* 9, no. 6 (November 1920): 193–195.

42. "Leon Durocher Explains Break," *Turtle Mountain Star* (Rolla, ND), October 6, 1921; Leon Durocher to C. A. Christensen, February 25, 1920, pp. 5–23 folder, box 1, Materials Relating to the Nonpartisan League, comp. Asher Howard, MNHS; Leon Durocher to A. C. Townley, May 13, 1919, and Anthony Walton to A. C. Townley, May 12, 1919, both reel 5, National Nonpartisan League Papers, MNHS.

43. Glenn H. Smith, "William Langer and the Art of Personal Politics," in Thomas Howard, *The North Dakota Political Tradition* (Ames: Iowa State University Press, 1981), 127–130; "Big Leaguers against Bills," *Bismarck Tribune*, March 14, 1919.

44. "Townley Damns Kositzky Who Hands It Back," *Bismarck Tribune*, March 1, 1919; "Kositzky Much in Demand with League Farmers," *Bismarck Tribune*, April 12, 1919; William Langer, *The Nonpartisan League: Its Birth, Activities, and Leaders* (Mandan, ND: Morton County Farmers Press, 1920), 34–36; William Langer to L. P. Zubrod, April 4, 1919, folder 1, box 15, William Langer Papers, UND; Buttree, *The Despoilers*, 180.

45. Terrence J. Tully to William Langer, May 1, 1919, and T. T. Jorstad to William Langer, May 3, 1919, folder 2, box 15, William Langer Papers, UND.

46. D. Jerome Tweton, "The Anti-League Movement: The IVA," in *The North Dakota Political Tradition*, ed. Thomas W. Howard (Ames: Iowa State University Press, 1981), 102; Richard Christen, "Personal Reflections on the Non-Partisan League" (senior thesis, Minot State College, 1975), 10; Arthur Warner, "The Farmer Butts Back," *Nation* 111, no. 2878 (August 28, 1920): 241.

47. W. E. H. Porter, Henry A. Wibert, and Chris Orton to William Langer, March 23, 1919, folder 1, box 15, William Langer Papers, UND; S. A. Olsness to John G. Johnson, October 13, 1920, folder 6, box 1, S. A. Olsness Papers, NDIRS.

48. Thomas Goebel, "'A Case of Democratic Contagion: Direct Democracy in the American West, 1890–1920," *Pacific Historical Review* 66, no. 2 (May 1997): 213–230.

49. Charles Edward Merriam and Louise Overacker, *Primary Elections* (Chicago: University of Chicago Press, 1928); Ralph S. Boots, "The Trend of the Direct Primary," *American Political Science Review* 16, no. 3 (August 1922): 412–431.

50. Samuel Huntington, "Election Tactics of the Nonpartisan League," *Mississippi Valley Historical Review* 36, no. 4 (March 1950): 616.

51. Ray McKaig to J. C. Kelly, February 4, 1918, NPL Correspondence folder K, box 1, Ray McKaig Papers, Idaho State Historical Society, Boise, ID (hereafter ISHS).

52. "'Save the Primary' Slogan in Montana," *Nonpartisan Leader*, June 9, 1919.

53. Clarence Berdahl, "The Richards Primary," *American Political Science Review* 14, no. 1 (February 1920): 93–105; Kathryn Otto, "Dakota Resources: The Richard Olsen Richards Papers at the South Dakota Historical Resource Center," *South Dakota History* 9, no. 2 (Spring 1979): 153–156.

54. W. N. Van Camp to W. H. King, February 22, 1918, 1917 King 1918 folder, box 82, Peter Norbeck Papers, University of South Dakota, Vermillion, SD (hereafter USD).

55. Ibid.; Berdahl, "The Richards Primary," 95; "Nonpartisan League to Act as Independent Political Party," *South Dakota Leader* (Mitchell), September 27, 1919; "Don't Fail to Attend Caucus at Your Polling Place," *South Dakota Leader*, October 11, 1919.

56. E. B. Russell, "Idaho Gang on Wild Political Spree," *Nonpartisan Leader*, February 10, 1919; Boyd A. Martin, *The Direct Primary in Idaho* (Palo Alto, CA: Stanford University Press, 1947), 66–75; "Idaho First State to Go Back to Corrupt Convention System," *Idaho Leader*, March 8, 1919.

57. Judson King, "Nation-Wide Attack on Primaries," *Nonpartisan Leader*, March 24, 1919; "'Save the Primary' Slogan in Montana"; "The Tide Is Rising," *Nonpartisan Leader*, April 14, 1919; "1919 Legislature Comes to an End," *North Platte (NE) Semi-Weekly Tribune*, April 29, 1919; Vance Monroe, "The Legislative Session in Colorado," *Nonpartisan Leader*, June 2, 1919; Joseph M. Dixon to Mark Sullivan, January 21, 1920, folder 7, box 21, Joseph M. Dixon Papers, University of Montana, Missoula, MT (hereafter UM).

58. Quote in Joseph M. Dixon to A. Howe, July 11, 1919, folder 5, box 21, Joseph M. Dixon Papers, UM; D. C. Dorman to Jesse R. Johnson, May 9, 1919, and W. Clavier to C. A. Sorensen, May 20, 1919, Correspondence May 1919 folder, box 2, C. A. Sorensen Papers, Nebraska State Historical Society, Lincoln, NE (hereafter NSHS); S. C. Ford to Joseph M. Dixon, June 7, 1919, folder 5, box 21, and Joseph M. Dixon to William Clavier, June 20, 1919, folder 7, box 24, Joseph M. Dixon Papers, UM; C. A. Sorensen to "Friend of the Direct Primary," August 1, 1919, Correspondence August 1919 folder, box 2, C. A. Sorensen Papers, NSHS.

59. "Nebraska Farmers Save Primary," *Nebraska Leader* (Lincoln), July 26, 1919; "League Fights Hard to Save Primary," *Nebraska Leader*, October 4, 1919; Joseph M. Dixon to Mark Sullivan, January 21, 1920, folder 7, box 21, Joseph M. Dixon Papers, UM; "The Disintegrating Party Primary," *New York Times*, September 29, 1920, 8; Oliver S. Morris, "The Nonpartisan League," *Nation* 111, no. 2894 (December 22, 1920): 733.

60. *Inaugural Address of Governor J. A. O. Preus to the Legislature of Minnesota, January 5, 1921* (Saint Paul: Syndicate, 1921), 3; "The Legislature," *Willmar (MN) Tribune*, March 16, 1921; "With the Minnesota Lawmakers," *Warren (MN) Sheaf*, April 27, 1921; Carl Chrislock, "The Drama of a Party's History," typescript, 1960, MNHS.

61. Warren G. Harding, "First Annual Message," December 6, 1921, American Presidency Project, http://www.presidency.ucsb.edu/ws/?pid=29562, accessed June 7, 2013.

62. William MacDonald, "North Dakota's Experiment," *Nation* 108, no. 2803 (March 22, 1919): 421; and Huntington, "The Election Tactics of the Nonpartisan League," 631–632.

63. Robert K. Murray, *Red Scare: A Study in National Hysteria, 1919–1920* (Minneapolis: University of Minnesota Press, 1955); Richard Gid Powers, *Not without Honor: The History of American Anticommunism* (New Haven, CT: Yale University Press, 1996).

64. Maurice McAuliffe, editorial, *Kansas Farmers' Union* (Salina), April 11, 1918.

65. Judson King, "Banking and Steel Interests and the Townley Trial," *Public* 22, no. 1123 (November 22, 1919): 1089–1090; Luke W. Boyce to J. F. Gould, January 29, 1920, Americas Committee, Jan.–May 1920 folder, Northern Information Bureau Records, MNHS; Carl H. Chrislock, *Watchdog of Loyalty: The Minnesota Commission of Public Safety during World War I* (Saint Paul: Minnesota Historical Society Press, 1991), 272–273.

66. William Pencak, *For God and Country: The American Legion, 1919–1941* (Boston: Northeastern University Press, 1989), 49–77; Jennifer D. Keene, *Doughboys, the Great War, and the Remaking of America* (Baltimore: Johns Hopkins University Press, 2001); "Wilson Declares Opponents Twist Treaty Meaning," *New York Times*, September 11, 1919, 7; "League Kills Attempt to Discredit Nonpartisans," *Ellsworth (KS) County Leader*, November 27, 1919; "Kansas Legion Post Defies Convention," *Nonpartisan Leader*, December 8, 1919.

67. "The American Legion Being Used by Gang as Political Buffer," *South Dakota Leader*, September 27, 1919; Keene, *Doughboys, the Great War, and the Remaking of America*, 161–178.

68. "North Dakota Employing Soldiers," *Nonpartisan Leader*, June 16, 1919; "Returned Soldier Organizer for League Writes Letter," *Idaho Leader*, July 12, 1919; "Returned Soldier Takes Up Challenge of 'New' Federation," *Nebraska Leader*, August 23, 1919.

69. "Help Fight Anarchy," *Ellsworth County Leader*, December 11, 1919; Luke W. Boyce to J. F. Gould, November 11, 1919, Americas Committee of Minneapolis—J. F. Gould, Nov. 1919 folder, Northern Information Bureau Records, MNHS; Harrison Fuller to Frank B. O'Connell, March 15, 1920, Department of Justice / Bureau of Information Investigative files, file 207238, reel 943, Collection M-1085, National Archives, http://www.marxisthistory.org/history/usa/groups /amlegion/1920/0315-fuller-tooconnell.pdf, accessed June 11, 2013.

70. "North Dakota, the Ex-Soldier and Taxes," *Nonpartisan Leader*, January 5, 1920.

71. Newspaper clipping, *Nebraska State Journal* (Lincoln), August 11, 1919, Non-Partisan League Scrapbook, NSHS; "Old Gang Attempts to Break Up League Meeting at Hartford," *South Dakota Leader*, October 18, 1919; "League Meeting at Monroe Broken Up by Mobsters," *South Dakota Leader*, October 25, 1919; "Mobsters Foiled at Monroe Meeting," *South Dakota Leader*, November 8, 1919.

72. "Farmer and Banker Kidnapped by Kansas Mob," *Ellsworth County Leader*, November 13, 1919; "Editor Wants to Know about the League, Etc., Etc.," *Ellsworth County Leader*, April 15, 1920; "Ensign Meeting Abruptly Ended," *Ellsworth County Leader*, May 13, 1920; "Mobbing Continues," *Ellsworth County Leader*, June 17, 1920; "Ormsbee, W. S. C. Grad, in Bad at Walla Walla," *Pullman (WA) Herald*, July 23, 1920.

73. "Farmers' Picnic Broke [*sic*] Up and Citizens of State Mob'd as Well as Speaker," *Ellsworth County Leader*, June 17, 1920; "H. T. B. Kelley to George Klein, May 28, 1920," reproduced in *Ellsworth County Leader*, June 17, 1920; "Egg Splattered Leaguers Snort at Government," *Ellsworth County Leader*, June 10, 1920; "Justifies Acts of Legion," *Hays (KS) Free Press*, July 1, 1920.

74. Frank W. Blackmar, ed., *History of the Kansas State Council of Defense* (Topeka: Kansas State Printing Plant, 1920), 3; P. E. Zimmerman to Clyde Reed, June 3, 1919, and P. E. Zimmerman to Clyde Reed, September 29, 1919, both folder 8, box 2, Henry J. Allen Papers, Kansas State Historical Society, Topeka, KS (hereafter KSHS); P. E. Zimmerman to "Gentlemen," October 6, 1919, *Ellsworth County Leader*, October 16, 1919.

75. Peter Norbeck to Henry J. Allen, May 7, 1919, folder 8: Anti Bolshevik, box 2, Henry J. Allen Papers, KSHS; "N. P. L. to Kansas," *Topeka (KS) State Journal*, November 5, 1919; Elmer Peterson to Henry J. Allen, November 4, 1919, folder 19, box 17, Henry J. Allen Papers, KSHS.

76. P. E. Zimmerman to Henry J. Allen, November 24, 1919, and P. E. Zimmerman, press release, December 20, 1919, both folder 19, box 17, Henry J. Allen Papers, KSHS; P. E. Zimmerman to Henry J. Allen, April 15, 1920, reel 1, Philip Zimmerman Papers, KSHS.

77. Elmer Peterson to Henry J. Allen, October 20, 1919, folder 19, box 17, Henry J. Allen Papers, KSHS.

78. G. B. Jacobson to editor, April 11, 1919, Oscar J. Sorlie Papers, NDIRS.

79. Theodore G. Nelson to Nicolay Grevstad, April 5, 1919, and Nicolay Grevstad to Will H. Hayes, April 6, 1919, both folder 4, box 2, Nicolay Grevstad Papers, Norwegian American Historical Association, Saint Olaf College, Northfield, MN.

80. Burleigh F. Spaulding, "Burleigh Folsom Spaulding," folder 1, box 1, UND; "Meeting of Joint Advisory Committee of Twenty One," April 8, 1920, folder 4, box 1, Theodore Nelson Papers, SHSND.

81. D. Jerome Tweton, "The Anti-League Movement: The IVA," in *The North Dakota Political Tradition*, ed. Thomas W. Howard (Ames: Iowa State University Press, 1981), 99–100; Nelson, *Scrapbook Memoirs*, 75; Charles James Haug, "The Industrial Workers of the World in North Dakota, 1918–1925," *North Dakota Quarterly* 41, no. 3 (Summer 1973): 15.

82. "A Plea for Fair Play," *Red Flame* 1, no. 5 (March 1920): 1.

83. Daniel Francis, *Seeing Reds: The Red Scare of 1918–1919, Canada's First War on Terror* (Vancouver: Arsenal Pulp, 2011).

84. "The Nonpartisan Political League," *Grain Growers' Guide* (Winnipeg, MB), September 5, 1917.

85. Ibid.; George Heal to John Ford, February 20, 1918, and John Ford to George Heal, March 9, 1918, both file 40, series 4, John Ford Fonds, Glenbow Archives, Calgary, AB (hereafter GA); Roy S. Weaver, *The Nonpartisan League in North Dakota* (Toronto: Canadian Reconstruction Association, 1921), 78.

86. "Saskatchewan League Prosperous," *Nonpartisan Leader*, December 22, 1919; W. M. Martin to Norman McKay, November 23, 1920, folder 25, General Correspondence, box 3270-4406, and W. M. Martin to Bishop Nykyta Budka, December 28, 1920, folder 23, General Correspondence, box 3270-4406, William Melville Martin Papers, Saskatchewan Archives Board, Saskatoon, SK; Weaver, *The Nonpartisan League in North Dakota*, 79–80; *Mind Your Own Business* (Nonpartisan League: Saskatoon, 1920).

87. "Non-Partisan Politics: The League's Opportunity," *Alberta Non-Partisan* (Calgary), January 18, 1918.

88. J. H. Ford to J. B. Erickson, January 3, 1918, file 35, series 4, John Hooper Ford Fonds, GA; "Non-Partisan League Annual Convention," *Alberta Non-Partisan*, March 29, 1918.

89. "Get Control of Government," *Alberta Non-Partisan*, February 8, 1918, 10; "The Democratic Non-Partisans," *Calgary News Telegram*, March 30, 1918, reprinted in *Alberta Non-Partisan*, April 12, 1918.

90. H. Higgonbotham to J. H. Ford, July 6, 1918, file 40, series 4, John Hooper Ford Fonds, GA; "Political Action," *Alberta Non-Partisan*, December 4, 1918.

91. J. H. Ford to John Glambeck, June 26, 1918, file 38, J. H. Ford to Henry Ross, May 28, 1918, file 63, J. W. Leedy to J. H. Ford, March 18, 1918, file 43, all series 4, John Hooper Ford Fonds, GA.

92. J. H. Ford to *Nonpartisan Leader*, April 19, 1918, file 52, series 4, John Hooper Ford Fonds,

GA; " 'Wild Farmers of Canada' on Warpath," *Nonpartisan Leader*, December 9, 1918; Weaver, *The Nonpartisan League in North Dakota*, 78, 80.

93. "The Logical Course," *Alberta Non-Partisan*, January 15, 1919; Anthony Mardiros, *William Irvine: The Life of a Prairie Radical* (Toronto: James Lorimer, 1979), 89; Bradford James Rennie, *The Rise of Agrarian Democracy: The United Farmers and Farm Women of Alberta, 1909–1921* (Toronto: University of Toronto Press, 2000), 89.

94. Rennie, *The Rise of Agrarian Democracy*,181; "Alberta Farmers Parliament," *Grain Growers' Guide*, January 29, 1919.

95. "The U.F.A. Method," *Alberta Non-Partisan*, January 30, 1919; "Declaration of Political Independence," *Alberta Non-Partisan*, May 22, 1919; H. W. Wood, "Political Action in Alberta," *Grain Growers' Guide*, May 7, 1919.

96. "Declaration of Political Independence," *Alberta Non-Partisan*, May 22, 1919.

97. "Alberta Farmers Enter Politics," *Nonpartisan Leader*, May 12, 1919.

98. "Alberta Farmers Hold Big Political Conventions," *Grain Growers' Guide*, June 25, 1919; "U. F. A. Political Conventions," *Alberta Non-Partisan*, July 7, 1919.

99. "The Provincial Convention," *Alberta Non-Partisan*, August 7, 1919; Rennie, *The Rise of Agrarian Democracy*, 204.

100. Charles Wooster, "Some Reasons Why I Join the Non-Partisan League," January 22, 1920, 1920—Nonpartisan League folder, box 12, Charles Wooster Papers, NSHS. Wooster's daughter Margaret pursued a distinguished career as a psychologist and professor and married historian (and fellow Nebraskan) Merle Curti in 1925.

101. Warner, "The Farmer Butts Back," 240; S. A. Olsness to L. B. Garnaas, September 27, 1920, folder 6, box 1, NDIRS.

102. "County Organization for Nonpartisans," *Nonpartisan Leader*, December 29, 1919; Gaston, *The Nonpartisan League*, 318.

103. Warner, "The Farmer Butts Back," 241.

104. Fred A. Harding, "Minnesota's Plan for New Campaign," *Nonpartisan Leader*, July 26, 1920.

105. James H. Shideler, *Farm Crisis, 1919–1923* (Berkeley: University of California Press, 1957); "Your Magazine on a New Basis," *Nonpartisan Leader*, July 26, 1920.

106. "A Farmer Governor—Why Not a Farmer President?," *Nonpartisan Leader*, January 26, 1920; lapel pin in possession of author; "League Announcement," *Nonpartisan Leader*, August 23, 1920.

107. "The Third Party," *Nonpartisan Leader*, August 9, 1920.

108. Warner, "The Farmer Butts Back,", 241; R. Todd Laugen, *The Gospel of Progressivism: Moral Reform and Labor War in Colorado, 1900–1930* (Boulder: University of Colorado Press, 2011), 119–120.

109. I. P. MacDowell, "Idaho Labor Joins Hands with League," *Nonpartisan Leader*, March 8, 1920; "Prospects Bright for League Success in South Dakota," *South Dakota Leader*, October 4, 1919.

110. Hamilton Cravens, "The Emergence of the Farmer-Labor Party in Washington Politics, 1919–20," *Pacific Northwest Quarterly* 57, no. 4 (October 1966): 151–153.

111. "Non-Partisans to Vote G.O.P.," *Sunnyside (WA) Times*, July 22, 1920, and "Triple Alliance Will Be Fought," *Yakima (WA) Republic*, July 22, 1920.

112. "Alliance and Nonpartizans Compromise; Split Up Votes," *Spokane (WA) Daily Chronicle*, August 1, 1920; Cravens, "The Emergence of the Farmer-Labor Party in Washington Politics," 154–155.

113. "Nebraska," *Nonpartisan Leader*, March 22, 1920; C. A. Sorensen to Arthur Weatherly, July 22, 1920, Correspondence—July 1920 folder, box 2, C. A. Sorensen Papers, NSHS; Elmo Bryant Phillips, "The Non-Partisan League in Nebraska" (master's thesis, University of Nebraska, 1931), 88–91; "Arthur G. Wray Will Lead the Progressives of Nebraska," *Nebraska Leader*, May 8, 1920.

114. "Montana," *Nonpartisan Leader*, March 15, 1920; "Labor and Farmers Convention at Emporia," *Ellsworth County Leader*, March 25, 1920; "Colorado," *Nonpartisan Leader*, April 5, 1920, 9; "Special Convention Colorado State Federation of Labor, Railroad Brotherhood, and Farmers' Non-Partisan League, Denver, June 12, 1920," photograph, National Farmers Union Papers, series 16, "Oversize Photographs," Special Collections, University of Colorado at Boulder, Boulder, CO; "Interests Seek to Halt Effort of Farmer Labor Groups in Colorado," *Colorado Leader* (Denver), July 2, 1920; "United Americans Seeking to Monopolize Patriotism," *Colorado Leader*, July 9, 1920.

115. F. H. Shoemaker to Henry Teigan, December 21, 1918, reel 4, National Nonpartisan League Papers, MNHS; Donald B. Marti, "Answering the Agrarian Question: Socialists, Farmers, and Algie Martin Simons," *Agricultural History* 65, no. 3 (Summer 1991): 53–69.

116. Henry Teigan to F. H. Shoemaker, December 27, 1918, reel 4, National Nonpartisan League Papers, MNHS; W. C. Zumach to C. A. Christenson, April 23, 1920, pp. 118–127 folder, box 2, Materials Relating to the Nonpartisan League, comp. Asher Howard, MNHS; "Wisconsin," *Nonpartisan Leader*, September 6, 1920.

117. "League Forces Put a Dent in the 'Solid South,'" *Nonpartisan Leader*, March 29, 1920, 11; "With the Organized Farmers," *Nonpartisan Leader*, June 28, 1920; John W. Canada, "A Plain People's Campaign: The New Force in Texas Politics," *Southland Farmer* (Houston, TX), March 15, 1920.

118. Warner, "The Farmer Butts Back," 240; "Statements by the President and the Secretary of the North Dakota WCTU," broadside, ca. November 1920, folder 2, box 1, Theodore Nelson Papers, SHSND; Steven L. Piott, *Giving Voters a Voice: The Origins of the Initiative and Referendum in America* (Columbia: University of Missouri Press, 2003), 211.

119. Advertisement, *Colorado Leader*, July 30, 1920; "League Precinct Meetings to Be Held Early Part of Next Month," *Nebraska Leader*, November 8, 1919.

120. "Members Take Your Wives . . . ," *Colorado Leader*, July 30, 1920; "See That She Votes!," *Nonpartisan Leader*, November 1, 1920; C. A. Sorenson to "Dear League Member," October 29, 1920, Correspondence—October 1920 folder, box 2, C. A. Sorenson Papers, NSHS; "We Cannot Win without the Woman Vote!," *Colorado Leader*, October 30, 1920; Lavern Schoeder, "Women and the Nonpartisan League in Adams and Hettinger Counties," in *Women on the Move*, ed. Pearl Andre (Bismarck: North Dakota Democratic-NPL Women, 1975), 48.

121. "Nebraska Makes Gains," *Nonpartisan Leader*, May 17, 1920; "Primary Election Returns of Election Held June 21, 1920," in *The Legislative Manual of the State of Minnesota* (Minneapolis: Harrison and Smith, 1921), 100–109; "How Minnesota Voted," *Nonpartisan Leader*, July 12, 1920.

122. Thomas A. Contois, "A Triumph of American Politics: Subduing Popular Democracy on the Northern Plains," manuscript in possession of author, 139–140.

123. "Let's Take Our Hats Off to Montana," *Nonpartisan Leader*, September 20, 1920; "Colorado 'Over the Top,'" *Nonpartisan Leader*, October 4, 1920.

124. "Great Victory in Wisconsin," *Nonpartisan Leader*, September 20, 1920.

125. "The Coming of a People's New Daily," *Nonpartisan Leader*, August 9, 1920.

126. "Townley Classifies Every Voter in the State," *Red Flame* 1, no. 11 (September 1920): 22.

127. "Cox Carries Fight into North Dakota," *New York Times*, September 8, 1920, 1–2; "Peach Day Celebrators Hear Democratic Leader," *Deseret News* (Salt Lake City, UT), September 16,

1920; "Harding Points to New Era for American Farmer as Census Gains," *Minneapolis Tribune*, September 9, 1920.

128. "Taft Sees Hope of League Downfall in North Dakota," *Minneapolis Tribune*, July 19, 1920; James F. Vivian, "'Not a Patriotic American Party': William Howard Taft's Campaign against the Nonpartisan League," *North Dakota History* 50, no. 4 (Fall 1983): 4–10.

129. Burton K. Wheeler with Paul F. Healy, *Yankee from the West: The Candid, Turbulent Life Story of the Yankee-Born U.S. Senator from Montana* (New York: Doubleday, 1962) 173–174.

130. Herbert E. Gaston, "Where Merchants and Farmers Agree," *Nonpartisan Leader*, February 2, 1920.

131. "Tells Non-Partisan Gains," *New York Times*, November 9, 1920, 1; C. R. Johnson, "The Nonpartisan League Defeated," *Nation*, 111, no. 2891 (December 1, 1920): 614.

132. "A Comparison," *Nonpartisan Leader*, November 15, 1920; Johnson, "The Nonpartisan League Defeated," 614.

133. "Leaguers Gain in Election," *Nonpartisan Leader*, November 15, 1920 and "News of League States," *Nonpartisan Leader*, November 29, 1920.

134. "Leaguers Gain in Election," *Nonpartisan Leader*, November 15, 1920; Kurt Wetzel, "The Defeat of Bill Dunne: An Episode in the Montana Red Scare," *Pacific Northwest Quarterly* 64, no. 1 (January 1973): 18; Henry G. Teigan, "The National Nonpartisan League," *The American Labor Year Book*, ed. Alexander Trachtenberg (New York: Rand School of Social Science, 1922), 426.

135. "Leaguers Gain in Election," *Nonpartisan Leader*, November 15, 1920; Teigan, "The National Nonpartisan League," 426; Gilbert C. Fite, "Peter Norbeck and the Defeat of the Nonpartisan League in South Dakota," *Mississippi Valley Historical Review* 33, no. 2 (September 1946): 217–236.

136. "Meet John J. Blaine of Wisconsin," *Nonpartisan Leader*, November 29, 1920; "Leaguers Gain in Election," *Nonpartisan Leader*, November 15, 1920; Contois, "A Triumph of American Politics," 145.

137. "Leaguers Gain in Election"; Warner, "The Farmer Butts Back," 241; Norris C. Hagen, *Vikings of the Prairie: Three North Dakota Settlers Reminisce* (New York: Exposition, 1958), 160–161.

138. "The Election Just Past—and the Next One," *Nonpartisan Leader*, November 29, 1920; "A Good Fighter," *Nonpartisan Leader*, February 7, 1921.

139. "From Montana," *Nonpartisan Leader*, February 7, 1921; and Shideler, *Farm Crisis, 1919–1923*.

140. "A Century of Progress," *Nonpartisan Leader*, January 24, 1921.

141. "Leaguers Ready for Winning Campaign," *Nonpartisan Leader*, March 21, 1921.

142. Charles Merz, "The Nonpartisan League: A Survey," *New Republic* 22, no. 284 (May 12, 1920): 336; "Leaguers Ready for Winning Campaign."

143. "Leaguers Ready for Winning Campaign," *Nonpartisan Leader*, March 21, 1921; Minutes of the State Executive Committee of the Nonpartisan League of North Dakota, May 1, 1921, and June 24, 1921, folder 15, box 1, North Dakota Nonpartisan League Collection, UND.

144. "Club Work in Summer," *Nonpartisan Leader*, May 30, 1921; "Urges Women to Be Alert during 1921," *Nebraska Leader*, January 15, 1921.

145. "A Fourth of July," *Nonpartisan Leader*, August 8, 1921; "Miss Keller on the Nonpartisan League," *Nonpartisan Leader*, October 3, 1921.

146. "Worse Than Wall Street," *Minneapolis Tribune*, March 1, 1921; Henry Raymond Mussey, "The Farmers and Congress," *Nation* 112, no. 2896 (January 5, 1921): 12.

147. Oliver S. Morris, "It Is Time the Truth Be Known and the Facts Faced: A Frank Statement

to Our Readers," *Nonpartisan Leader*, July 25, 1921, 16; Oliver S. Morris, "Extraordinary Offer to Leader Readers," *Nonpartisan Leader*, May 16, 1921; Oliver S. Morris, "A Word by the Editor," *Nonpartisan Leader*, June 13, 1921; "'Hard Times' and Your Organization," *Nonpartisan Leader*, September 19, 1921, 4; "Comment by the Editor," *National Leader* (Minneapolis) 16, no. 6 (July 1923): 3.

148. Orville Merton Kile, *The Farm Bureau through Three Decades* (Baltimore: Waverly, 1948), 62–63; "Farm Bureau Gaining Ground in This State," *Milwaukee Journal*, May 23, 1920, 1; "War Breaks Out in State among Farm Organizations," *Minneapolis Tribune*, September 9, 1921, 1; "Nonpartisan League News," *Nonpartisan Leader*, September 19, 1921; Nonpartisan County Executive Committee, Morton County, ND, to "Fellow Worker," February 8, 1921, folder 1, box 1, UND.

149. Nicolay Grevstad to Medill McCormick, April 7, 1921, folder 5, box 2, Nicolay A. Grevstad Papers, Norwegian-American Historical Association Archives, Saint Olaf College, Northfield, MN.

150. "Non-Partisan Men Tarred in Kansas," *New York Times*, March 14, 1921, 1; Bruce L. Larson, "Kansas and the Nonpartisan League: The Response to the Affair at Great Bend, 1921," *Kansas Historical Quarterly* 34, no. 1 (Spring 1968): 51–76.

151. Asher Howard, *The Leaders of the Nonpartisan League: Their Aims, Purposes and Records . . .* (Minneapolis: published by author, 1920); "Records Showing Socialist Hold on Nonpartisan League Filed," *Minneapolis Journal*, October 10, 1920; "Minnesota Historical Society Notes," *Minnesota History* 3, no. 8 (August 1920): 527.

152. Oliver S. Morris, "What Is Happening in North Dakota," *Nation* 112, no. 2905 (March 9, 1921): 367–368; Alvin S. Tostlebee, "The Bank of North Dakota: An Experiment in Agrarian Banking," *Studies in History, Economic, and Public Law* 64, no. 1 (1924): 112; George E. Akerson, "Nonparty Surrenders in North Dakota," *Minneapolis Tribune*, February 14, 1921, 1; "North Dakota and the Banks," *Nation* 112, no. 2904 (March 2, 1921): 330.

153. "The Bank of North Dakota," *Colorado Leader*, August 6, 1920; "$6,000,000 Bond Issue: The State of North Dakota," advertisement, *Nation* 112, no. 2910 (April 21, 1921): 536–537, quote on 536; Lynn J. Frazier, "A Message to the Farmers of America," *Nonpartisan Leader*, May 2, 1921; Rozanne Enerson Junker, *The Bank of North Dakota: An Experiment in State Ownership* (Santa Barbara, CA: Fithian, 1989), 68–69; "North Dakota and the Banks," *Nation* 112, no. 2904 (March 2, 1921): 330; "North Dakota Goes to the People," *Nation* 112, no. 2910 (April 13, 1921): 530–531; Tostlebee, "The Bank of North Dakota," 145–147.

154. "Guess They'll Sell the Bonds!" *Nonpartisan Leader*, June 27, 1921; "North Dakota Bond Campaign Winning," *Nonpartisan Leader*, May 30, 1921.

155. *Laws Passed at the Sixteenth Session of Legislative Assembly of the State of North Dakota* (Bismarck, ND: Secretary of State, 1919), 111.

156. Hilton Hornaday, "'Bitter Ender' Townley Foes Combat Plan," *Minneapolis Tribune*, February 16, 1921; Theodore Nelson, "Suggestions and Outline for Talk to Be Made by Organizers (Fieldmen and Speakers) for the IVA," folder 6, box 1, Theodore Nelson Papers, SHSND; D. Jerome Tweton, "The Anti-League Movement: The IVA," in *The North Dakota Political Tradition*, ed. Thomas W. Howard (Ames: Iowa State University Press, 1981), 115; Contois, "A Triumph of American Politics," 180–183.

157. "Committee of 21, February 24, 1921," folder 6, box 1, Theodore Nelson Papers, SHSND; "North Dakota Recall Faction Claims Control," *Minneapolis Tribune*, March 31, 1921; "Fusionists Vote Nonparty Recall in North Dakota," *Minneapolis Tribune*, April 1, 1921.

158. "Why the North Dakota I. V. A. Is Insincere," *Nonpartisan Leader*, October 17, 1921; J. H. Voracheck to Theodore G. Nelson, March 7, 1921, folder 7, box 1, Theodore Nelson Papers, SHSND.

159. O. S. Aaker to Theodore G. Nelson, March 4, 1921, and W. T. Cummins to Theodore G. Nelson, March 7, 1921, folder 7, box 1, Theodore Nelson Papers, SHSND.

160. "Why the North Dakota I. V. A. Is Insincere"; "Save the State or Get the Jobs?" *Bismarck Tribune*, February 17, 1921; "Nonpartisan League News," *Nonpartisan Leader*, September 19, 1921; "Recall Petitions Filed in North Dakota," *Nonpartisan Leader*, October 3, 1921.

161. Thomas M. Contois, *A Triumph of American Politics: Subduing Popular Democracy on the Northern Plains, Sample Election Returns, 1916–1921,* pamphlet from presentation at Nonpartisan League at 90 Conference, October 9, 2008, University of North Dakota, Grand Forks, ND, table 8, "Reducing the League's Agrarian Majority."

162. Theodore Nelson to J. T. Nelson, July 22, 1921, folder 8, box 1, Theodore Nelson Papers, SHSND. Thomas Contois speculates that the Minneapolis Chamber of Commerce provided the IVA with funding as early as December 1920. See Contois, "A Triumph of American Politics," 185–188. Given the antirecall stance of Alex McKenzie, the IVA's rejection of a February 1921 deal between the NPL and Minneapolis financiers to save the state bank, and Nelson's ongoing financial concerns, such claims remain dubious.

163. *The Independent Program* (1921), Theodore Nelson to Peter Norbeck, June 24, 1921, and Theodore Nelson to Peter Norbeck, September 23, 1921, all Hall–Outdoor Life, 1921 folder, box 86, Peter Norbeck Papers, USD; Theodore Nelson, "Election Prospective," folder 8, box 1, Theodore Nelson Papers, SHSND. Contois makes the deterioration of pro-League voters in North Dakota's rural districts between 1916 and 1921 clear. "A Triumph of American Politics," 221.

164. *Standard Atlas of Walsh County, North Dakota: Including a Plat Book of the Villages, Cities, and Townships of the County* (Chicago: Alden, 1910), 39; *Lankin (ND) Reporter,* November 30, 1916; G. K. Ness, *History of Fordville and Surrounding Area* (Fordville, ND: Ness, 1973), 190–191; "Howard, Geo.," list of subscribers to the *Independent,* ca. 1921, folder 4, box 2, Theodore Nelson Papers, SHSND; "Post Office: Fordville," Petition for the Recall of Lynn J. Frazier, Governor, box 1, series 30911, North Dakota State Archives, SHSND.

165. Public Statement on Recall, County Executive Committee of NPL, Griggs County Branch, ND, ca. 1921, and A. A. Liederbach to County Victory Campaign Committee and Precinct Crews, October 1, 1921, Political Organization Papers folder, box 3, John L. Miklethun Papers, MNHS.

166. "Townley Ousted by Nonparty in North Dakota," *Minneapolis Tribune,* April 2, 1921; Morlan, *Political Prairie Fire: The Nonpartisan League, 1915–1922* (1955; repr., Saint Paul: Minnesota Historical Society Press, 1986), 319; "Retract Charges," *Nonpartisan Leader,* July 25, 1921.

167. "North Dakota Faces Crucial Test in Battle for Recall," *Minneapolis Tribune,* September 18, 1921; *Stringent League Laws against Immorality Is One Reason for Bitter Hatred of Leading I.V.A.'s* (Fargo, ND, 1921), folder 1, box 6, Martin O. Thompson Papers NDIRS.

168. "High Court Denies Townley's Petition," *New York Times,* October 25, 1921; "A Talk with Townley," *Nonpartisan Leader,* September 5, 1921; "Townley Ready to Go to Jail," *Minneapolis Tribune,* October 26, 1921.

169. C. R. Johnson, "The Nonpartisan League Defeated," *Nation,* 111, no. 2891 (December 1, 1920): 614.

Chapter Six

1. "Nestos Leads Frazier by 17,000," *Bismarck (ND) Tribune,* October 29, 1921.

2. "Anti-League Candidates Leading in North Dakota," *New York Tribune,* October 29, 1921, 1; "Nonpartisan League Admits Defeat," *Tulsa World,* October 30, 1921, 1; "Recall Election in North

Dakota," *Albuquerque Evening Herald*, October 28, 1921, 1; "North Dakota Shows Anti-League Gain in Early Returns," *Washington (DC) Herald*, October 29, 1921, 1; "Victory Recalls Armistice Day to Fargo," *Minneapolis Tribune*, October 30, 1921, 1; "Thousands Honor Gov.-Elect Nestos in Big Celebration," *Ward County Independent* (Minot, ND), November 3, 1921.

3. Lorena A. Hickok, "Townley Begins Jail Term," *Minneapolis Tribune*, November 3, 1921.

4. "Triumph in North Dakota—League Program Victorious," *National Leader* (Minneapolis), November 14, 1921; Oliver S. Morris, "The Vote of the North Dakota Farmers," *Nation* 113, no. 2940 (November 9, 1921): 536.

5. John N. Hagan to H. F. Samuels, November 14, 1921, folder 1, box 1, John Hagan Papers, State Historical Society of North Dakota, Bismarck, ND (hereafter SHSND); "Nonpartisan Rule Broken," *Minneapolis Tribune*, October 30, 1921.

6. Michael Kazin's *The Populist Persuasion: An American History* (New York: Basic Books, 1995), Elizabeth Sanders's *Roots of Reform: Farmers, Workers, and the American State, 1877–1917* (Chicago: University of Chicago Press, 1999)—a book that has done much to restore farmers to early twentieth-century political history—and Doug Rossinow's *Visions of Progress: The Left-Liberal Tradition in America* (Philadelphia: University of Pennsylvania Press, 2008) only briefly mention the Nonpartisan League. Kazin's *American Dreamers: How the Left Changed a Nation* (New York: Knopf, 2011) ignores rural movements of every sort.

7. "Fight Started on Recall in North Dakota," *Minneapolis Tribune*, November 15, 1921; resolution, November 9, 1921, Ransom County Nonpartisan League, folder 21, box 6, Martin O. Thompson Papers, North Dakota Institute for Regional Studies, North Dakota State University, Fargo, ND (hereafter NDIRS); "Recall Contest Is Wrong," *National Leader*, November 28, 1921; "Leaguers Urge Townley Again to Head Fight," *Minneapolis Tribune*, November 21, 1921; "North Dakota's Nonparty Rule Brought to an End," *Minneapolis Tribune*, November 24, 1921.

8. "Townley Chafes at Confinement; Life behind Bars Doesn't Suit Him," *Minneapolis Tribune*, December 11, 1921.

9. "Townley Chides Followers for Deserting Cause," *Minneapolis Tribune*, February 4, 1922; "Townley Denies Split Possible on New Plans," *Minneapolis Tribune*, February 12, 1922.

10. "League Ignore's [*sic*] Townley's Plea, Calls Caucuses," *Minneapolis Tribune*, January 10, 1921; "Mr. Townley and the Dissenters," *Minneapolis Tribune*, February 15, 1922; "Townley Plan Splits League Ranks in Fight to the Finish," *Minneapolis Tribune*, February 19, 1922; "Nonparty Letters Call Halt on Pilot Townley," *Minneapolis Tribune*, March 4, 1922; "Retreat but Not Surrender," *Minneapolis Tribune*, March 22, 1922.

11. "Letter of Resignation," *Minneapolis Star*, May 12, 1922; "Townley, Vindicated in Court, Resigns," *National Leader*, April 17, 1922.

12. "Keep Your Eyes on Washington State," *Nonpartisan Leader* (Fargo, ND), August 5, 1918; Louis Levine, "Politics in Montana," *Nation* 107, no. 2783 (November 2, 1918): 507; Burton W. Folsom, "Immigrant Voters and the Nonpartisan League in Nebraska," *Great Plains Quarterly* 1 (Summer 1981): 162. See also Paula Baker, *The Moral Frameworks of Political Life: Gender, Politics, and the State in Rural New York, 1870–1930* (New York: Oxford University Press, 1991).

13. "What Our Clubs Are Doing," *Minnesota Herald* (Mora), May 15, 1922.

14. "Women Pleased at Recognition," *Minnesota Leader* (Saint Paul), April 10, 1920; "Real Farm Girl to Be State Treasurer," *Minnesota Leader*, April 10, 1920; "Suffragists Are Jubilant," *Minnesota Leader*, April 17, 1920.

15. "Wheeler, Arnold, Haste, Ford Endorsed with Others to Run on Democratic Ticket," *Montana Nonpartisan* (Great Falls), June 26, 1920.

16. "Nonpartisan Political League Volunteer Workers, Please Notice," *Colorado Leader* (Denver), September 3, 1920; "About Mrs. Ventzke," *Nonpartisan Leader*, October 3, 1921.

17. "Mrs. Minnie D. Craig—Genealogical Data (Mrs. E. O. Craig)," autobiography, no date, folder 3, box 1, Minnie D. Craig Papers, NDIRS, 28; Barbara Stuhler and Gretchen Kreuter, eds., *Women of Minnesota: Selected Biographical Essays* (Saint Paul: Minnesota Historical Society Press, 1977), 262–263; "Nonpartisan League News," *National Leader* 15, no. 2 (August 1922): 8.

18. "Not to Marry Senator," *New York Times*, May 7, 1913, 1; "Woman Leads People's Battles in South Dakota," *Minnesota Herald*, April 15, 1922; "Egg Volley Greets Woman Campaigner," *New York Times*, July 23, 1920, 3.

19. "Woman Is League Candidate in South Dakota," *Nonpartisan Leader*, March 22, 1920; "Woman Leads People's Battles in South Dakota," *Minnesota Herald*, April 15, 1922; Kristi Andersen, *After Suffrage: Women in Partisan and Electoral Politics before the New Deal* (Chicago: University of Chicago Press, 1996), 138.

20. "Woman Leads People's Battles in South Dakota"; "Convention of South Dakota Federation Indorses Principle of Industrial Organization," *Railroad Worker* 20, no. 8 (November 1922): 42; Carl J. Hofland, "The Nonpartisan League in South Dakota" (master's thesis, University of South Dakota, 1940), 67; William C. Pratt, "Another South Dakota; or, The Road Not Taken: The Left and the Shaping of South Dakota Political Culture," in *The Plains Political Tradition: Essays on South Dakota Political Culture*, ed. Jon Lauck, John E. Miller, and Donald C. Simmons (Pierre: South Dakota State Historical Society Press, 2011), 113.

21. "Readers on Birth Control, Pro or Con," *National Leader* 14, no. 7 (April 3, 1922): 9; William C. Pratt, "Women and the Farm Revolt of the 1930s," *Agricultural History* 67, no. 2 (Spring 1993): 214–223; Jenny Barker Devine, *On Behalf of the Family Farm: Iowa Farm Women's Activism since 1945* (Iowa City: University of Iowa Press, 2013).

22. "Oregon Editors in Session," *Pacific Printer and Publisher* 24, no. 3 (September 1920): 160–161; "Oregon and N.P.L.," *Oregon Voter* 27, no. 7 (November 12, 1921): 27; "Oregon Leaguers Enter Campaign," *National Leader*, December 12, 1921; "Spence Addresses Board," *Oregon Grower* 5, no. 1 (August 1923): 14; Lawrence M. Lipin and William Lunch, "Moralistic Direct Democracy: Political Insurgents, Religion, and the State in Twentieth-Century Oregon," *Oregon Historical Quarterly* 110, no. 4 (Winter 2009): 514–545.

23. "Political Bunk in Large Chunks," *Coconino Sun* (Flagstaff, AZ), January 27, 1922; "Attacking Farmers," *National Leader*, May 1, 1922; T. A. Larson, *History of Wyoming* (Lincoln: University of Nebraska Press, 1965), 454–455; "Independents to Support," *Wind River Mountaineer* (Lander, WY), July 21, 1922; "Non-Partisaners Take Kendrick as Their Candidate," *Wyoming State Tribune* (Cheyenne), July 14, 1922; "Odds in Wyoming against Mondell," *New York Times*, October 30, 1922, 14.

24. "Representative Hoare Doubts Wisdom of the New Party," *New State* (Lincoln, NE), December 3, 1921; "Farmers Get Ready for 1922 Campaign," *National Leader* 14, no. 6 (March 20, 1922): 6; Christian Sorensen to A. C. Townley, June 2, 1922, Correspondence June 1922 folder, box 3, C. A. Sorensen Papers, Nebraska State Historical Society, Lincoln, NE; "League News," *National Leader* 15, no. 3 (September 1922): 8–9; Verlaine Stoner McDonald, *The Red Corner: The Rise and Fall of Communism in Northeastern Montana* (Helena: Montana Historical Society Press, 2010), 69–75.

25. M. H. Hedges, "The Liberal Sweep in the West," *Nation* 115, no. 2994 (November 22, 1922), 543.

26. "League News," *National Leader* 15, no. 3 (September 1922): 8–9; A. C. Townley, "Members

Mean More Than Candidates Elected," *National Leader* 15, no. 4 (October 1922): 5; "Montana Comes Back," *Nation* 115, no. 2994 (November 22, 1922): 545.

27. "League Farmers Score Great Victories," *National Leader* 15, no. 5 (November 1922): 7; Wheeler, *Yankee from the West*, 189.

28. "Farmer Advance on the United States Senate," *National Leader* 16, no. 1 (January 1923): 6; Robert Kingsley, "Recent Variations from the Two-Party System as Evidenced by the Nonpartisan League and the Agricultural Bloc" (master's thesis, University of Minnesota, 1923).

29. "Courier-News Sold," *National Leader* 16, no. 5 (May 1923): 4; Paul May, "Star, 'Co-op' Daily, Wanes," *New York Times*, March 19, 1922, sect. 8, 15; "Comment by the Editor," *National Leader* 16, no. 6 (July 1923): 3.

30. "The Founder of the League" and "Comment by the Editor," *National Leader* 16, no. 6 (July 1923): 2–3.

31. John Lord, "The Future of the Nonpartisan League," *National Leader* 16, no. 6 (July 1923): 4.

32. "The Third Party Is Born," *Nation* 115, 2994 (November 22, 1922): 541.

33. Ibid.

34. Ibid.

35. Henry Teigan to H. H. Stallard, June 15, 1921, and Henry Teigan to Carl D. Thompson, July 8, 1921, roll 9, National Nonpartisan League Papers, Minnesota Historical Society, Saint Paul, MN (hereafter MNHS); A. E. Bowen to Knute Wefaldt [*sic*], telegram, ca. November 1922, January 1922–March 1922 folder, box 9, Knud Wefald Papers, MNHS.

36. A. C. Townley, "Should We Restore a Party or Build a New One?" *National Leader* 15, no. 5 (November 1922): 5.

37. A. C. Townley, "Why the Nonpartisan Plan Does the Work," *National Leader* 16, no. 2 (February 1923): 5.

38. Richard M. Valelly, *Radicalism in the States: The Minnesota Farmer-Labor Party and the American Political Economy* (Chicago: University of Chicago Press, 1989), 1–16.

39. Hugh Lovin, "The Farmer Revolt in Idaho, 1914–1922," *Idaho Yesterdays* 20, no. 3 (Fall 1976): 20–33.

40. Hugh Lovin, "The Nonpartisan League and Progressive Renascence in Idaho, 1919–1924," *Idaho Yesterdays* 32, no. 3 (Fall 1988): 1–15; Claudius O. Johnson, "William E. Borah: The People's Choice," *Pacific Northwest Quarterly* 44, no. 1 (January 1953): 15–22.

41. Annie Pike Greenwood, "Bill Borah and Other Home Folks," *Nation* 116, no. 3008 (February 28, 1923): 236.

42. Millard L. Gieske, *Minnesota Farmer-Laborism: The Third-Party Alternative* (Minneapolis: University of Minnesota Press, 1979), 69–71.

43. "Progressive Conferences Propose National Convention," *Minnesota Union Advocate* (Saint Paul), November 22, 1923, 1, 5; James Weinstein, "Radicalism in the Midst of Normalcy," *Journal of American History* 52, no. 4 (March 1966): 773–790; David P. Thelan, *Robert La Follette and the Insurgent Spirit* (Boston: Little, Brown, 1976), 190.

44. Murray E. King, "The Farmer-Labor Federation," *New Republic* 38, no. 487 (April 2, 1924): 145–147; Gieske, *Minnesota Farmer-Laborism*, 84.

45. Gieske, *Minnesota Farmer-Laborism*, 89.

46. George H. Mayer, *The Political Career of Floyd B. Olson* (Minneapolis: University of Minnesota Press, 1951); Gieske, *Minnesota Farmer-Laborism*; Valelly, *Radicalism in the States*.

47. Gieske, *Minnesota Farmer-Laborism*, 229; Elmer A. Benson, "Politics in My Lifetime," *Minnesota History* 47, no. 4 (Winter 1980): 154–161.

48. Gilbert Fite, "The Nonpartisan League in Oklahoma," *Chronicles of Oklahoma* 24, no. 2 (Summer 1946): 146–157; Gilbert Fite, "Oklahoma's Reconstruction League: An Experiment in Farmer-Labor Politics," *Journal of Southern History* 13, no. 4 (November 1947): 535–555; Garin Burbank, *When Farmers Voted Red: The Gospel of Socialism in the Oklahoma Countryside, 1910–1924* (Westport, CT: Greenwood, 1976), 157–183.

49. "Texas Nonpartisan Conference Declares for Union Labor," *Brotherhood of Locomotive Firemen and Enginemen's Magazine* 72, no. 4 (February 5, 1922): 4–5; Robert Edward Anderson, "The History of the Farm Labor Union in Texas" (master's thesis, University of Texas–Austin, 1928), 14–19, 51–65; James R. Green, *Grass-Roots Socialism: Radical Movements in the Southwest, 1985–1943* (Baton Rouge: Louisiana State University Press, 1978), 399–400.

50. Anthony Mardiros, *William Irvine: The Life of a Prairie Radical* (Toronto: James Lorimer, 1979), 102–103, 109–114, 132–133.

51. Benjamin Stolberg, "Third Party Chances," *Nation* 118, no. 3065 (April 2, 1924): 364–365.

52. Alan Dawley, *Struggles for Justice: Social Responsibility and the Liberal State* (Cambridge, MA: Harvard University Press, 1991), 297–333; Julia Ott, *When Wall Street Met Main Street: The Quest for an Investors' Democracy* (Cambridge, MA: Harvard University Press, 2011), 130–131.

53. Marc Stears, *Demanding Democracy: American Radicals in Search of a New Politics* (Princeton, NJ: Princeton University Press, 2010), 56–84.

54. Waldo Frank, *The Re-discovery of America: An Introduction to a Philosophy of American Life* (New York: Charles Scribner's Sons, 1929), 94, 208, 223.

55. Clark Davis, "The Corporate Reconstruction of Middle-Class Manhood," and Jeffrey M. Hornstein, "The Rise of the Realtor: Professionalism, Gender, and Middle-Class Identity," in *The Middling Sort: Explorations in the History of the Middle Class*, ed. Burton J. Bledstein and Robert D. Johnston (New York: Routledge, 2001), 201–216 and 217–234; Catherine McNichol Stock, *Main Street in Crisis: The Great Depression and the Old Middle Class on the Northern Plains* (Chapel Hill: University of North Carolina Press, 1992), 1–7; Dawley, *Struggles for Justice*, 297–333.

56. Deborah Fitzgerald, *Every Farm a Factory: The Industrial Ideal in American Agriculture* (New Haven, CT: Yale University Press, 2003).

57. Ronald R. Kline, *Consumers in the Country: Technology and Social Change in Rural America* (Baltimore: Johns Hopkins University Press, 2000).

58. John Miklethun, manuscript, ca. 1928, Political Organization Papers, undated and 1918–1921 folder, box 3, John Miklethun Papers, MNHS; Glenn H. Smith, "William Langer and the Art of Personal Politics," in *The North Dakota Political Tradition*, ed. Thomas W. Howard (Ames: Iowa State University Press, 1981), 123–150.

59. Theodore Saloutos and John D. Hicks, *Agricultural Discontent in the Middle West, 1900–1939* (Madison: University of Wisconsin Press, 1951), 145–148; Jason McCollom, " 'We Love You People Better Than We Love Ourselves': North Dakota, Western Canada, and the International Wheat Pool Movement in the 1920s," paper presented at the Association for Canadian Studies in the US Conference, Tampa, FL, November 20, 2013; Theodore Saloutos, "The National Producers' Alliance," *Minnesota History* 28, no. 1 (March 1947): 37–44.

60. Charles Conrad and Joyce Conrad, *50 Years: North Dakota Farmers Union* (n.p., 1976), 11; Saloutos and Hicks, *Agricultural Discontent*, 229–253; William C. Pratt, "Rural Radicalism on the Northern Plains, 1912–1950," *Montana: The Magazine of Western History* 42, no. 1 (Winter 1992): 47–48.

61. Lowell K. Dyson, *Red Harvest: The Communist Party and American Farmers* (Lincoln: University of Nebraska Press, 1982), 1–50.

62. Walter F. Bell, "Farm Bloc," in *Encyclopedia of the United States Congress*, ed. Robert E. Dewhirst (New York: Facts on File, 2007), 206–207; Saloutos and Hicks, *Agricultural Discontent*, 321–341; Patrick G. O'Brien, "A Reexamination of the Senate Farm Bloc, 1921–1933," *Agricultural History* 47, no. 3 (July 1973): 248–263; Gilbert C. Fite, *George N. Peek and the Fight for Farm Parity* (Norman: University of Oklahoma Press, 1954); Kimberly Porter, "Embracing the Pluralist Perspective: The Iowa Farm Labor Bureau Federation and the McNary-Haugen Movement," *Agricultural History* 74, no. 2 (Spring 2000): 381–392; John Phillip Gleason, "The Attitude of the Business Community toward Agriculture during the McNary-Haugen Period," *Agricultural History* 32, no. 2 (April 1958): 127–138.

63. David A. Horowitz, "The Perils of Western Farm Politics: Herbert Hoover, Gerald P. Nye, and Agricultural Reform, 1926–1932," *North Dakota Quarterly* 53, no. 4 (Fall 1985): 97.

64. Darrel LeRoy Ashby, "Progressivism against Itself: The Senate Western Bloc in the 1920s," *Mid-America: An Historical Review* 50 (October 1968): 291–304; and O'Brien, "A Reexamination of the Senate Farm Bloc," 262.

65. Washington Correspondent, "The Progressives of the Senate," *American Mercury* 16, no. 64 (April 1929): 385.

66. C. Fred Williams, "William M. Jardine and the Foundations for Republican Farm Policy, 1925–1929," *Agricultural History* 70, no. 2 (Spring 1996): 216–232; David B. Danbom, *The Resisted Revolution: Urban America and the Industrialization of Agriculture, 1900–1930* (Ames: Iowa State University Press, 1979), 120–146.

67. Theodore Saloutos, *The American Farmer and the New Deal* (Ames: Iowa State University Press, 1982), 34–49; William D. Rowley, "M. L. Wilson: 'Believer' in the Domestic Allotment," *Agricultural History* 43, no. 2 (April 1969): 277–288; M. L. Wilson, *The Reminiscences of Milburn Lincoln Wilson* (1973; repr., New York: Columbia University Oral History Research Office, 1975), 159, 226, 233–234, 243–244.

68. Scot A. Stradley, "Senator Lynn Frazier and Federal Agricultural Policy," *North Dakota History* 66, no. 3 (Summer 1999): 36; Saloutos, *The American Farmer and the New Deal*, 46–47.

69. Stock, *Main Street in Crisis*, 128–147; William Pratt, "Rethinking the Farm Revolt of the 1930s," *Great Plains Quarterly* 8 (Summer 1988): 131–144.

70. Donald L. Miller, *The New American Radicalism: Alfred Bingham and Non-Marxian Insurgency in the New Deal Era* (Port Washington, NY: Kennikat, 1979).

71. Stradley, "Senator Lynn Frazier and Federal Agricultural Policy," 30–40; David A. Horowitz, *Beyond Left & Right: Insurgency and the Establishment* (Urbana: University of Illinois Press, 1997), 95–98, 104–105; Conrad and Conrad, *50 Years*, 23.

72. Ernest Feder, "Farm Debt Adjustments during the Depression—the Other Side of the Coin," *Agricultural History* 35, no. 2 (April 1961): 78–81; James A. Munger and Ernest Feder, *The Frazier-Lemke Act: Its Impact on Farmers and Lenders in the Northern Great Plains*, (Washington, DC: USDA, 1957), 1–5.

73. Walter Lippman, "Today and Tomorrow: Borah versus Roosevelt," *New York Herald Tribune*, April 18, 1936, 15.

74. John M. Jordan, *Machine-Age Ideology: Social Engineering and American Liberalism, 1911–1939* (Chapel Hill: University of North Carolina Press, 1994); Richard S. Kirkendall, *Social Scientists and Farm Politics in the Age of Roosevelt* (Columbia: University of Missouri Press, 1966); Jess Gilbert, "Eastern Urban Liberals and Midwestern Agrarian Intellectuals: Two Group Portraits of Progressives in the New Deal Department of Agriculture," *Agricultural History* 74, no. 2 (Spring 2000): 162–180; Saloutos, *The American Farmer and the New Deal*, xiii.

75. Jeffrey M. Hornstein, *A Nation of Realtors: A Cultural History of the Twentieth-Century American Middle Class* (Durham, NC: Duke University Press, 2005).

76. Saloutos, *The American Farmer and the New Deal*, 244–246.

77. Stock, *Main Street in Crisis*, 86–127; Clifford B. Anderson, "The Metamorphosis of American Agrarian Idealism in the 1920s and 1930s," *Agricultural History* 35, no. 4 (October 1961): 182–188.

78. Alan Brinkley, *Voices of Protest: Huey Long, Father Coughlin, and the Great Depression* (New York: Vintage, 1983); Edwin Amenta, *When Movements Matter: The Townsend Plan and the Rise of Social Security* (Princeton, NJ: Princeton University Press, 2006).

79. Edward C. Blackorby, "William Lemke: Agrarian Radical and Union Party Presidential Candidate," *Mississippi Valley Historical Review* 49, no. 1 (June 1962): 67–84; Ronald L. Feinman, *Twilight of Progressivism: The Western Republican Senators and the New Deal* (Baltimore: Johns Hopkins University Press, 1981), 102, 45–47; David A. Horowitz, *America's Political Class under Fire: The Twentieth Century's Great Culture War* (New York: Routledge, 2003), 45–47.

80. Blackorby, "William Lemke," 67–84.

81. Horowitz, *Beyond Left & Right*, 148–149.

82. Feinman, *Twilight of Progressivism*, 117–135; Horowitz, *Beyond Left & Right*, 140–144.

83. Wayne S. Cole, *Senator Gerald P. Nye and American Foreign Relations* (Minneapolis: University of Minnesota Press, 1962), 60–96, quote on 76. See also Nels Erickson, "Prairie Pacifist: Senator Lynn J. Frazier and America's Global Mission, 1927–1940," *North Dakota History* 52, no. 4 (Fall 1985): 26–32.

84. Cole, *Senator Gerald P. Nye*, 124–201; Feinman, *Twilight of Progressivism*,136–202; Horowitz, *Beyond Left & Right*, 179–180; Wayne S. Cole, *America First: The Battle against Intervention, 1940–1941* (Madison: University of Wisconsin Press, 1953), 131–154.

85. Cole, *Senator Gerald P. Nye*, 227–235; Horowitz, *Beyond Left & Right*, 160.

86. James T. Sparrow, *Warfare State: World War II Americans and the Age of Big Government* (New York: Oxford University Press, 2011); Alan Brinkley, *The End of Reform: New Deal Liberalism in Recession and War* (New York: Vintage, 1995); Michael Sandel, *Democracy's Discontent: America in Search of a Public Philosophy* (Cambridge, MA: Harvard University Press, 1998), 250–273.

87. Wendy L. Wall, *Inventing the "American Way": The Politics of Consensus from the New Deal to the Civil Rights Movement* (New York: Oxford University Press, 2008).

88. John Earl Haynes, *Dubious Alliance: The Making of Minnesota's DFL Party* (Minneapolis: University of Minnesota Press, 1984); Jennifer A. Delton, *Making Minnesota Liberal: Civil Rights and the Transformation of the Democratic Party* (Minneapolis: University of Minnesota Press, 2002); Rossinow, *Visions of Progress.*

89. Delton, *Making Minnesota Liberal*, 158–169.

90. Ibid., 31–33.

91. Robert Loren Morlan, "The Rise of the Nonpartisan League" (master's thesis, University of Minnesota, 1947); Morlan, "The Political History of the Nonpartisan League, 1915–1922" (PhD diss., University of Minnesota, 1949); Larry Remele, "Introduction to the Reprint Edition," in *Political Prairie Fire: The Nonpartisan League, 1915–1922*, by Robert L. Morlan (1955; repr., Saint Paul: Minnesota Historical Society Press, 1985), ix–x.

92. Morlan, *Political Prairie Fire*, preface; Remele, "Introduction to the Reprint Edition," xxi.

93. Remele, "Introduction to the Reprint Edition," x–xi; Theodore G. Nelson, *Scrapbook Memoirs* (Salem, OR: Your Town, 1957), 102.

94. Lloyd B. Omdahl, *Insurgents* (Brainerd, MN: Lakeland Color, 1961).

95. Larry Remele, "The Lost Years of A. C. Townley (after the Nonpartisan League)," North Dakota Humanities Council Occasional Paper no. 1 (1988): 1–27.

96. Walter D. Young, *The Anatomy of a Party: The National CCF, 1932–1961* (Toronto: University of Toronto Press, 1970), 12–33; Mardiros, *William Irvine*, 180–183; Richard Allen, *The Social Passion: Religion and Social Reform in Canada, 1914–1928* (Toronto: University of Toronto Press, 1971), 204.

97. Seymour M. Lipset, *Agrarian Socialism: The Cooperative Commonwealth Federation in Saskatchewan* (Berkeley: University of California Press, 1959), 73–87, 103–109; Young, *The Anatomy of a Party*, 39–100; Mardiros, *William Irvine*, 184–188; George Hoffman, "Frank Eliason: A Forgotten Founder of the CCF," *Saskatchewan History* 58 (Spring 2006): 18–31; Nelson Wiseman, "The Socialist Imprint on Saskatchewan Politics," *Saskatchewan History* 65 (Fall–Winter 2013): 26–33, 43–44.

98. Edward A. Bell, *Social Classes and Social Credit in Alberta* (Montreal: McGill-Queen's University Press, 1994); Nelson Wiseman, "The Pattern of Prairie Politics," in *The Prairie West: Historical Readings*, ed. R. Douglas Francis and Howard Palmer (1985; repr., Edmonton: Pica Pica, 1992), 640–660; Alvin Finkel, *The Social Credit Phenomena in Alberta* (Toronto: University of Toronto Press, 1989).

99. Young, *The Anatomy of a Party*, 103–137; Nelson Wiseman, *In Search of Canadian Political Culture* (Vancouver: University of British Columbia Press, 2007), 211–236; Antonia Maioni, *Parting at the Crossroads: The Emergence of Health Insurance in the United States and Canada* (Princeton, NJ: Princeton University Press, 1998).

100. Michael Dempsey, "Northern Lights: An Interview with John Hanson and Rob Nilsson," *Film Quarterly* 32, no. 4 (July 1979): 3, 5–6, 8; *Cine Manifest*, directed by Judy Irola, New York: New Video Group, 2006.

101. Dempsey, "Northern Lights," 2–10; Clay Jenkinson, "The Making of *Northern Lights*," in *A Humanities Guide to Northern Lights* (Bismarck: North Dakota Humanities Council, 1981).

102. Henry R. Martinson, " 'Comes the Revolution . . .': A Personal Memoir," *North Dakota History* 36, no. 1 (Winter 1969): 40–109.

103. Michael Kernan, "Eyeing the Reels of Fortune; Banking on the Reels of Fortune; John Hanson's 'Northern Lights,'" *Washington Post*, November 16, 1979, F1; Emanuel Levy, *Cinema of Outsiders: The Rise of American Independent Film* (New York: New York University Press, 1999).

104. Vincent Canby, " 'Northern Lights,' Story of Early Labor Wars," *New York Times*, September 19, 1979, C19; Roger Ebert, "Northern Lights," *Chicago Sun-Times*, August 27, 1980, http://www.rogerebert.com/reviews/northern-lights-1980, accessed September 13, 2013.

105. "*Northern Lights* Transcript of Dialogue and Subtitles," in *A Humanities Guide to Northern Lights*. Just three years before the movie, Martinson complained that, in fact, the insurgent farmers "couldn't understand Karl Marx." Robert Carlson, "An Interview with Henry R. Martinson," *North Dakota History* 43, no. 2 (Spring 1976): 20. See also Michael Anderegg, "The Fiction Film as Artifact: History, Image, and Meaning in 'Northern Lights,'" *North Dakota History* 57, no. 3 (Summer 1990): 14–23.

106. Martinson, " 'Comes the Revolution,'" 107–109.

107. Dempsey, "Northern Lights," 10; Ann Markusen, "Who Were Your Grandmothers, John Hanson?," *Quest: A Feminist Quarterly* 5, no. 2 (1980): 25–35.

108. Michael Moore, "Michael Moore's Action Plan: 15 Things Every American Can Do Right Now," October 22, 2009, http://michaelmoore.com/words/mikes-letter/michael-moores-action-plan-15-things-every-american-can-do-right-now, accessed August 22, 2013; David A. Graham, "Socialism Thrives in North Dakota," *Newsweek*, April 22, 2010, http://www.thedaily

beast.com/newsweek/2010/04/22/socialism-thrives-in-north-dakota.html, accessed August 22, 2013; "Bank of North Dakota: America's Only 'Socialist' Bank Is Thriving during Downturn," *Huffington Post*, April 18, 2010, http://www.huffingtonpost.com/2010/02/16/bank-of-north -dakotasocia_n_463522.html, accessed August 22, 2013; Josh Harkinson, "How the Nation's Only State-Owned Bank Became the Envy of Wall Street," *Mother Jones*, March 27, 2009, http://www .motherjones.com/mojo/2009/03/how-nation%E2%80%99s-only-state-owned-bank-became -envy-wall-street, accessed August 22, 2013; Les Leopold, "North Dakota, Socialist Haven?," *Salon*, March 29, 2013, http://www.salon.com/2013/03/29/north_dakota_is_bringing_socialism_back _partner/, accessed August 22, 2013; Gretchen Dykstra, "Pragmatism on the Prairie," *New York Times*, March 30, 2012, A19; Mark Stephen Jendrysik and Dana Michael Harsell, "Egalitarian Populism on the High Plains, or, Why Are There No Parking Meters in North Dakota?" *Journal of Popular Culture* 46, no. 2 (April 2013): 396.

109. Horowitz, *Beyond Left & Right*, 310.

110. *Midland Cooperative* (Minneapolis), October 19, 1953, Newspaper Clippings 1919–1953 folder, Joseph Gilbert Papers, MNHS.

Epilogue

1. Robert George Paterson, "North Dakota: A Twentieth-Century Valley Forge," *Nation* 117, no. 3031 (August 8, 1923): 134.

Index

Page numbers in italics refer to figures.

Made in the USA
Columbia, SC
02 September 2022

66501299R00219